Tourism Ethics

ASPECTS OF TOURISM

***Series Editors*:** Chris Cooper, *Oxford Brookes University, UK*, C. Michael Hall *University of Canterbury, New Zealand* and Dallen J. Timothy *Arizona State University, USA*

Aspects of Tourism is an innovative, multifaceted series, which comprises authoritative reference handbooks on global tourism regions, research volumes, texts and monographs. It is designed to provide readers with the latest thinking on tourism worldwide and in so doing will push back the frontiers of tourism knowledge. The series also introduces a new generation of international tourism authors writing on leading edge topics.

The volumes are authoritative, readable and user-friendly, providing accessible sources for further research. Books in the series are commissioned to probe the relationship between tourism and cognate subject areas such as strategy, development, retailing, sport and environmental studies. The publisher and series editors welcome proposals from writers with projects on the above topics.

Full details of all the books in this series and of all our other publications can be found on http://www.channelviewpublications.com, or by writing to Channel View Publications, St Nicholas House, 31–34 High Street, Bristol BS1 2AW, UK.

ASPECTS OF TOURISM: 81

Tourism Ethics

2nd Edition

David A. Fennell

CHANNEL VIEW PUBLICATIONS
Bristol • Blue Ridge Summit

DOI https://doi.org/10.21832/FENNEL6355
Library of Congress Cataloging in Publication Data
Names: Fennell, David A., 1963- editor.
Title: Tourism Ethics/David A. Fennell.
Description: Second Edition, reworked and updated. | Blue Ridge Summit, Pennsylvania: Channel View Publications, [2018] | Series: Aspects of Tourism: 81 | Includes bibliographical references and index.
Identifiers: LCCN 2017035560| ISBN 9781845416355 (hardback : alk. paper) | ISBN 9781845416348 (paperback : alk. paper) | ISBN 9781845416362 (pdf) | ISBN 9781845416379 (epub) | ISBN 9781845416386 (kindle)
Subjects: LCSH: Tourism—Moral and ethical aspects.
Classification: LCC G155.A1 T591747 2018 | DDC 174/.991—dc23 LC record available at https://lccn.loc.gov/2017035560

British Library Cataloguing in Publication Data
A catalogue entry for this book is available from the British Library.

ISBN-13: 978-1-84541-635-5 (hbk)
ISBN-13: 978-1-84541-634-8 (pbk)

Channel View Publications
UK: St Nicholas House, 31–34 High Street, Bristol BS1 2AW, UK.
USA: NBN, Blue Ridge Summit, PA, USA.

Website: www.channelviewpublications.com
Twitter: Channel_View
Facebook: https://www.facebook.com/channelviewpublications
Blog: www.channelviewpublications.wordpress.com

The policy of Multilingual Matters/Channel View Publications is to use papers that are natural, renewable and recyclable products, made from wood grown in sustainable forests. In the manufacturing process of our books, and to further support our policy, preference is given to printers that have FSC and PEFC Chain of Custody certification. The FSC and/or PEFC logos will appear on those books where full certification has been granted to the printer concerned.

Typeset by Nova Techset Private Limited, Bengaluru and Chennai, India.
Printed and bound in the UK by Short Run Press Ltd.
Printed and bound in the US by Edwards Brothers Malloy, Inc.

To my parents, John and Nancy Fennell for inspiring me to live your philosophy of 'roots and wings.' We miss you.

'Nothing we do, however virtuous, can be accomplished alone; therefore, we are saved by love.'
(Reinhold Niebuhr)

Contents

Boxes, Figures and Tables ix
Preface xi
Acknowledgements xvii

1 Introduction 1

2 Human Nature 16

3 The Basis of Ethical Discourse 51

4 Applications of Ethics 86

5 The Nature of Politics and Economics 115

6 The Business Side of Ethics 133

7 Ethics and the Natural World 166

8 Broad-based Concepts and Issues in Tourism 200

9 Codes of Ethics 224

10 Models and Methods of Moral Reasoning 251

11 Case Study Analyses 283

12 A Moral Tourism Industry? 319

Appendix: WTO Global Code of Ethics for Tourism 353
References 360
Index 393

Boxes, Figures and Tables

Boxes

Box 1.1	Tourism impacts in Goa	3
Box 2.1	The trapped tourists	40
Box 4.1	The consequences of indirect reciprocity	93
Box 6.1	Tropic Ecological Adventures, Ecuador	163
Box 6.2	Pro-poor tourism around South Africa's Addo Elephant National Park	164
Box 7.1	Environmental justice problems and solutions	176
Box 7.2	The myth of ecotourism	188
Box 7.3	The self-interest or egocentric approach to environmental ethics	193
Box 7.4	The homocentric approach to environmental ethics	194
Box 7.5	The ecocentric approach to environmental ethics	195
Box 7.6	Beyond Leave No Trace	198
Box 8.1	The Tragedy of the Commons	205
Box 8.2	Certification	217
Box 9.1	The Country Code	225
Box 9.2	Code of ethics for tourists: Sustainable tourism	229
Box 9.3	Code of ethics for the industry: Sustainable tourism	229
Box 9.4	The Gwaii Haanas Watchman Program	234
Box 9.5	Pros and cons of codes of ethics	241
Box 10.1	PricewaterhouseCooper's code of conduct for ethical decision-making	253
Box 10.2	Kohlberg's stages of moral development (*e.g. in running your business what is considered to be morally right*)	256
Box 10.3	The hockey game	272
Box 10.4	Ethical scenarios	281

Figures

Figure 1.1	Tourism interactions	4
Figure 2.1	Payoffs from the Prisoner's Dilemma game	31
Figure 9.1	Levels of moral discourse	245
Figure 9.2	Code development process	248
Figure 10.1	Interactions model of ethical decision-making in organisations	259
Figure 10.2	Framework for ethical conduct	262
Figure 10.3	Social intuitionist model of moral judgement	265
Figure 10.4	General theory of marketing ethics	266
Figure 10.5	Ethical decision-making approach (comprehensive)	270
Figure 10.6	Buddhist ethics: Balancing problems and benefits	274
Figure 11.1	A model of ethical triangulation	296
Figure 11.2	Moral development in tourism organisational cultures	314
Figure 12.1	Increasing cooperation and coordination in research	322
Figure 12.2	The knowledge value chain	330
Figure 12.3	Framework for tourism ethics	331
Figure 12.4	Framework for the interdisciplinary exchange and diffusion of ethical knowledge in tourism	339

Tables

Table 4.1	Interactions of motivations and outcomes in determining morality and immorality of social acts	94
Table 6.1	Ethical orientations: A comparison	141
Table 6.2	Special ethical considerations for tourism	142
Table 6.3	Theoretical ethical climate types	147
Table 7.1	Contrasting paradigms	191
Table 7.2	Definitions of environmental ethics	192
Table 10.1	Schumann's moral principles framework	269
Table 10.2	Risk and response options for wrongdoing	277
Table 11.1	Categories of Holy Land tour operators	301

Preface

In writing a book on tourism ethics it would seem appropriate, if not a bit embarrassing, to come clean on certain matters. In a previous publication I reported observing tourists pilfer small pieces of sandstone on the surface of Australia's Ayers Rock. Lest we think that I occupy some lofty moral position as the author of this book, we can easily put this nonsense to bed by my admittance that one of the culprits was me. Can I take it back? No. But what is most encouraging about the whole event is that it has been bothering me since 1987, the year in which I committed the act. In reflection, I justified my actions from the apparent need to savour the memory of Ayers Rock forever by having a tangible bit of this renowned attraction for my mantle. In reality, however, the act was more than just the need to savour a memory. As I think back, it was also about a novice traveller trying to demonstrate to others how extensively he had travelled. The act was made easier, in my opinion, as it was based on what might be termed a collective, consensual and temporary state of self-interest, demonstrating that ethics is very much situational. A sense of mischief and 'adventure', however, soon turned to disappointment as I didn't have the sense to protect the rock (sandstone) over the course of my six-month trip, thus demonstrating my limited knowledge of geology. In a matter of weeks, the rock was reduced to sand, showing me how erosion works through 'natural' processes, rendering it both useless to me as well as to anyone else who might have been presented with the opportunity to view it.

I make light of the situation but it is serious enough to demonstrate that: (1) the natural resources we use for our pleasure are threatened in innumerable ways; and (2) the limits of such resources are tested daily by the Golden Hordes, the majority of whom often have little knowledge of their actions. This has been proven time and time again in decades of tourism research. But where has this research taken us? We know the result, indeed we can easily predict it, but we still have trouble addressing, or worse yet, stemming the tides of negative change. In this regard, we need to be vigilant in heeding the words of McKercher who has noted that our field has been 'entrenched in an intellectual time warp that is up to 30 years old' (McKercher, 1999: 425). The friction (impacts) of this time warp (indeed, new books continue to surface on impacts), has pulled us far behind many other disciplines that have progressed further both conceptually and theoretically through an interdisciplinary agenda based on ethics. Ethics is serious stuff. It has been on the minds of people for thousands of years. Not so in tourism, however, where ethics is barely a decade old.

In recognising the immense void in ethics in our field, I have attempted to take the reader to a new destination; one where the waters are just now being tested. This has necessitated the development of a book that delves heavily into theory. Even those aspects that are of an applied nature are tied to theoretical models and methods. This renders the book less applicable to the younger undergraduate crowd and more applicable to the senior group, as well as graduate students and researchers in tourism. Having said this, every effort was made to be as accessible as possible with the theoretical content within. I should also note that, although it is a book for students and researchers, it is as much a book for the author, as selfish as this may seem. Taken from the existential domain, in order to be free we need to be personally authentic. This book is a tangible result of this authenticity; about my past, the present, and perhaps where I would like to see myself in the future. In being free to make this leap, however, we must also be willing to take responsibility for our actions. I am fully prepared to take responsibility (hence, the apparent need to come clean on the Ayers Rock situation) for what might be too much of a leap in faith in the traditional sense. With this in mind, it is the purpose of this book to:

(1) Introduce moral concepts and issues in tourism in a comprehensive yet accessible fashion. This will entail an analysis of ethical systems and theories as well as a link to the many ethical situations that exist in the field of tourism.
(2) Examine ethics from the perspective of many more established disciplines for the purpose of diversifying tourism scholarship. This is critical if we are to emerge from the present intellectual stagnancy, as noted earlier.
(3) Develop a theoretical and conceptual template for tourism that provides a foundation for research, planning, development and management of tourism from an ethics based perspective. In this context, the book points to the fact that there is an absence of an underlying ethical basis for critical thought in tourism.

The intent is to formulate these ideas, theories and issues for a readership that might not be well versed in ethics or moral philosophy; two terms that to some theorists are synonymous and to others distinct (the latter being the more general), as discussed in Chapter 3. Indeed, this is an area of thought that can be extremely deep. The only requirement I ask of the reader is a willingness to look upon the many positive and negative aspects of the tourism industry from the side of ethics. In doing so we may yet uncover more intriguing ways to study a field that continues to challenge us in theory and practice, and in time and space. It should also be noted that the approach used here afforded me the opportunity to study ethics with an open mind and not as a slave to any one discipline. This method was critical in seeking a diversity of explanations that would allow for an enhanced view of the place of ethics in tourism, in efforts to help pull us out of our theoretical cul-de-sac.

The book is organised into 12 chapters, which move from the theoretical to the applied, and back. Chapter 1 provides an overview of the tourism industry, with a particular focus on sustainable tourism, alternative tourism, impacts and the current research on tourism and ethics. It follows with an examination of some new realities and moralities in tourism. Chapter 2 briefly summarises the rich foundation of literature on the cultural and biological basis of human nature. The argument carried forward is that we must have a firm grasp of human nature in order to better understand the role of ethics as a fundamental aspect of our natures. The chapter ends with a brief discussion of the evolution of ethics. Chapter 3 focuses on the basis of ethical discourse, including a discussion of classical antiquity and philosophical terminology (i.e. philosophy, morality and ethics, values and norms), absolutist theories, such as deontology and teleogy, as well as existentialism as one of the predominant subjectivist ethical theories. Chapter 4 takes a more applied look at the circle of morality, justice, rights, responsibility and free will. The bulk of the work in Chapter 5 focuses on an illustration of the history of trade as a fundamental aspect of our human natures, including aspects of self-interest and cooperation. Politics, power and capitalism are examined, along with implications for tourism and development in lesser developed countries. The chapter ends with an examination of social status and the culture of consumption, laying the foundation for a more comprehensive treatment of business and ethics in Chapter 6. Here, corporatism is compared with individualism, and corporate responsibility is examined along with trust and culture within organisations. The chapter also looks at business ethics, and ends with some ethical responses that have taken place in tourism, including marketing, fair trade and pro-poor tourism.

Chapter 7 centres on ethics and the natural world through an examination of ecosystems and ecosystem services, stewardship, values and rights, and environmental ethics. Several definitions of environmental ethics are discussed, as well as a number of different models of human–environment relationships. In Chapter 8, the focus switches to what might be termed broad-based issues and concepts that have an effect on tourism. The discussion includes work on major ethical responses in tourism, including common pool resources, social traps and governance, as well as accreditation, best practice benchmarking and the precautionary principle. Chapter 9 deals specifically with codes of ethics in tourism, and includes work in other fields, mostly business, in an attempt to provide further theoretical and applied guidance for tourism studies. Pros and cons of codes of ethics are discussed, along with a code development process and a comprehensive ethical programme. The chapter ends with an examination of the World Tourism Organization's Global Code of Ethics. In Chapter 10, a number of models and methods of moral decision-making are introduced, which are later applied to a series of different ethical dilemmas in Chapter 11. The intent of both chapters is to show that ethical dilemmas in tourism can be examined in a number of different ways. The final chapter, 12, uses the concepts of interdisciplinarity, knowledge and complexity in the development of a comprehensive ethical framework for tourism. This framework builds upon a number of key themes that are introduced throughout the course of the book. These include: (1) current knowledge in tourism; (2) micro interactions (those that are indicative of the day-to-day and face-to-face interactions of people involved in tourism); (3) macro interactions (more broadly based issues in tourism); (4) the importance of knowledge in the humanities (e.g. ethics) in solving tourism-related problems; and (5) the theoretical contributions of biology in addressing tourism issues. Only through an enhanced understanding of these five different domains, it is argued, can we begin the task of assembling a base of knowledge for the purpose of more clearly addressing the various impacts that continue to shackle the tourism field.

In this book I have, in the words of Humphrey (1992), 'a big fish to fry', for much of what appears in the pages to follow is relatively new for students of tourism, at least by virtue of what does not appear in our journals. I make no claim as to the size or weight of the fish (as did Humphrey), only that it succeeds in getting students and researchers to adopt new techniques for effective angling. In striving to make this happen, I have had to angle in many different ponds, which has enlightened me in ways unimaginable. But I also subscribe to the words of the evolutionary biologist Ernst Mayr (1988), who, in acknowledging his own limitations as a non-expert in the field of ethics, chose to steer himself clear of finding definitive answers on ethics by electing to ask open questions. In this regard, it is worthwhile to acknowledge Rawls, who observed that we learn about moral philosophy by studying the noted figures who have gone before us: Kant, Aristotle, Nietzsche and so on (Rawls, 2000). If we are lucky, we find a way to go beyond them. While Rawls had the tools to do this, I make no such claim. My aim here is to better understand an area that is both fascinating and still lacking in tourism, especially as we are in the midst of this moral turn in tourism (Caton, 2012).

But why the need for a second edition? Around about the time that the first edition of this book was in production, Macbeth (2005) made a compelling argument for a new platform in tourism research: ethics. His perspective was indeed timely as it sustains, and moves forward, several changes taking place in tourism research and practice over the years designed to curb negative impacts. The first edition of *Tourism Ethics*, was crafted in the same spirit of what Macbeth was after: to place ethics in a central position in tourism research. Given the steady flow of tourism ethics research over the intervening years, ethics is just starting to find its stride. Is it a platform unto itself? The jury is still out. Books that have been added to the literature include work by Hall and Brown (2006); Fennell and Malloy (2007) on codes of ethics in tourism, Mowforth and Munt (2008) on tourism and responsibility, Tribe (2009) on philosophical issues in tourism, Fennell (2012a) on tourism and animal ethics, Frenzel *et al.* (2012) on slum tourism and ethics, Lovelock and Lovelock (2013) on ethics in tourism, Mostafanezhad and Hannam (2014) on moral encounters in tourism, and Weeden and Boluk (2014) on the management of ethical consumption in tourism. Routledge has also recently published a four volume set of classic papers on ethics, entitled *Tourism Ethics* (Fennell, 2016), the title not to be confused with the present book. Routledge has also developed a book series devoted specifically to tourism ethics, and the *Journal of Ecotourism* as of 2018 will be dedicating one edition per year solely to tourism ethics. The consistency in publication of these volumes and initiatives on tourism ethics over the years serves notice that the waters are no longer *just* being tested, as was the case in the first edition, but rather that ethics in tourism is not going away. We are a moral species by nature. This book, more than any other in the tourism studies field, makes this point abundantly clear. It is the task of tourism scholars, practitioners, and policy makers to take this leap together in the development of a tourism industry that values more than just profit and pleasure at the expense of so many.

Acknowledgements

Thanks are extended to Channel View for inspiring me to take on this second edition, especially Elinor Robertson, who was the first point of contact and Sarah Williams for keeping it all on track. Thanks are also extended to an anonymous reviewer who did an outstanding job identifying the many changes required to bring this edition up-to-date. I wish to thank my wife, Julie, and children Sam, Jessie and Lauren, for their continued patience, support, and understanding.

1 Introduction

Tourism cannot be explained unless we understand man, the human being
Przeclawski, 1996: 239

Introduction

This chapter discusses the background behind the tendency of tourism researchers to examine impacts as the traditional root of ethical issues in tourism. The chapter also analyses alternative tourism and sustainable tourism paradigms as the field's most frequently used means by which to alleviate the negative impacts of the industry. A brief summary of work on tourism and ethics provides a generalised snapshot of the range of studies undertaken to date in addressing ethical issues in tourism. The chapter further discusses the negative backlash that has come about regarding the so-called 'new tourism', and sets the stage for the discussion in later chapters on human nature and ethics, and how these relate more specifically to tourism.

Tourism Impacts

One of the longest-standing traditions in tourism research, which is almost universal in our books and academic papers, is the necessity of discussing at the outset the idea that tourism is the world's foremost economic engine. This is natural from at least two perspectives. The first is that it seems to legitimise the importance of tourism through an approximation of its overall magnitude regarding foreign receipts, employment and other such indicators. Second, it demonstrates that, apart from its position as the formidable economic giant, there are associated costs, which have been discussed almost universally as sociocultural, economic and ecological impacts.

The concern over tourism impacts originates from the 1950s, when the International Union of Official Travel Organizations' (the precursor to the WTO) Commission for Travel Development first initiated discussions on how to minimise destinational impacts (Shackleford, 1985). During the 1960s, publications such as *National Geographic* and *Geography* picked up on the negative impacts from tourism in places that were at the leading edge of the mass tourism phenomenon, including Acapulco (Cerruti, 1964) and the Balearic Islands, Spain (Naylon, 1967). The pace of international tourism intensified during the 1970s, and impacts were discovered in many more of the sea, sun, sand and sex destinations, such as Gozo (Jones, 1972), as well as in city environments, including London, where Harrington (1971) observed how unregulated hotel development led to a lower quality of life. Tourism research on impacts hit its stride during the 1970s on the strength of work from scholars such as Budowski (1976), whose classic paper on the interactions between tourism and environmental conservation were explained as: (1) conflict; (2) coexistence; or (3) symbiosis. In the majority of cases he felt that the relationship was one of coexistence, moving towards conflict. Such were the conclusions of other esteemed authors, who felt that poorly planned tourism development had many serious effects on the integrity of the natural world (Cohen, 1978; Krippendorf, 1977).

The conflict so often identified by these and many other successive papers is no more clearly articulated than in the following case study on tourism impacts in Goa, India (see Box 1.1). The maturity of the tourism product in Goa has created a level of competition and fractioning within society that is extreme – conflict in the words of Budowski (see the work of Lea, 1993, who discusses both the impacts of tourism on Goa as well as the beginnings of responsible tourism). This fact has been supported by literally dozens of academic reports, which identify the polarisation of socioeconomic conditions, usually between the lesser developed countries and the most developed countries, leading to a number of interaction problems between tourists and resentful hosts (see, for example, Ahmed *et al.*, 1994). The intersection of many competing interests from a number of different stakeholder groups frames the basis for the impacts that we experience in tourism. The main groups involved in these interactions include tourists, inhabitants of the destination and tourism brokers, as illustrated in Figure 1.1 (Fennell & Przeclawski, 2003). The combinations of these interactions are extensive, and include: (1) the tourist's own personal or existential experiences, and interactions with other tourists, residents of the destination, tourism brokers and the ecology of the region; (2) residents' interactions with tourists, brokers, the community in general and ecology; and (3) brokers' interactions with tourists, residents, other brokers and the natural world. These interactions range along a continuum from negative (hostile) to positive (symbiotic), as noted above, and are moderated by time, space, situational factors, resource allocation and a whole host of other elements.

Box 1.1 Tourism impacts in Goa

India's smallest state, Goa, is indicative of the extent to which tourism can transform a region. Noronha examined over 50 newspaper articles from just 1995 to 1997, documenting much of the uneasiness that tourism has created in the region, which was identified as problematic 10 years earlier when German tourists were pelted with cow dung on their arrival. A decade later, articles were suggesting the following: land values had skyrocketed due to tourism; very few of the lush green hills that once were prevalent remain; agricultural land has been lost to tourism development; government officials have been targeted through allegations of misappropriation and corruption in the name of tourism; building regulations have been violated, especially by the large hotels; no proper scientific or economic assessments have been undertaken to plan tourism; no priorities have been established; rapid urbanisation has transformed the region; age-old storm water drains have been turned into sewage conduits; the beach plays host to drug dealers whom the police turn a blind eye to; folk art has been eroded with great loss to authenticity; water shortages and electricity shortages have occurred because of the demand placed on the infrastructure from large hotels; waste disposal systems are overrun; transport systems are inadequate; the water pipeline meant for locals has been taken over by hotels; tourists on several occasions have been beaten up by villagers; strong-armed tactics (gangsters) have been used to displace local people for the development of hotels; local residents have protested plans to have hawking zones established in certain areas; beach shacks and temporary restaurants have been shut down by authorities, because they charge lower prices than the hotels; beaches are overly crowded; hotels often do not pay staff for up to three months, because of their own slim margins; the cost of living for locals has gone up markedly; apartment blocks are turned into makeshift hotels, which undercut the cost of rooms in hotels; child sex abuse is rampant; AIDS is becoming a problem; paedophiles have been identified in Goa; the police have been known to extort money and frame people on drug-related charges; tourists have been raped; the density of hotel establishments per km in some areas is excessive (43 establishments per km); firms have pushed local authorities to privatise Goa's old and historic forts; politicians push for more tourism as visitation begins to fall; scarcely 10% of Goans have benefited from tourism; beaches that cater to up to 10,000 per day remain without toilets; and malaria is spreading throughout Goa. Given the magnitude of the problem in Goa, it is easy to see how mismanagement and greed have dictated the levels of growth in this region.

Source: Noronha (1999).

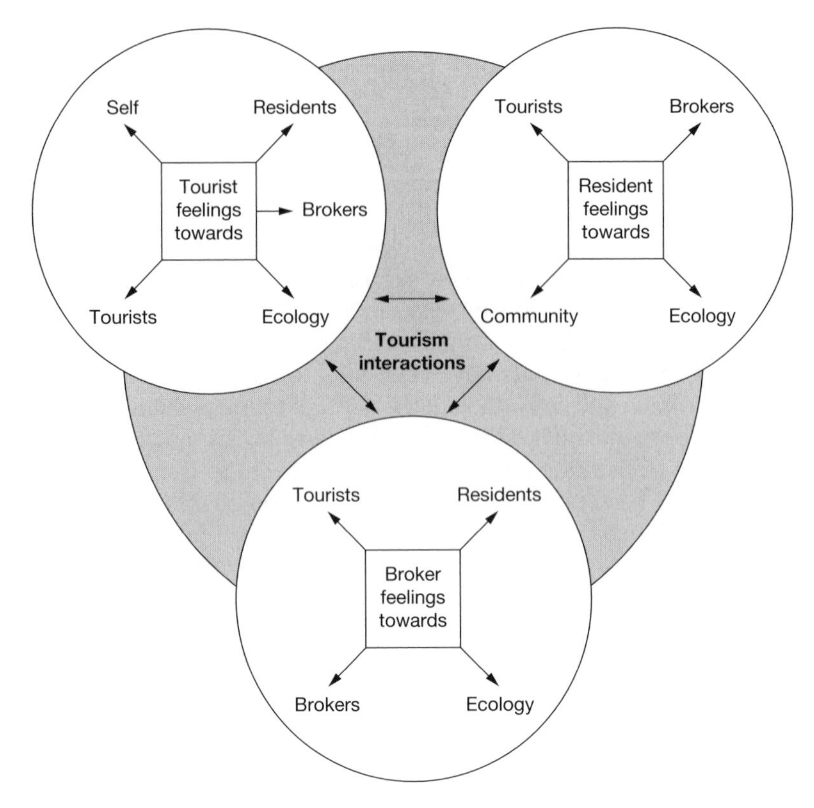

Figure 1.1 Tourism interactions
Source: Fennell and Przeclawski (2003).

Alternative tourism (AT) and sustainable tourism (ST)

It is not the purpose of this section to fully elaborate on the development and impact of AT and ST, which can be found in many recent tourism publications, but rather to briefly provide an historical context that emphasises certain paradigmatic changes in tourism that were borne out of efforts to both understand and mitigate tourism impacts, and thus to implicitly strive to become ethical.

The intensity of moral concern in tourism intensified during the late 1970s and early 1980s through the AT paradigm, which emerged through its potency in providing an alternative to mass tourism. The tenets of the ecodevelopment paradigm of the 1970s, including enlarging the capacity of individuals, self-sufficiency of communities, and social and environmental justice, were articulated through AT, which was meant to be both a softer and gentler form of tourism (Riddell, 1981; Weaver, 1998). This meant that 'small scale' was thus better than 'large scale'; locally oriented was better than externally oriented; low impact better still than high impact. These polarised options where recognised early in the work of Dernoi, who observed that AT would: (1) provide economic benefits for individuals and families (e.g. through accommodation provision); (2) allow the local community as a whole to benefit; (3) allow the host country to benefit through the avoidance of leakages and the reduction of social tensions; (4) provide an option for cost-conscious travellers coming from the 'north'; and (5) realise cultural and international benefits across countries and continents (Dernoi, 1981).

There is little question that AT provided a needed backdrop from which to gain perspective on the often disingenuous side of mass tourism. However, the dichotomous positions that are inherent in the mass-alternative perspectives are rarely encountered in their purest forms because of the sheer complexity of different attractions, accommodations, transportation and facilities that the traveller encounters on a day-to-day basis (thus minimising the true alternative nature of the trip) (Weaver, 1998). This was identified early by Butler, who quite effectively observed that, while mass tourism has a callous side, it may be just as destructive to promote AT without being confident of what it can achieve for the community, socially, environmentally and economically (Butler, 1990). This sentiment has led theorists to conclude that it is perhaps best to view AT not as a replacement for mass tourism (this will surely never take place), but rather as a model in helping to amend some of the problems that are inherent in mass tourism (Butler, 1990; Cohen, 1987). So where AT is perhaps most beneficial is in defining the range of the continuum regarding tourism development. And as development economists might suggest, it is perhaps better to have a balanced approach to development within a region, including a number of active sectors in the economy for the purpose of achieving balanced growth, including mass tourism and AT.

Alternative tourism articulated many of the tenets supported by the sustainable development (SD) platform, which emerged late in the 1980s. SD subsumed ecodevelopment, but also intensified at a broader scale in its application to poverty, limits on technology and unfettered growth, cross-cultural applications, its ability to be integrative, and its use at broader scales (Redclift, 1987). For tourism this meant that if the industry was to become sustainable, it would do so by adhering to a number of basic principles, including: (1) reduction of tension between stakeholders; (2) long-term viability and quality of resources; (3) limits to growth; (4) the value of tourism as a form of development; and (5) visitor satisfaction (Bramwell & Lane, 1993), and through the realisation that ST is a process and an ethic (Fennell, 2002). The need to articulate such criteria in ST is underscored by González Bernáldez, who notes that benefits and costs must be weighed equally in a better understanding of the impacts of tourism, as follows (González Bernáldez, 1994):

Benefits

- Increases and complements financial income.
- Improves facilities and infrastructures.

- Allows greater investment for the preservation of natural and cultural enclaves.
- Avoids or stabilises emigration of the local population.
- Makes tourists and local populations aware of the need to protect the environment and cultural and social values.
- Raises the sociocultural level of the local population.
- Facilitates the commercialisation of local products and quality.
- Allows for the exchange of ideas, customs and ways of life.

Costs

- Increases the consumption of resources and can, in the case of mass tourism, exhaust them.
- Takes up space and destroys the countryside by creating new infrastructure and buildings.
- Increases waste and litter production.
- Upsets natural ecosystems, and introduces exotic species of animals and plants.
- Leads to population movement towards areas of tourist concentration.
- Encourages purchase of souvenirs that are sometimes rare natural elements.
- Leads to a loss of traditional values and a uniformity of cultures.
- Increases prices and the local population loses ownership of land, houses, trade and services.

But how is it that we determine what is a tourism benefit and what is a cost? To whom, and at what scale? McKercher (1999) has noted that the most unfortunate reality confronting tourism is that its plans and models have been mostly ineffective at controlling the adverse effects of the tourism industry. If traditional models explained tourism fully, he suggests, then they would also be able to offer insights into how best to control such impacts. Traditional models in tourism are ineffective because they imply strongly that: (1) tourism can be controlled; (2) its players are formally coordinated; (3) it is organised easily in a top-down fashion; (4) service providers achieve common, mutually agreed-upon goals; (5) it is the sum of its parts; and (6) an understanding of each of these parts will allow us to understand the whole. Therefore, by nature, tourism is far too complex to be explained by linear, deterministic models. This presumably includes sustainable tourism too. So, if ST is more about development than conservation, it is because the former reflects more of who we are and what we represent. We can demand from science all we want regarding a more ecocentric lifestyle. It does not mean that change will be easily attained or socially desired, despite the new morality that has emerged regarding more ethical attitudes about a number of different social and ecological issues (Fox & DeMarco, 1986).

What has come about, along with AT and ST, to ameliorate the various dysfunctions that characterise the industry, are a series of codes of ethics as well as a range of policies and regulations. For example, the World Tourism Organization (WTO) has published a global code of ethics for tourism (see Chapter 9 for more of a discussion on this document, as well as the Appendix), outlining a vast array of directives that need to be followed in generating good behaviour and positive experiences (WTO, 2001). In the minds of many, however, such a cookbook of guidelines is an example of the leading edge of tourism ethics. But we must be careful that (1) identifying these impacts and prescribing guidelines for their control; and (2) rectifying them, are two very different mindsets and actions. In the latter, we have largely been unsuccessful.

Our propensity to investigate impacts has drawn us into a circuitous loop of reactance, preventing us from focusing on the underlying nature of these disturbances. So, with all due respect to the WTO and others who have attempted to wrestle with these difficult long-standing social and ecological issues, we have not yet committed ourselves to an examination of the broader underlying questions that create these impacts. That is, we have not made the leap from recognising impacts and attempting to ameliorate them beyond that which has been deemed acceptable to the industry. This is very much akin to setting standards for the industry on the basis of what is deemed 'right' or 'good', without fully understanding the meaning of right or good. We can do this intuitively and anecdotally, but what we have not yet effectively determined is right and good from an ethical standpoint.

Tourism and Ethics

In asking if the multi-billion-dollar tourism industry can 'put aside its dirty tricks and become ethical', Boyd (1999) strikes to the heart of an important issue: although tourism is often touted as being a saviour in many regions, experts, including the UN, suggest that it has failed because of the 'displacement of local and indigenous people, unfair labor practices, corruption of or disrespect for culture and a myriad of other human rights abuses, along with environmental contamination' (Boyd, 1999: 1).

In India, for example, women must walk miles to get water because hotels siphon it off from the groundwater for their own excessive uses; while, in Burma, thousands of Burmese are forced from their homes to make way for huge new tourism developments (Wheat, 1999). Both are cases where economic priorities in the name of tourism have given way to significant, and unthinkable, human rights abuses (see Lovelock, 2008). These are not isolated cases. Increasingly communities are losing their cultural integrity because of tourism – as a force of globalisation – which will only intensify, based on the forecasts for huge increases in international tourism over the next three decades.

It is not just those peripheral, marginalised places that are hit by the ugly side of tourism, but places that are part of the mainstream. In one case of rickshaw rip-offs, a Detroit policeman on vacation in Toronto was charged CAD$240 for two rickshaws that took him and his wife and daughter five blocks (McGran, 2003). After a debate in which the police were summoned, the drivers and customer agreed on CAD$30 for the driver, still above the 'regular' rate. The same trip for the family by bus would have been CAD$6.75, CAD$5.50 by taxi or CAD$11.86 by car, including parking. This is also not an isolated case. Rickshaw drivers have been disrupting life in Toronto for some time by misquoting fares, leaving customers out in the middle of intersections, crowding sidewalks, blocking streetcars and not providing exchange on US dollars. The hotel sector and the Toronto Tourism Board have both been alerted to the problem and have in turn informed city council. Although divided on the issue, many in council feel that rickshaws should be banned from the city altogether. While this does not appear to be a legal option, councillors have been asked to look into stronger bylaws, stricter enforcement or stricter licensing. And, while some companies appear to be playing by the 'rules', others have not been so forthcoming – which has placed all who are involved in a precarious position. These examples illustrate that ethics for tourism is not restricted to the frequent structural inequalities between the North and South, but rather pervades all aspects of the industry in time, space and circumstance. Unfortunately, however, we have been slow to recognise the depth of the disparity.

Tourism research on ethics

The genesis of ethics in tourism appears to have developed in hospitality management owing to the emphasis of hospitality's relationship to service and business (Wheeller, 1994). This research provided the foundation for the move to establish the International Institute for Quality and Ethics in Service and Tourism (IIQUEST), which was designed to bridge the gap between ethics and issues related to community relations, sexual harassment, the rights of guests and so on (Hall, 1993). For example, in one of the earlier papers on hospitality and ethics, Whitney found that the value of a company's code of ethics, and whether profit should be the sole factor in influencing business decision, was based on traditional values rather than those that violate traditions. These latter situations create ethical dissonance (ethical conflict) between stronger ideological aspects (those that they believe in) versus operational ones (those that they practice) (Whitney, 1990).

Although tourism studies researchers were largely inactive in the area of ethics in the early 1990s (D'Amore, 1993; Payne & Dimanche, 1996), there was a burgeoning recognition of the importance of ethics in tourism (Hughes, 1995), especially in regard to the lesser developed world (Lea, 1993). Two vehicles were seen to be instrumental in the development of an interest in tourism studies and ethics during the 1990s. One of these was the International Association of Scientific Experts in Tourism (AIEST) Congress in Paris (1992), which proposed the creation of a commission to deal with the ethical problems in tourism (Przeclawski, 1996). The other was the Rio Earth summit of 1992, whose attendees committed themselves to Agenda 21. Chapter 30 of this plan is as follows:

> Business and Industry, including transnational corporations, should be encouraged to adopt and report on the implementation of codes of conduct promoting best environmental practice, such as the International Chamber of Commerce's Business Charter on Sustainable Development and the chemical industry's responsible care initiative. (Genot, 1995: 166)

By the mid-1990s, a series of articles appeared in the literature on ethics and tourism in general (Ahmed *et al.*, 1994; Upchurch & Ruhland, 1995; Walle, 1995), which were accompanied by research that emphasised specific forms of tourism, such as ecotourism (Duenkel & Scott, 1994; Karwacki & Boyd, 1995; Kutay, 1989; Wight, 1993a, 1993b). Typical of much of the early research on ethics is the following example, which describes 'principles fundamental to sustainable ecotourism [that] can be listed from an ethics based perspective'. Some of these include:

- It should not degrade the resource and should be developed in an environmentally sound manner.
- It should provide first-hand, participatory, and enlightening experiences.
- It should involve education among all parties – local communities, government, non-governmental organizations, industry, and tourists (before, during, and after the trip) (Wight, 1993a: 56).

An analysis of these fundamental principles, however, begs the following questions: which ethics-based perspective? Where do these come from? Why are they important? How can we serve these up and in what form? And can these principles not be grounded in research or theory which demonstrates their importance as foundations? Some of this work began the task of looking deeper into how the theoretical aspects of ethics may prove beneficial to the tourism industry. Ecotourism has frequently been used as a barometer in tourism studies that sought to justify its position as one of the most ethical forms of tourism (see Fennell & Malloy, 1995, 1999; Karwacki & Boyd, 1995; Malloy & Fennell, 1998a, 1998b; Stark, 2002). For example, Karwacki and Boyd (1995) charged that ecotourism is unethical because those who stand to gain the most from it (political figures and service providers) do so at the expense of the poverty-stricken citizenry. Under utilitarian scrutiny, ecotourism fails because its economic benefits (to a few) do not come close to its externalities (e.g. pollution, loss of culture and local resources). The authors also noted that tourism can be unjust based on process (top-down development) and outcomes (shifts from agriculture to seasonal tourism). In this regard, one of the classic examples of the uneasiness over ecotourism is that put forward by Wheeller (1994), who notes that the marriage of travel (and the sophistication as something more lofty than tourism) and concern for the environment, in the most ostentatious way, has created a 'new' form of ethical travel (see also Munt, 1994). This so-called more respectable form of tourism is seen by Wheeller as an excuse to behave in much the same way, but from a higher moral platform. This is very much akin to Butcher's opinion of tourism, which he says used to be fun and adventurous. Instead, these aspects have been removed from tourism because of the new ethical imperative that currently pervades the industry. In this tourism world, pleasure-seeking has been regulated in the face of social and environmental concerns. So hedonism, once a virtue of tourism, has now become a sin.

> The moral baggage associated with travel now threatens to shackle a spirit of adventure for travellers young and old. As travel has become a focus for moral codes, something has been lost along the way. If travel is to really be a 'life-expanding activity', or a 'unique experience' of any kind, then it has to rely on the individual, be they reckless or sensitive, impulsive or well prepared. (Butcher, 2003: 141)

We do not really get a sense of what morality is, theoretically or conceptually, in Butcher's work. Instead we are left to make assumptions about the place of ethics in tourism. The fact is, we know very little about morality and its application to tourism. Furthermore, we cannot view ethics on the basis of individuality, as Butcher has done in the aforementioned quote. Such egoistic accounts fail to embrace the fact that the meaning that we so often crave in life often comes not by existing in isolation, but rather as a by-product from the experiences and relationships we have with others.

Other studies on ethics in tourism have used justice in emphasising the importance of fairness in the industry. For example, Hultsman (1995) used the concept of 'just' tourism as a metaphor to suggest that ethical tourism is that which is virtuous (e.g. fair and honourable) among a number of different choices, but also to suggest the fact that tourism is 'merely' or 'only' a 'small thing'; that tourism should be organised and delivered in a principled manner. He notes: 'Should tourism reach the point of being considered by service providers as first a business and second an experience, it is no longer "just tourism"; it is an industry' (Hultsman, 1995: 561). His notion of 'just' tourism is premised on the work of Aldo Leopold's land ethic, which is essentially a limitation on the notion of freedom in the interests of appropriate social conduct – do nothing that will harm the natural world. The underlying rationale of his ethical framework is subjective or intuitive in its orientation. In this regard it has been criticised on the basis of difficulty in defining virtue in the universe of differing cultures, because of varying interpretations of what is virtuous (Yaman, 2003). But, like the medical equivalent of 'Do no harm', his work stands as a first principle from which to venture into other ethical realms, and so provides a clear link to sustainable and resource management tools, such as the precautionary principle (as noted in Chapter 8). One of the most recent and most comprehensive treatises on tourism and ethics, which is decidedly sociological in its outlook, is a book by Smith and Duffy (2003). These authors quite effectively summarise a number of the most important ethical theories (e.g. utilitarianism, ethics of care) and how these apply in a tourism context. Their work stands as a tangible representation of the need for a more comprehensive look at how ethics can aid in addressing the innumerable conflicts that exist in tourism.

If we can gauge demand on the basis of conferences (or perhaps it is conferences that generate demand), then a further recognition of the importance of ethics to tourism came about from the first internet conference on tourism ethics in 1998, sponsored by the *International Journal of Contemporary Hospitality Management*, and facilitated by MCB University Press. One of the chief aims of the conference was to explore many of the key ethical issues faced by those who promote and market tourism. The conference included papers on a draft code of ethics, ethical tourism, tourists with a social conscious, the ethics of destination promotion, the ethical challenges of managing pilgrimages to the Holy Land, the sustainability of Indian religious traditions, the impacts of tourism in Goa, ethical dimensions of rural tourism in Estonia and factors of intrusion in homes and castles (MCB University Press, 1998). While few of these papers examined the theoretical aspects of tourism, many represent the 'ethics as impact' perspective that continues to prevail in the literature (see Lea, 1993 for a discussion of two other ethical perspectives in tourism, including Third World and traveller ethics).

Encouraging, however, is the fact that recent research has taken to examining ethics in the context of tourism types. Examples include Third World tourism, as above (Ahmed *et al.*, 1994; Lea, 1993; see also Hudson, 2007); social tourism (Higgins-Desbiolles, 2006; Minnaert *et al.*, 2006; see also Hall & Brown's 2006 detailed work on welfare tourism); poverty or slum tourism (Dyson, 2012; Frenzel *et al.*, 2012; Selinger, 2009); responsible tourism (Grimwood, 2013; Weeden, 2011); ecotourism (Karwacki & Boyd, 1995); sex tourism (Eades, 2009; Wonders & Michalowski, 2001); medical tourism (Connell, 2006; Meghani, 2011); and backpacking (Speed, 2008). In this latter case, Speed found that although backpackers have been characterised as good tourists, survey research shows that this group does not necessarily demonstrate ethical behaviour as the behaviour corresponds to current ethical models. Fair trade tourism and pro-poor tourism are other examples of ethics and tourism types, which will be given further weight, below.

Overall, however, we can conclude, as noted previously, that tourism research is not driven by an ethics agenda. This is unfortunate because many other disciplines, such as business, the environment, medicine, law, sport and marketing, are active in this area. The reasons for the dearth are open to debate. Perhaps these other disciplines have longer and more rigorous histories, allowing them to arrive at ethics as a natural progression in their research agendas. Perhaps still the ethical transgressions in tourism do not have the same social implications as medicine or business. It may also be that tourism research has not embraced knowledge from other fields to the extent that it might. Tourism by nature is said to be interdisciplinary (if we take, for example, the mission of the journal *Annals of Tourism Research*). Whether it is truly interdisciplinary is subject to debate. A strong case can be made for too much insularity in tourism research (e.g. our focus on impacts) as opposed to a receptiveness that would welcome the theory on ethics from other magisteria in attempts to strengthen our field. Tourism is not so far removed from these disciplines to not warrant a commitment to ethics. But, at the same time, if we attempt to construct a world view on tourism and ethics at all, we often do so from the perspective of environmental ethics alone. This cannot prove to be a fruitful avenue because it ignores who we are as a species, as we shall see in coming chapters. Tourism ethics can only be operant through the acceptance and integration of knowledge from other more established disciplines, such as biology, anthropology, psychology and business. This leads to the belief that, although we are left with impacts and sustainable development as the most dominant conceptual bases from which to find meaning in tourism, we have steadfastly chosen to ignore ethics, which, arguably, is a vacant niche that we can no longer afford to ignore.

Tourism's new moralities and realities

Writing on the value of responsibility in tourism, Goodwin (2003) notes that the Ethical Purchasing Index, which measures the growth in the ethical market place, moved from 100 in 1999 to 115 in 2000. In some sectors, he notes, ethical purchasing by consumers rose 18.2% between 1999 and 2000, compared with a total market growth of 2.8%, showing growth six times faster than the overall market. (Overall, however, the ethical market represents only 1.6% of the total market.) In his review of the work of Tearfund (2002) on ethical travel, Goodwin reports that 52% of tourists would book a trip with a company if they had a written code of ethics to guarantee good working conditions, protect the environment and support local charities. This was up 7% from their 1999 study, which asked the same question. Another Tearfund study reported that 27% of UK tourists felt that a service provider's ethical policies were important to them in choosing who to travel with (Tearfund, 2000a, as reported in Weeden, 2001). The conclusion that Tearfund has come to and the priorities and recommendations they are amassing to effect change in the tourism industry (Tearfund, 2001) are based on the realisation that tourism is an ethical beast. In this they observe that:

> Tourism is not just an economic transaction or a series of activities which can be isolated from everyday life or from their impact on people. The very fact that we travel to another culture and come into direct contact with the people there raises a number of ethical issues. Do local people want tourists visiting them? What are the working conditions in the tourism industry? What change does tourism make to family relationships and values? Where does the money go – who benefits? What are the environmental consequences of travel? Does travel to a particular place support democracy and human rights, or undermine them? (Tearfund, 2000b: 5)

To many, however, ethics is uncomfortable and disturbing as a focus of study because it invades our behavioural tendencies, because people are uneasy with being told what to do. Because of this there are few who are sensitive to the moral or ethical environment, namely the climate of ideas about how to live a good life (Blackburn, 2001). But to be fair, ethics too can be wrong in its support of ideologies and utopias that have more to do with the agendas of a few at the expense of the many. But, like it or not, we are ethical animals. Although we often fail to behave properly, we have a propensity to tell each other what to do, and to grade, evaluate and compare (Blackburn, 2001).

Added to this is the notion that an ethical tourism industry is constrained by an unwillingness to discuss morality without being accused of moralising (Denhardt, 1991; see also the previous discussion on the work of Wheeller, 1994 and Butcher, 2003) or, more bluntly, of being a moral crusader. Different stakeholder groups have vastly different ideas of the place of ethics in tourism. Perhaps the academics are perceived as being too idealistic and moralistic, where industry has the perception that an ethical approach will be bad for business. The self-righteousness of the moral crusader cuts into the notion that we cannot always be angelic in our actions: there is a side to human behaviour that allows us to be more self-interested, especially in the realm of business or perhaps after having spent thousands of dollars on a yearly vacation. But does the moral high ground offer us only one way to act? Does it promote moral indoctrination? Does it mean that those who champion ethics seek only authority for themselves? Obviously we must be careful not to impose our own moral principles on others. It is an objective approach to morality that we are interested in, and those who go down the moral path must be clear on this.

In contrast to the promising conclusions of Tearfund (2002), above, Travel Wire News (2004) reports on a study conducted by Norwich Union on the travel habits of British holidaymakers. The study found that about one-third of the sample travel abroad without knowing anything about the country they are visiting. Furthermore, 'experiencing another culture' was reported to be the least important consideration when booking a holiday.

A further 29% travel without knowing anything about the local customs of the destination. The study also found that 25% of travellers spend less than two hours selecting their travel destination.

Where To Start?

In preparing to write this book I was convinced that, in order to better understand ethics as an essential component of the human condition, I was compelled to look deeper than tourism, and deeper still than the conventional material on ethics. My reading led me to research on the biological and cultural basis of human nature, providing a holistic base from which to understand human nature.

But the main question remained: how to 'sell' this material to an audience that is either unaware of this subject matter or unappreciative of its fit in a field that is arguably unidirectional. That is, it was important for me to look beyond my own world view on tourism. Thankfully, this came from the work of Przeclawski, who suggested that in our attempts to understand tourists and tourism as a phenomenon we must first recognise that tourism is a form of human behaviour. More precisely, he wrote, as illustrated in the quote that leads into this chapter, that: 'Tourism cannot be explained unless we understand man, the human being' (Przeclawski, 1996: 236; see also Wheeller, 1994). We have not attempted to do this in tourism studies, and so I include aspects of human nature as an essential point of departure for our discussion in this book. I do this in order to provide a firm foundation of why ethics is such an important aspect of our human nature. Its importance has been underlined by Kagan, who wrote that 'more philosophical works have been written on morality than on any other human quality because it is a unique and distinctive characteristic of our species' (Kagan, 1998: 7).

The need to travel also seems to be an important part of who we are: to escape and to experience new places. For example, even before 100 years had passed after the death of Jesus Christ, travellers were visiting Jerusalem for penance, for thanksgiving, or simply to walk the streets that Christ had (Boorstin, 1985). There were over 200 monasteries and hospices near Jerusalem by the early 5th century, and the traveller, on his or her pilgrimage, was aided by travel guides and lodgings all along the way. Perhaps the most noteworthy of these pilgrims, Boorstin notes, was the Muslim Ibn Battuta (1304–1374), who at the age of 21 travelled over 75,000 miles in visiting every Muslim country, including four pilgrimages to Mecca. So, if we wish to understand why people went (and still go) on pilgrimages, or to Thailand as sex tourists, or why ecotourism is the only type of travel that some people will take part in, then we have to know something about human nature: the basic drives, intuitions and processes that affect behaviour. In doing so, it is hoped that the theories and approaches cited act as a springboard for further development of tourism studies. But a note of caution. In embracing ethics we must be ready to grasp the oft-quoted contention that philosophy is eternally unsettled and only occasionally stirred by new facts; that philosophy is often unlike the sciences in that 'disagreement is its essence; settled opinions are its stagnation; and in philosophy, newer is not always better' (Garofalo & Geuras, 1999: 21).

In concluding this opening chapter I quote Ayn Rand who, ironically, given her political persuasion, wrote:

> Yes, this *is* an age of moral crisis. . . . Your moral code has reached its climax, the blind alley at the end of its course. And if you wish to go on living, what you now need is not to *return* to morality . . . but to *discover* it. (Rand, 1957: 13)

In tourism we have but one option, if the options are return or discovery. I would suggest the latter, since a return would mean that there has been some prior concerted effort to examine tourism and ethics in the first place. In moving the agenda forward we need to take the first steps in assembling a base of knowledge on this relevant and timely discipline of study.

2 Human Nature

Whether there is despair or love or friendship – or indeed benevolence, ambition, tolerance, greed, sincerity, curiosity, a belief in truth or in new clothes – what is helpful is to be able to turn that door-handle with some sense as to what inner forces motivate us. To believe that forces exist and to be able to assume them – that is the beginning.

Saul, 2001: 2

Introduction

Although this is foremost a book about tourism and ethics, especially in the context of environment and business, it is also a book about human nature. In understanding tourists or tourism in general, we must first understand what it is to be human, as noted by Przeclawski in Chapter 1, and, in particular, what stimulates us towards behaviour deemed good and bad. Increasingly we hear of police officers and judges who are corrupt, taxpayers who are dishonest, leaders of corporations who swindle and government officials on the take. These pillars of society, the people who we traditionally place the greatest amount of faith and trust in, have been shown to let us down, time and time again. Human behaviour is a complicated matter. It has challenged scholars for millennia in much the same way as it challenges us today.

The debate over human nature has traditionally boiled down to a discussion on nature versus nurture, or the level of genetic (biology) or environmental (cultural) factors that influence the decisions we make. And because there are two schools of thought it follows that there are two errors in attempts to understand the human condition (Wilson, 1993). The first is to assume that culture is everything, and the second is to assume it is nothing. Both arms are essential in our attempts to understand ethics in the context of tourism. In doing so, this chapter will examine culture, sociobiology, cooperation, commitment, new sciences and the evolution of ethics, in gaining deeper insight into who we are.

Culture

In the earliest stages of his *The Ascent of Man*, Bronowski (1981) writes that, while animals evolved a set of adaptations that allow them to occupy rather specific niches in the natural world, humanity has not limited itself to any such role. In fact it is quite the reverse. Humans are free, through technology, innovation, imagination, reason and so on, to change the world in innumerable ways. The reshaping of the natural world through human ingenuity is referred to not as a biological evolution, but rather a cultural one. The focus on culture as a driver for change within the human species has been a consistent theme over time. We are constantly in a state of cultural change. This is emphasised in Bronowski's example of Laplanders who are contrasted with the reindeer they so heavily rely on for food, clothing and comfort:

> And yet the Lapps are freer than the reindeer, because their mode of life is a cultural adaptation and not a biological one. The adaptation that the Lapps have made, the transhumance life on the move in a landscape of ice, is a choice that they can change; it is not irreversible, as biological mutations are. For a biological adaptation is an inborn form of behaviour; but a culture is a learned form of behaviour – a communally preferred form, which (like other inventions) has been adopted by a whole society. (Bronowski, 1981: 48)

Culture can be defined as shared sets of symbols and their definitions (Hagedorn, 1981), and it includes technological and social innovations that have accumulated over time and that have helped individuals, collectively, to live, much as the Lapps have. These innovations are often specific to a group and have both psychological and physiological utility. Culture is also the predominant means by which to widen the human imagination – in Bronowski's terms – to move our minds through space and time and to see ourselves not only in the past but to think beyond our current situation in visualising the future. This cultural phenomenon, which has taken place over just a few thousand years, contains almost the whole ascent of humankind at a rate of over 100 times faster than biological evolution.

Our understanding of how or why change has come about has been the subject of great debate, particularly among anthropologists. The early pioneer anthropologist, Edward Tylor, felt that, although societies differ in many ways, they all descend from a common evolutionary basis. His student, Franz Boas, however, took the opposite stance in suggesting that all ethnic groups are born with the same basic cognitive abilities, regardless of environment and upbringing (Fox, 1989). Today's science recognises this fact, but in the years following Boaz's tenure, his students began to interpret and recast this critical finding in a different light. Cultures were thought to be an outgrowth of experience and the sum total of the ideas of the unit, rather than the individual. Individual enterprise was thus suppressed in favour of the collective nature of the group. This group-centred or superorganic view of humanity fails to recognise the importance of the individual in shaping social processes within groups and society (Kroeber, as cited in Degler, 1991). The group concept was emphasised by other notable social scientists in the early part of the 20th century, including Margaret Mead, who wrote that 'we are forced to conclude that human nature is almost unbelievably malleable, responding accurately and contrastingly to contrasting cultural conditions' (Mead, 1935/1963: 280). The sense that the history of humanity, its culture and its civilisations was somehow intricately tied to nature was very quickly and quietly erased by anthropologists, setting the stage for the time-tested debate on nature versus nurture. The extent of this debate has been summarised by Pinker in three areas: the 'blank slate', the 'noble savage' and the 'ghost in the machine' (Pinker, 2002).

Slates, savages and ghosts

'Blank slate' is a term used to refer to the notion that humans are born with a *tabula rasa*, or a scraped tablet. This perspective was championed by many scholars, including John Locke, David Hume, John Stuart Mill and John Watson. Learning was thought to be solely the function of experiences that individuals gathered throughout their lives. As a doctrine grounded in experience, it has opened the door for a tremendous variety of ethical and political ideologies, including differentiation between human beings on the basis of a number of characteristics. Science could now explain why some races were smarter than others, why one gender was better than another, and why some children excelled while others did not. It was all based on the experiences that these individuals had throughout their lives. As such, it led to the belief that if you changed the experience through media, education or rewards, you could change the person (Pinker, 2002). But it also meant that all had equal opportunity for success, which made the role of the teacher critical. The founding father of behaviourism, John Watson, wrote:

> Give me a dozen healthy infants, well-formed, and my own specified world to bring them up in and I'll guarantee to take any one of them at random and train him to become any type of specialist I might select – doctor, lawyer, artist, merchant-chief, and yes, even beggar-man and thief, regardless of his talents, penchants, tendencies, abilities, vocations and race of his ancestors. (Watson, 1924/1998, as cited in Ridley, 2003: 185)

Watson was convinced that people were conditioned by their environment and what ensued was a tumultuous period of behaviouralism championed further into the 1900s by one of Watson's disciples, B.F. Skinner, who felt that people were essentially devoid of all instinct, i.e. all the tools that people needed to live their lives were generated by what they learned in the environment. But the main lesson we have learned, in part from psychology, is that we often refuse to look outside our own often restricted world view. Although specialisation serves us well in giving us credibility as a distinct science, it often does little to blow open the doors of insularity. And when we are wrong, a whole house of cards can come down. The 'blank slate' held intellectual court for many years.

Columbus found, much to his dismay and disappointment, that the people that he encountered on his travels were not horribly deformed and grotesque, but rather well built and handsome (Boorstin, 1985). These were not monsters as originally thought, but the sensationalisation of Columbus's accounts laid the groundwork for the new science of anthropology – a science that sought to show how such groups progressed from a state of savagery through to more advanced and civilised conditions. Even more instrumental in painting the 'noble savage' in romantic colours was Jean Jacques Rousseau (1712–1778) in the second of his two essays published under the title *Discourse on the Origin and Foundations of Inequality Among Men.* Rousseau felt that natural man, the 'noble savage', was more virtuous than Western man because his state within nature was simplistic and beautiful (1755/1964). The savage was thus more peaceful and selfless than his Western counterparts, because traditional societies were devoid of greed and violence, which were seen as by-products of the civilised world. Rousseau's main thesis was based on the notion that 'Everything that comes from nature will be true' (Rousseau, 1755/1964: 104). This premise naturally spilled over into humanity. Those who were immersed in nature were naturally more virtuous.

This perspective stood in stark contrast to that proposed by Thomas Hobbes (1588–1679), who wrote '*Homo homini lupus*' ('Man is a wolf to man'). His most famous work, *Leviathan, or the Matter, Form, and Power of a Commonwealth, Ecclesiastical and Civil*, published in 1651, was a commentary on political theory and supported the belief that, because people were inherently self-interested, humanity is inevitably destined for war and hostility. (This was tacitly or perhaps remotely supported by scholars like Kenneth Clark, who, in his *Civilisation*, observed that all great civilisations in their early stages are based on success in war (Clark, 1969).) In fact, Rousseau's work can be seen as the antithesis of Hobbes's, and he makes regular contrast to the Hobbesian perspective in his *Second Discourse*: 'There is, besides, another principle which Hobbes did not notice, and which . . . tempers the ardor he has for his own well-being by an innate repugnance to see his fellow man suffer' (Rousseau, 1755/1964: 130).

The concept of the noble savage had a profound effect on intellectuals in the 19th and 20th centuries and was further compounded by explorers and intellectuals of the era. But more recently this view of human nature has been refuted by science. The noble savages of Tahiti, for example, were in fact not virtuous in the least, and practised all manner of sacrifice, incest and infanticide.

The doctrine of the 'ghost in the machine' was attributed to the scientist René Descartes (1596–1650), who wrote of a mind–body dualism in humans, thus rejecting the popular notion of the time, one reinforced by Hobbes, that the workings of the mind could only be explained in mechanical terms. Descartes suggested instead that behaviour was freely chosen and not subject to the laws of the universe. He noted that, although we can doubt the existence of our bodies as we imagine ourselves and our soul from an incarnate perspective, we cannot do so of our minds as the very fact that we can think presupposes that our minds exist. The dualism constructed from this mindset put in motion questions regarding the legitimacy of a separation between mind and body:

> There is a doctrine about the nature and place of minds which is so prevalent among theorists and even among laymen that it deserves to be described as the official theory. . . . The official doctrine, which hails chiefly from Descartes, is something like this. With the doubtful exception of idiots and infants in arms every human being has both a body and a mind. Some would prefer to say that every human being is both a body and a mind. His body and his mind are ordinarily harnessed together, but after the death of the body his mind may continue to exist and function. Human bodies are in space and are subject to mechanical laws which govern all other bodies in space. . . . But minds are not in space, nor are their operations subject to mechanical laws. . . . . . . Such in outline is the official theory. I shall often speak of it, with deliberate abusiveness, as 'the dogma of the Ghost in the Machine'. (Ryle, 1949, as cited in Pinker, 2002: 9)

It follows that, if people are machines, how is it that we are able to be responsible, to love and to feel compassion? These are emotions that surely cannot be controlled by a set of levers or gears activated in order to elicit certain outcomes, but yet it is still normal to suggest, rather to hope, that, even though the body has long decomposed, the soul continues to exist in some unaltered state. Dualism was the theoretical construct that said that the universe is made up of both physical things, such as bodies and brains, and mental stuff, which includes subjective feelings. Both of these exist semi-independently of one another. This doctrine has given way to monism, which asserts that there is just one entity that is responsible for our feelings and the things, such as brains, that generate these feelings (Humphrey, 1992).

The inherent weaknesses of these three doctrines illustrate that cultural evolution has not been well understood. With no firm answers the door has stood open for other interpretations of human nature, principally by biologists, but also by anthropologists themselves. For example, Brown (1991) observes that his field far too frequently elects to dwell on the differences instead of the universals that exist between the multiplicity of cultures in existence. He suggests that: (1) universals not only exist but are important to any broad conception of the task of anthropology; (2) universals form a heterogeneous set; (3) the study of universals has been tabooed as an unintended consequence of assumptions that have predominated in anthropology (and other social sciences) throughout the 20th century; (4) human biology is a key to understanding many human universals; and (5) evolutionary psychology is a key to understanding many of the universals that are of greatest interest to anthropology. Brown sums up the agency of human behaviour through evolutionary theory:

> If we assume that society and culture are products of human action, or that society and culture (including language) are evolved characteristics of humans, and that humans themselves are products of organic evolution, then evolutionary theory offers the only explanatory framework for universals that is potentially all-inclusive. (Brown, 1991: 99)

These thoughts were echoed by Fox, also an anthropologist, who noted:

> And the 'message' (lord help us) is the same: the unity of anthropology, its uniqueness among the social sciences, lies in its devotion to the study of the human species as a product of evolution. . . . All I'm saying is that in the end, the master paradigm that holds all this effort together must be the theory of natural selection and must connect us to our evolutionary past via the theory of the evolution of social behavior, or what we are doing will not be anthropology but some branch of some other discipline masquerading as such. (Fox, 1994: xi–xii)

Biological Antagonism

Thomas Malthus (1766–1834) wrote in his *Essay on Population* that population multiplies faster than food. Shortages of key resources, such as land and minerals, were prime causes of economic downturns, poverty and malnutrition (see Furedi, 1997). Such thinking in the hands of Darwin meant that animals must also be subject to competition in order to survive, with the most successful of these best able to adapt to the conditions of their surroundings. Variations that were favourable (the concept of natural selection) were preserved, leading to the creation of new species, while those less favourable were not. The linkages to free enterprise are savoury, with the inference that natural laws slip with ease into the spirit of commerce. As Fox (1997) asserts, Darwinism is simply the transposition of classical economics to the natural world, Malthus being the ancestor of both.

The concept of natural selection, the fundamental premise driving the theory of evolution, is a process that takes several generations. Those individuals who are better adapted – more fit – outbreed others who are less well adapted, allowing a population to undergo slight changes in their appearance and habits that render them more successful. Adaptations are also important in the event of environmental change, allowing the species to maintain a stable relationship with its surroundings. Adaptation is also gauged as a function of reproductive success. Most important to our discussion here is that natural selection also favours the individual organism that leaves the most surviving offspring. In his summary of the basis of Darwin's theory, Trivers (1985) illustrates that natural selection: (1) refers to individuals, not groups within populations; (2) favours individuals that maximise the number of their surviving offspring; (3) selects individuals to maximise the number of surviving offspring they produce; and (4) emphasises the production of surviving offspring, such that reproduction is a process that is biologically meaningful for the passing on of the individual's genes for the future.

Although the foregoing are widely accepted principles for the biological world, the perspective of a gene-centred view of human agency has been at the centre of debate for some time. This has led biologists such as Ehrlich (2000) to argue that genes do not shout commands at us in determining behaviour, but they do whisper suggestions. That is, genes do not absolutely determine one's destiny, but they do define a range of possibilities for the individual. But, taken too far, the 'biology as all' focus takes us down a slippery slope. The notion that we are slaves to our genes – the biological determinism of eugenics – drove us to the point where we felt compelled to manipulate genes in making a superior class of people. Other doctrines of science, such as Lamarkism, were founded on the belief that life experiences or acquired characteristics were instrumental in the evolutionary process. This means that, using the example of physical fitness, we cannot build up our muscles through exercise so that our offspring will be just as fit.

Self-interest, altruism and inclusive fitness

Perhaps one of the most enduring themes over time in the efforts to understand human nature has been the struggle to determine whether we are inherently self-interested as a species or cooperative. But no matter which side of the debate one is on, there is little argument that the primacy of the tradition of self-interest in philosophy and science makes it difficult to ignore as a shaper of human nature (Griffin, 1997). Science has backed this up. Even in the sanctimonious confines of the womb have we found selfishness. Researchers have shown that, while traditionally the relationship between mother and foetus has been viewed as a cooperative one, there is conflict between both in the form of a genetic tug-of-war (Haig, 1993). Foetal genes have been found to invade the maternal endometrium and remodel the arteries into low-resistance vessels. The foetus (which contains only half of the mother's genes) can overtax the blood sugar equation with at times quite devastating effects on the mother. In the struggle for blood sugar, the mother, although selfless in her efforts to nurture the child, also must ensure her own survival and ability to reproduce further. The will to survive through self-interest appears to be ingrained at a very early age, at least physiologically.

During the 1960s, evolutionary biology was revolutionised by the work of W.D. Hamilton (1964; see also Maynard Smith, 1964), whose theory of kin selection (otherwise known as inclusive fitness theory) explains that altruism takes place among those who are related genetically – altruism as a function of natural selection. Hamilton observed that: 'Selection operates when carriers of some genes out-reproduce carriers of other genes. If altruistic behaviour were inherited, it would be more likely to spread if the altruism were directed at close relatives, because relatives share genes' (Hamilton, 1964: 38). (Altruism has been defined by Kitcher (1993), as behaviour that promotes the fitness of another organism at costs in fitness to the agent.) Hamilton realised that, in the social animals, such as bees and wasps, it was not the queen who manipulated workers through chemical means, but rather it was the workers who were more closely related to the larvae than the queen herself. Workers are thus seen to nurture the larvae, and to manipulate them to increase their productivity in propagating replicas of the workers' genes. By helping their sisters breed, bees and wasps are leaving more copies of their genes in the next generation than by trying to breed themselves. Simple observation of a bee colony shows what appears to be unbridled altruistic behaviour on the part of the worker. Closer observation, however, shows each worker – who is sterile – striving for genetic eternity through the queen's offspring, its only hope for immortality, rather than through its own. As Ridley (1998) notes, it does so with just as much gene selfishness as the person elbowing aside his or her rivals up the corporate ladder. So, donating organs to a brother or sister to aid in their survival, as opposed to donating to someone who was not a family member, increases the representation of the donor's genes in the next generation. This relates to a stronger form of kin selection (helping one's brother) than a group selection process (helping someone from outside the family).

Following from Hamilton, G.C. Williams (1966) argued that animals and plants are designed not to do things for themselves or their species, but rather for their genes, since it is the gene that is responsible for passing along the characteristics of that individual. Indeed, the truly individualistic and egocentric person wishes to reproduce so that his or her genetic self is transmitted to an offspring, given that he or she has a rather fixed lifespan. Genes, Williams wrote, include any portion of chromosomal material that lasts for enough generations to serve as a unit of natural selection. Genes therefore are replicators that are characterised by their longevity as demonstrated by their appearance in many successive generations. As such, individuals will behave with the purpose of enhancing the possibility of their genes surviving and replicating. This means that, if one animal is seen doing something to benefit another, the observer can conclude that it is being manipulated by the other or being subtly selfish (Buss, 1987; Paradis & Williams, 1989; Wilson & Sober, 1994). The popular writer, Bill Bryson, nicely encapsulates the role of genes to the lay person in suggesting that they can be likened to the

> keys of a piano, each playing a single note and nothing else, which is obviously a trifle monotonous. But combine the genes, as you would combine piano keys, and you can create cords and melodies of infinite variety. Put all these genes together, and you have (to continue the metaphor) the great symphony of existence known as the human genome. (Bryson, 2003: 408)

Genes are not devoted to you, so the argument goes, but rather to themselves, where the desire to reproduce (i.e. spread one's genes) appears to be the most powerful force in nature, to the point where sex, according to many evolutionary biologists, is simply a reward that encourages us to pass on our genes. Perhaps the most widely recognised proponent of the genetic view of self-interest is Richard Dawkins, who wrote, in *The Selfish Gene*, originally published in 1976:

> The argument of this book is that we, and all other animals, are machines created by our genes. . . . I shall argue that a predominant quality to be expected in a successful gene is ruthless selfishness. This gene selfishness will usually give rise to selfishness in individual behaviour. However, as we shall see, there are special circumstances in which a gene can achieve its own selfish goals best by fostering a limited form of altruism at the level of individual animals. (Dawkins, 1999: 2)

Dawkins' metaphor is an interesting way to demonstrate the laws of natural selection, at the level of the gene. There really is no such thing as a selfish gene in the sense that the gene itself possesses its own motives for its own survival. It is simply, in a very mechanical sense, doing what it has evolutionarily been programmed to do: make copies of itself for its own longevity. These genes, far from being selfish and individualistic bandits, 'act' in the most altruistic manner possible by making us empathetic, caring and unselfish in our regard for other beings.

Although we appear to be 'hard-wired' by our genes, Dawkins, in his later edition, was careful to suggest that we can learn to be altruistic. We are not necessarily compelled to obey our genes throughout our lives. Culture, the great agitator against the selfish gene concept, is noted by Dawkins as an important aspect in our development over millennia. But even so, the amount of evidence supporting a gene-centred view of behaviour is ominous. For example, if a parent was asked by an outsider how much love he had for his biological child and his stepchild, the answer might surely be equal love. However, researchers have shown that this is not necessarily true (Daly & Wilson, 1999). Step-parents are far more likely to abuse their stepchildren than their own biological brood. The reason for this is that stepchildren do not carry the genes of parents, and therefore an over-expenditure in their upbringing of the stepchild, at the expense of their own children, does not make sense from a natural selection standpoint: selection typically favours one's own.

Even though education and learning about altruism can take place, Williams and Hamilton have shown us that selfish genes sometimes use selfless individuals to achieve their goals. Just because bees act according to selfish directives does not take away from the fact that they are acting in an altruistic manner by furthering the security and growth of the hive.

Reciprocal altruism

One of the other main theories that emerged out of the work in the 1960s on gene theory was proposed by Robert Trivers (1971), who noted that, although animals and people have shown to be self-interested, they periodically and/or regularly demonstrate a willingness to cooperate with others they are not related to.

He felt that cooperation might occur if there was a chance that the beneficiary might reciprocate in the future. However, far from being an altruistic gesture (altruism also defined as 'behavior that benefits the another organism, not closely related, while being apparently detrimental to the organism performing the behavior, benefit and detriment being defined in terms of contribution to inclusive fitness' (Trivers, 1971: 35)), Trivers argued that both individuals would be doing each other a favour, as long as the benefit of receiving the favour was not outweighed by the cost of returning it. We find these relationships among the primates, where grooming of neighbours is a common practice. One of the most quoted examples of reciprocity is in the work of Wilkinson on reciprocal food-sharing (the regurgitation of blood to others at roosting times) in the vampire bat, based on his observations of this species in Costa Rica. He reported that blood-sharing depends both equally and independently on the degree of relatedness (kinship) as well as on an index of reciprocation (with non-kin). According to Wilkinson, reciprocity persists in this group under three conditions:

> (1) enough repeated pairwise interactions must occur to permit role exchanges and ensure that a net benefit accrues to all donors; (2) the benefit of receiving aid must exceed, on average, the cost of donating; and (3) donors must be able to recognise and not feed previous recipients that fail to reciprocate. (Wilkinson, 1984: 182)

Regarding the third condition, bats roost in the same place, live for up to 18 years, and get to know one another quite well, and thus are able to play the game repeatedly. As such, bats who donate get something back later on if they have had an unsuccessful hunt; bats that refused to donate in the past will not benefit from a gift of blood by another. This means that bats are well aware of each other and are quite adept at knowing the score. The basis of the cooperation in Trivers' and Wilkinson's work was founded on the notion that altruism for its own sake was not the prime motivation of animals, but rather was to secure selfishly desired favours from another member of the troop. The more frequently members engaged in these acts of reciprocity, the greater the opportunity for cooperation (see Sober, 1992).

Trivers, however, was also interested in gaining a better understanding of human social behaviour. He wanted to know why we act the way we do at the most fundamental level, and how this could be linked to pre-existing knowledge. In his book *Social Evolution*, Trivers (1985) examined the literature in psychology, economics, history and political science, but none of these had a secure foundation in pre-existing knowledge. For example, he wanted to know the scientific justification for why humans are self-interested and seek to maximise economic gains. Psychology, he notes, could not provide these explanations as it was grounded in the theory of learning, as we saw earlier in this chapter. Pinker corroborates what Trivers found in the 1970s, in observing that social psychology is often a mishmash of phenomena that are oftentimes unexplained by theory. Missing, he notes, is the deductive foundation of other disciplines, where 'a few deep principles can generate a wealth of subtle predictions – the kind of theory that scientists praise as "beautiful" or "elegant"', (Pinker, 2002: 241). Trivers, according to Pinker, developed the first theory applicable to social psychology that could be called elegant. By following the genes, one could understand the psychological system that regulates his or her altruism. This system is characterised by the following examples of reciprocal altruism (Trivers, 1971):

A complex, regulating system. Given the unstable nature of human altruism, natural selection will favour a complex psychological system where people regulate their own altruism and cheating tendencies as well as the responses to these tendencies in others.

Friendship and the emotions of liking and disliking. Tendencies to like others and to form friendships will be selected as the immediate emotional rewards motivating altruistic behaviour.

Moralistic aggression. Injustice, unfairness and lack of reciprocity motivate human aggression and indignation.

Gratitude, sympathy and the cost/benefit ratio of an altruistic act. Humans are selected to be sensitive to the costs and benefits of altruistic acts. The greater the need state the greater the tendency for an individual to reciprocate.

Guilt and reparative altruism. Cheaters (defined as those who fail to reciprocate) pay for their transgressions by being cut off from all future acts of aid. The cheater will be selected to make up for his misdeed and to show that he will avoid cheating in the future.

Subtle cheating: the evolution of mimics. Once friendship, moralistic aggression, guilt, sympathy and gratitude have evolved to regulate the system, selection will favour mimicking these traits in order to influence the behaviour of others to one's advantage.

Detection of the subtle cheater: trustworthiness, trust and suspicion. Selection will favour the ability to detect subtle cheaters as well as the ability to distrust those who perform altruistic acts without the emotional basis of generosity (i.e. those who are insincere).

Setting up altruistic partnerships. Selection will favour a mechanism for establishing reciprocal relationships. This might include performing altruistic acts towards strangers in inducing friendship.

Multiparty interactions. Selection may favour the formation of norms of reciprocal conduct. This would stem from learning how to deal with cheaters, and developing rules of exchange.

Developmental plasticity. Selection would favour the ability for reciprocal altruism to grow, learn and be adaptive under a number of different circumstances.

Reciprocal altruism, therefore, provides the needed basis by which to explain how and why morality evolved in humans. Sympathy and trust prompt people to do favours for each other, while gratitude and loyalty compel them to repay such favours. On the other hand, anger and contempt motivate the individual to punish the cheater. Reputations are built on these transactions and this can be communicated among the group, leading to partnerships and friendships that can be bound over time. This is all possible because of the neural network that has been established in the brain, through natural selection, that provides the capacity for conscience (Pinker, 2002) – a topic I will deal with shortly.

Reciprocal altruism is thus a form of symbiosis in which each organism helps itself by helping others (Trivers, 1971). This symbiosis has a lag time however, because altruists must often wait for another to return a favour. The return may come directly, as in the case where an organism returns the favour of food-sharing, or it may be indirect, such as by sounding out a warning call in the event of the appearance of predators. Trivers also explained that reciprocal altruism can evolve as people have the ability to ostracise those who fail to help them, especially after they have been helped by the altruist. Reciprocators, especially when they cooperate, are able to reap the benefits of trade and therefore out-compete, over the long run, those who would otherwise choose to cheat.

The theory of reciprocal altruism has many applications to social science, as noted by Pinker above. Although the research surrounding biological needs provided the scientific basis from which to understand purposive behaviour, psychologists recognise that there are other needs that are either biologically or psychologically linked. These include the needs for activity, stimulation, novelty, competence, power, self-determination, security, social approval and predictability (Wallach & Wallach, 1983). In other research, Batson and his colleagues have found that self-interested egoism is not the only threat to the common good (Batson *et al.*, 1999). In experiments on the allocation of resources (raffle tickets) between respondents (to the group, other individuals or just themselves), when decisions were private and public, the authors discovered that empathy-induced altruism (actions which benefit another person for whom empathy is felt) can also be a threat to the collective. In situations where an individual's welfare and what is best for the group are found to coincide, resources can be spread equitably. However, in situations where altruism is expressed for another on the basis of empathy, the decision to act for that person only can be a threat to the common good. The authors note that this other form of altruism happens frequently in society, including when politicians give posts to friends over others who are more qualified, and when whalers, loggers and herders extract out of concern for their families. For this reason, the authors feel that the long-standing focus on egoism as a function of self-interest is too narrow to the exclusion of their findings on empathy-induced altruism. The application of this type of altruism is often expressed in game theory (as discussed below).

The implications for tourism are savoury regarding reciprocal altruism, especially where tourism involves a significant array of one-off interactions among agents (Fennell, 2006). Think of this: are you more likely to cooperate with an individual while on vacation, or someone you interact with on a regular basis at home, especially if that someone on vacation is not cooperative (i.e. altruistic towards you). As such, it may not be rational to cooperate as a tourist or tourism service provider in consideration of the limited amount of time to build trust through long-term stable associations (see Kurzban, 2003, for a general description of this perspective). If this is the case, then tourists and service providers have the potential to be viewed less as mechanisms of trust and altruism and more as a means to a profit end based on cost/benefit calculations and cheating. This line of thought has been touched on tangentially through social exchange theory, which may be viewed as the social science alternative to reciprocal altruism. For example, recreation theorists observe that social exchange theory is based on the notion that people enter into relationships (e.g. service provider and participant) for the purpose of securing rewards, which are sustained if rewards are valued and costs do not exceed benefits, and which evolve over time in a positive way (Searle & Brayley, 1993). We may choose to frequent the same restaurant time and time again if we can be sure that we get good food, the price is right and we continue to have positive interactions with the hosts. A fair exchange may also equate to the maximisation of experiences for tourists and the minimisation of ecological disturbance to the local ecology, which would provide a social benefit to the hosts. Social exchange thus has an element of social justice in which individuals seek transactions and relationships in society that are fair. Although this theory serves as a good explanation for business and marketing perspectives (it emerged from exchange theory in economics), it fails to fully explain the reasons why people behave in this manner in the same way that the theory of reciprocal altruism does, that is, from biological and social standpoints. It would thus prove useful to perhaps more fully investigate social exchange theory using the lessons learned from reciprocal altruism, as noted above.

Game theory

Another theory in evolutionary biology that substantiated the work of Hamilton (1964) and Williams (1966) was game theory, which detailed how cooperation occurs between non-relatives. Game theory came about through the work of two distinguished scholars. John Von Neumann developed the idea in the 1940s that real-life games are nothing like games that we think about today (e.g. chess), where there is a solution or a correct procedure in different situations. Real-life games consist of tactics, deception and bluffing, with an attempt to act according to what others might do, and in consideration of all sorts of other variables, including time, space and circumstance. John Nash (1950), the famous Princeton mathematician, first worked out the theory in the early 1950s, on the premise that a strategy adopted by a player needs to be an optimal one based on the strategies adopted by other players. Important in this is the fact that players do not have or consider incentives to deviate from these optimal strategies (this a concept we shall need to return to in Chapter 8). Game theory is based on how instrumentally rational individuals (i.e. those who act in their own self-interest) make decisions when they are mutually interdependent (the welfare of one player is partially determined by the actions of another) (Romp, 1997).

One of the first adaptations of game theory in a biological context was developed by Lewontin (1961), who argued that evolution of genetic mechanisms can be examined as a game between species and nature. Strategies adopted by a species should be designed to give it the best chance for survival. Maynard Smith (1974) argued that individuals employ evolutionary stable strategies (ESSs) – cost-benefit calculations in simplistic terms – that are based on pre-programmed behavioural directives within the individual. An ESS is the best strategy for an individual animal, for example, depending on what the majority of the population are doing. Perhaps more accurately stated, it is a strategy that, if most members of the population adopt it, cannot be bettered by an alternative strategy (Dawkins, 1999: 69).

In an analogy of aggressors (hawks) and pacifists (doves) of the same species, researchers have shown why animals of the same species do not typically fight to the death in, for example, efforts to secure a mate (Maynard Smith & Price, 1973). In a hypothetical game mirroring the wild, one cannot tell whether a member of the same species is a hawk or dove until a confrontation occurs. Hawks always fight hard and only retreat if seriously harmed. Doves, conversely, only threaten others and never hurt anyone else. If a hawk fights a dove, the dove retreats and does not get hurt. If a hawk fights a hawk they continue until one or either is seriously injured or one dies. If a dove fights a dove, there is a period of prolonged posturing and nobody gets injured. The key to the game was not to understand who could defeat whom (we know that hawks defeat doves in battle), but rather which was the most successful ESS in helping the species evolve in the fittest possible manner. These researchers demonstrated mathematically that the average pay-off to the dove may be higher because of the avoidance of injury, with a lower pay-off to the hawks because of the high rate of injury. Consequently, the more docile behaviour of doves became a more useful strategy, with the inference that its genes would more likely spread throughout the population over time. This research helped to: (1) demonstrate that animals, like humans, are involved in the creation of optimal solutions that benefit the genetic and long-term viability of the population; and (2) formulate a whole new realm, using computers, by which to explore the possibilities of cooperation using the theoretical basis of game theory. This was done chiefly through the Prisoner's Dilemma game.

Prisoner's Dilemma is an example of a non-zero-sum game which is characterised by mixed motives in conjunction with conflicting parties, where there are intrapersonal and interpersonal dilemmas. Here, the rational choice of strategy by both parties leads to an outcome that is worse for both players than if they had selected their strategies irrationally. The non-zero-sum game is an example of a non-cooperative game in which players cannot make agreements to choose strategies in a joint fashion (Rapaport & Chammah, 1965). This is different to a zero-sum game, which is a pure conflict of interest where its conduct departs from rational norms, and where the strategies can be cooperative.

The Prisoner's Dilemma game is derived from the example of two collaborators in a crime who have been caught and independently questioned by authorities. They have two options: to stay silent or to defect (betrayal). If Jones blames Smith, and Smith stays silent, Smith goes to jail and Jones goes free. If each betrays the other, they both go to jail and get a reduced sentence for owning up to it. If Jones and Smith cooperate with each other by saying nothing, they get a lesser offence, with not enough evidence to convict them. This latter option is a reward, as both do not receive a longer sentence. Using a hypothetical example, if both players cooperate, the reward for mutual cooperation is $300. If both defect, the punishment is a $10 fine. If one defects and the other cooperates, the cooperator gets nothing (sucker's payoff) and the defector gets $500

		Player 1: Cooperate	Player 1: Defect
Player 2	Cooperate	Fairly good REWARD (for mutual cooperation) $300	Very bad SUCKER'S PAYOFF $100 fine
	Defect	Very good TEMPTATION (to defect) $500	Fairly bad PUNISHMENT (for mutual defection) $10 fine

Figure 2.1 Payoffs from the Prisoner's Dilemma game
Source: Adapted from Dawkins (1999).

(temptation). In general, mutual defection is the best solution for all, because there is only a $10 fine for each given the other scenarios. But if your partner decides to cooperate, you are still better off defecting (see Figure 2.1). The outcomes for the four cells of the matrix are as follows (Dawkins, 1999):

> Outcome 1. We have both played COOPERATE. The banker pays each of us $300. This respectable sum is called the Reward for mutual cooperation.
> Outcome 2. We have both played DEFECT. The banker fines both of us $10. This is called the Punishment for mutual defection.
> Outcome 3. You have played COOPERATE; I have played DEFECT. The banker pays me $500 (the Temptation to defect) and fines you (the Sucker) $100.
> Outcome 4. You have played DEFECT; I have played COOPERATE. The banker pays you the Temptation payoff of $500 and fines me, the Sucker, $100.

These outcomes are further examined by Dawkins, who offers the following explanation as to the merits of one decision over another:

> I know there are only two cards you can play, COOPERATE and DEFECT. Let's consider them in order. If you played DEFECT (this means we have to look at the right hand column), the best card I could have played would have been DEFECT too. Admittedly I have suffered the penalty for mutual defection, but if I'd cooperated I'd have got the

> Sucker's payoff which is even worse. Now let's turn to the right thing you could have done (look at the left hand column), play the COOPERATE card. Once again DEFECT is the best thing I could have done. If I had cooperated we'd both have got the rather high score of $300. But if I'd have defected I'd have got even more – $500. The conclusion is that, regardless of which card you play, my best move is *Always Defect.* (Dawkins, 1999: 204–205)

As Ridley explains:

> Do not get misled by your morality. The fact that you are both being noble in cooperating is entirely irrelevant to the question. What we are seeking is the logically 'best' action in a moral vacuum, not the 'right' thing to do. And that is to defect. It is rational to be selfish. (Ridley, 1998: 54)

Although the optimal or best fit solution in a Prisoner's Dilemma game is to defect, it does not make sense to do so in games that are repeated or iterated (i.e. when players are faced with the same dilemmas time and time again). Research has shown this over the last couple of decades, and has linked very well with the findings of Maynard Smith and his evolutionary stable strategy. (Remember hawks and doves, and that doves that retaliate against doves prove more successful.) In particular, Axelrod (1984) and Axelrod and Hamilton (1981) developed a number of Prisoner Dilemma tournaments by computer in which a series of programmes were submitted by individuals for the purpose of winning the tournament. Of the 14 programmes submitted initially, many of those that were designed to be ruthless did poorly, and a relatively simple and nice programme or strategy established by Anatol Rapoport, called Tit-for-tat, which was designed to cooperate initially with other programmes and then mirror what another 'player' did next, was successful at winning tournaments again and again. As Axelrod explains:

> What accounts for Tit-for-tat's robust success is its combination of being nice, retaliatory, forgiving and clear. Its niceness prevents it from getting into unnecessary trouble. Its retaliation discourages the other side from persisting whenever defection is tried. Its forgiveness helps restore mutual cooperation. And its clarity makes it intelligible to the other player, thereby eliciting long-term cooperation. (Axelrod, 1984: 123)

One of the real-life examples used by Axelrod to confirm the viability of Tit-for-tat as a basis for cooperation was the First World War. At times, combatants developed unofficial truces over stalemates concerning pieces of territory, which they may have fought over repeatedly for weeks or months on end. Truces were governed by simple revenge if one side defected by firing at the other either voluntarily or through orders by superiors. In such cases, peace was restored through mutual cooperation and respect.

> What makes it possible for cooperation to emerge is the fact that the players might meet again. This possibility means that the choices made today not only determine the outcome of this move, but can also influence the later choices of the players. The future can therefore cast a shadow back upon the present and thereby affect the current strategic situation. (Axelrod, 1984: 12)

Despite the overriding success of Tit-for-tat, it has been criticised because it is based on recognition: recognition of bats at the roost, and recognition of an enemy across the line. Indeed, officials regularly changed the position of their troops, according to Axelrod, in order to disrupt the mutual cooperation built between the two sets of troops. So, time and recognition are important aspects of an emerging relationship, but these are not always achievable in the real world. If everyone cooperates by doing unto others as they would have done to them, there is always the chance that someone will enter into the community who advantages themselves by another set of principles. This throws off the prevailing strategy as the new cheater is advantaged to the disadvantage of all others. It is also criticised because it is vulnerable to mistakes. If one Tit-for-tat player starts to defect, either by mistake or intentionally, and thus take the duo out of a cooperative loop that has existed for a time, it generates a long period of unprofitable misery (Ridley, 1998).

Through a number of incarnations of different programmes, Ridley reports that a more successful Prisoner's Dilemma programme, called Firm-but-fair and developed by Marcus Frean at Cambridge, was based on the notion that players do not have to move simultaneously (e.g. bats do not feed each other at the same time, but rather take turns). Firm-but-fair was found to cooperate with cooperators, return to cooperating after a mutual defection, punish a sucker by further defection, and continue to cooperate after acting as a sucker in a previous round of play. The application of this to our understanding of human and animal cooperation is that, if you are placed in a position to act before another agent, it pays to be nice. That is, cooperation can be achieved through such an approach. Ridley's analogy is that you greet strangers with a smile if you are to avoid having the individual form a negative opinion of you.

The Prisoner's Dilemma is certainly no 'silver bullet' in providing a model to fully explain cooperation in different environments. It has been roundly criticised for its inability to ameliorate commons-related issues. Because of its focus on the individual, it is difficult to understand the choices of multiple agents (see Kimber, 1981; Peters, 1990). And as groups increase in magnitude, there is an associated erosion of the potential to secure cooperation as there are too many chances for defection and too many random or chance events.

There are at least two possible suggestions by which to address the aforementioned problem. The first is referred to as a moralistic strategy (Boyd, 1992). This entails the punishment of not only defectors, but also the institutions that fail to recognise and punish these defectors. The second is developed by Kitcher (1993), who suggests that the reputation of players and social ostracism (refusing to interact or play the game with an individual) are strategies that may limit interaction.

It is indeed interesting that the brain seems to be equipped with a way to understand cost-benefit analysis regarding exchanges. People are able to detect cheaters, or those who seek out the rewards without paying the costs (Cosmides *et al.*, 1992). The fact is, we are constantly involved in Prisoner's Dilemmas in informal day-to-day interactions with others, and more specifically in fisheries, forestry and soil conservation (McKay & Acheson, 1990). The classic example of the Prisoner's Dilemma is in fisheries management, where the right thing to do for all fishermen would be to catch fewer fish with diminishing stocks (e.g. the Atlantic cod fishery). But since doing so would advantage other anglers if they continue to catch more than their fair share, it would be foolish to forfeit one's share to other more selfish agents if the outcome will be a depleted stock anyway. Obviously the best action would be to limit the catch or impose a moratorium on a species, but in the absence of acceptable data on marine animals and with the need to earn a living on the resource, individual selfishness reigns over collective well-being.

Sociobiology, defined as the 'application of evolutionary biology to social behavior' (Barash, 1977: ix), is the sub-field that emerged in biology in response to the work of Hamilton, Trivers and others, through the detailed efforts of Wilson (1975/2000). We can say that sociobiology and many of the theories that it encompasses, such as inclusive fitness, reciprocal altruism and game theory, are thought by many to be 'closed systems'. By this it is meant that these theories are defendable by: (1) not allowing any evidence to count against the theory, i.e. always finding some way of explaining away putative counter-evidence; or (2) answering criticism by analysing the motivations of the critic in terms of the theory itself (Stevenson & Haberman, 1998: 13). This has led many to view the debate on the legitimacy of selfish genes and sociobiology as over (Alcock, 2001). These terms have emerged to be accepted as part of the new sciences in the new millennium.

The group

There is a well-cited example in biology involving ants, where, in their efforts to gain and guard territory, distinct colonies of an ant species will steal the eggs and larvae of other colonies. Weakened by the disappearance of their charges, the colony will simply fall apart. What the queen ants (the 'leaders' of separate colonies) do is band together in a temporary cooperative relationship for their own mutual benefit. That is, they recognise that there is safety in numbers, thereby preserving the life of their own colonies (Hölldobler & Wilson, 1990). But is the motivation to save the colony? Biologists have found that the formation of herds, packs and shoals is based, not on the welfare of the group, but rather on the safety in numbers concept. It is straightforward selfishness, therefore, that compels fish to swim in unison and frogs to push their neighbour closer to a predator. In this latter study, it was shown that frogs will move to situate themselves in gaps between others in avoiding predation. Frogs stay put only if their gap is smaller than neighbouring gaps. If it is larger, it moves into smaller gaps creating aggregations of frogs for the purpose of selfishly avoiding predation (Hamilton, 1971).

But do not humans forge temporary alliances at times of war or over social policy for their own benefit that compel people to strike up such arrangements for short periods of time? Alexander (1979) notes that human evolution has been guided by intergroup competition and aggression. This is because the viability of larger groups within a society or network involves higher costs to individuals. Hunting serves to illustrate this point. If too many members of a tribe involve themselves in the hunt, then there would be too little to distribute among the hunters. There appears to be an optimal number of participants that guarantees success at the hunt and distribution of rewards. In no other species do we find social competition so complex and diverse, which has created the need for greater social complexity, intelligence and cleverness in dealing with rival groups (Alexander, 1979). Morality, in these cases, is not necessarily designed to allow people to live in harmony, but rather to unite societies in a sufficiently binding form to deter enemies. Alexander's work on moral systems can be viewed as contractarian because: (1) individuals seek their own interests; (2) their interests are primarily reproductive; (3) interests of individuals can be furthered by cooperating with others; (4) the mechanisms to achieve this involve status and reputation; and (5) rules for self-interest involve restraints such that one's success does not negatively impact the success of others in attaining their own self-interest.

This has prompted Fox to observe that 'In any social animal, genetic self-interest often works best in a collective social situation. Thus once sociality has evolved, only the most social organisms will be likely to prosper genetically' (Fox, 1997: 152). So it then becomes a matter of organisms working in socially cooperative ways in order to get ahead, in areas of reproduction, work, family, and so on. The key is to envision a benefit that comes to the agent over the long term. So men will pursue sociality as an end in itself that may pay dividends through the attraction of a mate. Fox extends this premise to morality in suggesting that it will pay for us to invest in moral systems of the community because these will ultimately provide protection for us and the longevity of our genes. So systems that are comprised of praise, blame, responsibility, reward and punishment will continue – indeed there are no societies that have been devoid of these basic ethical elements. The major stumbling block to Fox is the dichotomy between individualism on the one hand and the necessary revival of civic responsibility on the other. The challenges are found in greed, egoism and cynicism, which themselves have, arguably, taken over as basic virtues (see Chapter 5).

But while we are compelled to live in communities with thousands and millions of others, we are not group selected, meaning that individuals act against their own self-interest for the benefit of the group, as noted by Ridley (1998). In contrast, we are designed to exploit the group for ourselves rather than sacrifice ourselves for the group. This, Ridley suggests, makes us the most collaborative and social of all the creatures, yet, at the same time, the most belligerent and violent.

I mention this only briefly because in tourism we are pulled together in temporary groups for a common purpose over a prescribed period of time (e.g. a seven-day tour). All members of the group are united over a shared experience, although their motives and outcomes may be different. If we are to coexist in small tour groups, what is it that allows this to take place? Coupled with this question is the temporary reciprocity that must go hand in hand with this 'groupishness'. Perhaps we need to rely on these other people to get us through these uncertain experiences. In this, I recall a six-month trip to the South Pacific that I took after my undergraduate days, where I met dozens of people in the youth hostel network, most of whom promised to keep in touch with each other in the future. But we never did, despite the very positive experiences we had. I have not seen a single one of these individuals and cannot see how this might change. What brought us together in the first instance was circumstance (e.g. living in youth hostels and experiencing new and unique areas), but might it also have been a shared sense of allegiance in a foreign environment? We sought each other out as a form of kinship; a substitution of the social group that we rely on at home. Of course I have no way of proving this, but it is interesting to speculate on these temporary relationships that exist and why we are compelled to develop them. I find it interesting that, although assembled as a group for convenience purposes, there may be very little apart from the experience itself that holds these groups of individuals together. But maybe it is more than that. Travel presents the opportunity for new friends and new alliances, and the security that goes along with this: the need to trust and the need to be trusted, which I shall examine in more detail below.

Emotion and Reason

The philosopher Thomas Hobbes (1651/1957) believed that human behaviour is driven completely by merciless self-interest. He said: 'I put for a general inclination of all mankind, a perpetual and restless desire for Power after power, that ceaseth only in Death.' But surely this is a limited, rational form of psychological egoism. Politically and economically it has some uses, but it cannot be used to fully explain the gamut of human emotions. If we were so coarse in our actions the world would be a sad place indeed. In contrast, Adam Smith (1759/1966), although most clearly linked to his book on *An Inquiry into the Nature and Causes of the Wealth of Nations* (1776), wrote another book on *The Theory of Moral Sentiments* (1759), which by many accounts is just as seminal. His thesis was based on the premise that sympathy was important in motivating people to put the needs of others ahead of theirs, and that all of us were richly endowed with these moral sentiments.

The importance of sentiment as an underlying mechanism for morals was recognised by the Scottish philosopher David Hume (1739/1978), who took forward the idea that knowledge was derived from experience in strongly influencing the schools of scepticism and empiricism. In the case of the former school, Hume took issue with the concept of causality – that nothing can occur without a cause. Although he recognised that events were linked, one could not always prove that one event caused the other. Hume's morality was premised on the notion of experience, rejecting the notion that reason could discern vice from virtue. He could not envision the mind as holding the capability to fuse reason with feeling (today we do know that emotional and cognitive aspects of empathy are inseparable), touching off a long and distinguished debate on reason and passion. The English philosopher John Locke (1690/1979), too, argued for a uniformity of reason, but rejected René Descartes' philosophy of innate desires and axioms. Instead Locke spoke of the mind as a 'blank slate', upon which experience writes through sensation (e.g. light, taste, hearing) and reflection – as discussed earlier in this chapter.

Descartes (1637/1998) believed that the deductive method of enquiry (i.e. using axioms or general principles and reasoning to deduce or arrive at conclusions based on principles) was the only means by which to arrive at objective truths in his attempts to understand reality. He was instrumental in elevating reason to a position of supreme authority, and independent of experience. This was explained through his *Cogito, ergo sum*, or 'I think therefore I am'. That is, whenever people thought, they were aware of this thinking and therefore of their existence as a thinking being; a perspective that he later used to prove the existence of God. Similarly, Immanuel Kant (1781) insisted that it was the rational judgements of people that provide the basis for knowledge, rather than simply experience. So, in examining a painting, although we sense colour (a passion or sentiment), it is the mind that rationally distinguishes between these in assembling colours and shapes accordingly.

We go back to the fundamentals of our earlier discussions in recognising that we have the ability to place ourselves in the shoes of others who may be either more fortunate than us or less fortunate. We are thus able to understand stress and anxiety in others, which elicits emotion. There are very few organisms that are able to do this. Apparently, chimpanzees and dolphins are able to accomplish this, and they are the closest of all the other higher-order animals to humans in intelligence and have the most highly organised social structures of these beings.

In a previous section we saw how game theory provided an evolutionary explanation of how to mediate between self-interest and cooperation. But if people are gene-selected for self-interest, what is the human capacity that allows us to cooperate? The clue to this, at least for me, is in the eyes of a toddler who has not been socialised in the ways of cooperation and partnerships. If, in my interactions with my one-year old, I feign crying in our play after she has in some way harmed me, I see compassion in her eyes and in her responses; how she steps back and thinks why she has done it, perhaps imagining how she would feel if this were to happen to her in the future. It is just barely there, but it is surfacing nevertheless and waiting to blossom. This has been noted by scientists who say that prosocial behaviours such as helping, sharing and provision of comfort start developing in children between the ages of one and two, with increasing frequency and variety throughout this time period (Zahn-Waxler *et al.*, 1992). This concern is not just for members of the child's immediate family, but is also extended to include unfamiliar individuals, thus supporting the contentions of Mayr (1988) on the innate ability of very young children to be ethical. Morality is said to be adaptive in the earliest years, before grammar and sexual reproduction, because it allows the human child to understand what is right and wrong in navigating the tumultuous waters of play and siblings, as well as a raft of environmental hazards (Kagan, 1998). So, if it is true that we are self-centred beasts, it must also be true that we are at some level hard-wired for the capability to balance this self-interest with compassion and love – the Yin and the Yang that rests within us all.

In his seminal work on altruism, Trivers (1971) noted that emotions played a key role in reciprocity by compelling us to favour those who are altruistic in their actions. We like people who are altruistic and dislike others who are not. The importance of emotion in ethics therefore plays a critical role. His model was premised on feedback from an altruistic action that paves the way to cooperation. In this sense there is an extrinsic motivation for being altruistic; a goal that must be achieved in short order. But what of the intrinsic motivation that must surely be built into kind acts? If we were all goal-oriented extrinsics, we might never be able to achieve long-term harmonious relationships.

Tourism researchers have started to make important inroads in the connection between emotion and consumption. Malone (2014) for example argues that tourists make purchase decisions for the purpose of eliciting feelings of pleasure and emotion. These choices are often not made from strictly a rational standpoint, but rather from the emotive perspective in efforts to satisfy hedonic needs (Fennell, 2009). In contrast, agents who make deliberate ethical choices in tourism may do so by overlooking non-rational, factors i.e. emotion, according to Malone, suggesting that an individual's choice for to be ethical in tourism is rational. But is this always the case? Soper (2008) argues that some consumers experience displeasure by following the path of hedonism. For these consumers, termed alternative hedonists, pleasure is gained by consuming products that are better for people and the environment.

The theory of commitment

In his book on *Passions within Reason*, Frank (1988) observed that, if we are to reach a state of cooperation, people must be prepared to forego unbridled self-interest. It is the emotions that keep us on a biological and cultural track. Love, for example, is an emotion that reduces infidelity and helps to secure monogamous relationships. Frank says that it is irrational, say, to feud with others or cheat a community member, because it does nothing in an effort to form long-term stable relationships. For example, I have a friend who has recently opened a fish and chip shop with his brother. He interacts with the customers, works the cash register and keeps the books; his brother is the chef. But in this arrangement both may cheat each other in the name of self-interest (acknowledging the influence of the theory of inclusive fitness). Rationally, the chef could exaggerate costs related to the food and the accountant could easily steal money from the till. It does not stop here, as each could cheat one another for fear of being cheated. Reason does not help the situation as the brothers might not be able to convince, within a shadow of a doubt, that the other is operating 'above board' (recall the Prisoner's Dilemma). Frank argues that we do not attempt to reason in this way. If we did, we would never be able to conduct business because relationships would simply fall apart. He says we bring irrational commitment to the table from our emotions. In this way we trust each other and we avoid guilt by refusing to steal and cheat. Consequently, we need our emotions because they keep us in check; they prevent us from doing the unthinkable, which we would have a very difficult time living with. Our emotions, much more than reason, guarantee our commitment and act as the basis for the moral choices we make.

Consider the practice of tipping waiters, cab drivers and other service providers. Why do we do this if, in all likelihood, we will never see these people again, i.e. no shadow of the future as explained by Axelrod (1984)? A possible explanation for this is that, in keeping with the theory of commitment, it gives humanity an opportunity to be trustful and open. And in the old game of reciprocal altruism this, somewhere down the road, will enhance your stock as a virtuous person who is trustworthy and capable of returning favours. As Frank notes in regards to tipping:

> The motive is not to avoid the possibility of being caught [not tipping], but to maintain and strengthen the predisposition to behave honestly. My failure to tip in the distant city will make it difficult to sustain the

> emotions that motivate me to behave honestly on other occasions. It is this change in my emotional makeup, not my failure to tip, that other people may apprehend. (Frank, 1988: 18–19)

As such, trustworthiness, provided it is recognisable, creates valuable opportunities that might not be otherwise available. The more one behaves selflessly, the more that agent will find that he or she will be able to cooperate with others and harvest the benefits that society has to offer. This is supported by research that has found that we leave tips partly out of fear of retribution, but mostly out of a sense of duty, a desire to please and a belief in fairness (see, for example, Smith, 1993).

Revisiting the work of Batson (1990) on empathy-induced altruism, it has been found that we are truly social animals; much more social than social psychologists would give us credit for. The argument by social psychologists en masse is that we are social egoists; the only people we care about are ourselves. While our thoughts and behaviour may at times be social, our hearts are individualistic (Walster *et al.*, 1973). Batson suggests that an understanding of one's altruism and caring can be a function of values. Do we value someone for their own sake (terminal value) or do we value them for our own sake (instrumental value). He also notes that, in order to examine the level of empathy that people have towards others, we must examine the motivations for this empathy. Empathy and motivation are tied together in his 'empathy–helping relationship' model – the 'relationship between feeling for and helping a person in need' (Batson, 1990: 339). The motivations for helping someone can be explained in three ways: (1) aversive arousal, whereby empathetically aroused individuals help another who is in distress simply to rid themselves of their empathetic arousal; (2) empathy-specific punishment, where we help another in distress because we would feel shame and be socially ostracised if we did not; and (3) empathy-specific reward, where empathy for someone who is suffering generates sadness, and the empathetically aroused person works to relieve this sadness. While Batson's research shows that there is a link between altruism as an ultimate goal through attempts to improve the welfare of

Box 2.1 The trapped tourists

On an outing, six tourists suddenly become trapped by a landslide in a seashore cave, and the cave is rapidly being flooded by a rising tide. There is a glimmer of light at the top of the cave, and the tourists rush towards it. An overweight tourist reaches the hole first, struggles to get through it, and becomes stuck. Those below scream for help as the waters engulf them. Rescuers outside are faced with a dilemma. They have drills that can be used to free the overweight tourist, but this will take so much time that the others will drown. However, the rescuers have a stick of dynamite that can be used to blow the overweight tourist out of the hole, killing him, but allowing the other five to be freed. The overweight tourist pleads for his life, and his fellow travellers scream for theirs. What is the best ethical course of action?

Source: Adapted from Griffin (1997).

others, he concludes by suggesting that the altruistic caring we have for others is tied up in the empathy that we have for these individuals. If empathy is low, there emerges a pattern of egoistic motivation on the part of the helper.

Where Batson's work can be challenged is through the argument that people have boundless reserves of empathy that they can use to care for others. Midgley (1994) illustrates that people have evolved a limit to how much empathy that they can arouse. If people cared too much for others they would not be able to care for themselves. In the case of the trapped tourists (Box 2.1), too much empathy on the part of those trapped would perhaps spell their end, while rescuers would not be able to bring themselves to use the dynamite to blow the hole open because it would tragically kill the trapped tourist.

We recognise that reason is not the source of truth, but rather simply a tool that helps us deal with the shades of goodness and badness. As such, reason without passion and passion without reason, provide no basis from which to build a cohesive understanding of morality. In this regard, researchers have pieced together four major families of emotions that comprise our morality (and in support of the theory of reciprocal altruism). These four include (Haidt, 2002):

Other-condemning emotions. Contempt, anger and disgust, which move people to punish cheaters (see Trivers (1971) in drawing comparisons to the work of Haidt).

Other-praising emotions. Gratitude, elevation or moral awe, which allow us to praise altruists.

Other-suffering emotions. Compassion, sympathy and empathy, which allow us to help someone in need.

Self-conscious emotions. Guilt, shame and embarrassment, which move us to avoid cheating or to right wrongdoing.

So the basis of morality must lie in the rational actions of people, but also in the motives for acting. For example, we often act out of respect for a law or a duty and thus do things for others to fulfil these obligations. We purchase a gift for a visitor, for example, out of respect for society's unwritten rules. Often these gifts are purchased at the last moment, or the gifts have been so large and awkward that one could scarcely get them home. Are these gifts of love or gratitude? As parents we know that by taking a child to an event we will surely see the pleasure in the child from our efforts. But taking the child because of one's duty takes away from the goodwill of the act. Doing so out of love, and the passion that forms the foundation of this love, we inherently know or feel that what we have done is good. This is the essence of goodwill.

The New Sciences of Nature via Nurture

The debate on nature vs. nurture in which we engaged in the earliest part of the chapter has been taken to task of late in response to new knowledge that has emerged from the human genome project. This knowledge suggests that there is a very intricate tie between nature and nurture that has only gained acceptance at the dawn of the new millennium. It follows from earlier work illustrating that the behavioural sciences are now widely accepting the theory and methodologies of genetics (Plomin *et al.*, 1990), that there is plasticity in the gene structure that allows information to flow in and out of the genome (Miller, 1991). Ridley (2003) writes that current research is leading us to the conclusion that it is no longer nature vs. nurture, but rather nature *via* nurture. What this means is that gene researchers are finding that genes often take their cues from what goes on in the environment. Research has shown that many different types of animals, including humans, can grow new cortical neurons in response to the experiences of the environment. Similarly, these neurons are lost without environmental stimulation:

> It is genes that allow the human mind to learn, to remember, to imitate, to imprint, to absorb culture and to express instincts. Genes are not puppet masters or blueprints. Nor are they just the carriers of heredity. They are active during life; they switch each other on and off; they respond to the environment. They may direct the construction of the body and brain in the womb, but then they set about dismantling and rebuilding what they have made almost at once – in response to experience. They are both cause and consequence of our actions. Somehow the adherents of the 'nurture' side of the argument have scared themselves silly at the power and inevitability of genes and missed the greatest lesson of all: the genes are on their side. (Ridley, 2003: 6)

At one time, it was thought that one gene only was responsible for one behaviour. When it was found that there were only 30,000 genes, as uncovered by the human genome project in 2001, the nurture crowd became very excited. However, it soon became common knowledge that there are innumerable ways in which genes can be used. To put the potential into perspective, just 33 genes, each with the ability of turning itself on and off (as twos) would be enough to make all of us in the world unique. There are more than

10 billion ways of flipping a coin 33 times. Although humans have about the same number and kinds of genes as the chimpanzee, the secret is not the number or type, but rather how the genes are used in different patterns and orders (Ridley, 2003).

As it happens, with us sharing so many of the same genes as other animals, this is the most logical explanation as to why species have evolved to be so different in body plan. Time appears to play a significant role in the development of these body plans. A chimpanzee has a different head from a human not because of a difference in the blueprint of genes, but rather because genes switch on and off at different times during the development of the jaw, cranium and so on. If the jaw of a chimp is larger than that of a human it is because the gene responsible for shutting down growth does so at a later time than in humans.

In reference to behaviour, as the central theme of our discussion, Ridley is quick to point out that genetics is not responsible for answering questions on what determines behaviour, but rather on what varies. In individuals, it is the genes that vary, and as much as 50% of the variation in a population is tied to genes. What this means is that personality is heritable. The parent with more than one child will know this to hold true by detecting personality differences in their children from an early age. For example, Tom is very different from Jane in the way that he is more introverted, more curious and less antagonistic than his sibling. Personality is thus the result of characteristics passed along by parents. In fact, research has shown that a propensity to cheat, steal, fight, destroy property and lie is heritable. This genetic link to behaviour demonstrates that antisocial disorders are in fact much more deeply ingrained than behaviour that might be the result of social dysfunction (Lykken, 1995, 2000).

Such a high degree of variation means that children's personalities are not identical to the personalities of their parents, nor are they shaped by child-rearing. The personalities of children who are reared in the same home are no more similar than those separated at birth (Pinker, 2002). Although children contain many traits of their parents, their behaviour is only marginally a function of parental influence. Indeed, Trivers predicted this based on his extensive observations of many other species, in writing that:

> When the parent imposes an arbitrary system of reinforcement (punishment and reward) in order to manipulate the offspring into action against its own best interests, selection will favor offspring who resist such schedules of reinforcement. They may comply initially, but at the same time search for alternative ways of expressing their self-interest. (Trivers, 1985: 159)

Variation can also be explained by group socialisation theory, which further helps us to understand that a large part of the personalities of groups, such as children, can be explained by group interaction, as noted by Harris (1998). As such, what has not come from the genes, in large part, comes from the interactions that children have with their peer group and not from their parents. Socialisation includes the norms and skills that allow people to operate successfully within society. We say 'successfully' because of the assimilation of norms and skills. Deviance is outside of what is normatively acceptable by society. Harris observes that the influence of the peer group can have far more weight than the influence of parents on children. What this suggests is that frequently children choose to model themselves after their friends rather than their parents, and so group approval depends more on assimilating with the group (the culture within the group) than conforming to parental influences. Researchers now believe that approximately 50% of our personalities can be explained by innate wiring, another 40% can be attributed to the peer group and the unique environmental experiences that accompany the individual through the different stages of development, while only approximately 10% of our personalities is actually a function of parental guidance (Pinker, 2002).

The sciences that have enabled us to make these new discoveries include evolutionary psychology, cognitive neuroscience and behavioural genetics, leading some researchers to posit that selection pressures are essential in generating hypotheses about the design of the human mind (Cosmides *et al.*, 1992). The human genome project has been important in allowing researchers to unravel many of the underlying genetic functions behind behaviour. For example, those individuals who have a longer D4DR dopamine receptor gene are more prone to participation in high-risk activities, such as sex with strangers, bungee jumping and ice-climbing (Benjamin *et al.*, 1996). There is now a physiological explanation as to why people behave in such ways, beyond psychological measures such as Zuckerman's (1979) sensation seeking scale. Correspondingly, the psychological and physiological research might in the future be combined to gain a more comprehensive analysis of the person, and the individual motives responsible for participation in these activities. This new knowledge has come about at a time when scientists have learned a great deal more about the brain's intricate influence on behaviour.

The work of the evolutionary psychologist David Buss has been important in casting light on the manner with which evolution can be applied to the mind (Buss, 1995). He notes that psychology is in disarray as a result of a number of mini-theories that cannot account for an understanding of the functional properties of the mind (i.e. what the mind is designed to do). Although evolutionary biology and psychology are founded on the same principles, the psychologist's view differs from that of the sociobiologist in the way they view 'fitness'. Often the sociobiologist views people as fitness maximisers (maximising their gene representation) in all endeavours. Buss notes that this is not always the case, as men do not always line up to donate sperm to sperm banks, nor do they try to strive to reproduce to their maximum. Indeed, many men choose not to reproduce at all. Instead of taking on the role of fitness strivers at all times, he would rather define people as adaptive executors, in reference to our ability to modify behaviour according to setting and circumstance. As such, while we can discuss human behaviour in the context of inclusive fitness and reciprocal altruism as foundational theories, it is important that we do not go directly from these foundations without an account of the psychological mechanisms on which these end products are founded (e.g. aspects of social and legal systems).

The implications of all of this knowledge on tourism are significant. This is particularly true, for example, in regards to the findings that new neurons are created through rich environmental experiences. It is not unrealistic, therefore, to suggest that the act of travel provides the foundation for the creation of new neurons through the experience of travel. This also underlines the importance of learning through travel. As part of my doctoral comprehensive exam, I was asked if virtual reality might not be an alternative to the act of travel: one might gain an appreciation for the destination from the lifelike features of a machine, without actually having to go there. My response was that we lose out on the full travel experience, in the tradition of Clawson and Knetsch (1966), who argued that trip-planning, getting there, being there and getting home, along with the recollection phase, were important parts of the overall package. Now it seems there may be another explanation. By choosing virtual reality instead of the authentic trip, the mind may miss out on some quite critical stimuli, which might create new neurons and in turn new avenues for learning. Fennell (2009) explored the dimension of the nature of pleasure in pleasure travel. In surveying the research on neuropsychology, he found that pleasure is a function of the neurotransmitter dopamine (particularly the anticipation of rewards and incentive), but also that the release of opioids is associated with consumption, satiation, bliss and rest. Interesting is the fact that dopamine transmission is altered after a few experiences of the same reward. This would explain why tourists seek out new destinations out of novelty alone.

The Evolution of Ethics

Although there is general agreement regarding the evolutionary basis of morality – the publication of Darwin's *Origin of Species* put an end to the belief that morality was a god-given gift (Mayr, 1988) – it has been slow to be universally accepted, first because of the failure to link human traits with evolutionary development, and second because biologists over the last 30 years have been vague about how to apply natural selection to social systems (MacIntyre, 1981). What has been accepted is that the *capacity* to develop ethics is widely recognised as a function of biological evolution, but the products of the ethical capacity result from culture and society. The ability to be moral, in a biological sense, is attributed to *Homo sapiens* only because of the intellectual capacities of humans to: (1) anticipate the consequences of their actions; (2) to make value judgements; and (3) to choose between different courses of action (Ayala, 1987; Ehrlich, 2000). This has been corroborated by other researchers who feel that ethics has evolved in humans the ability to: (1) infer the thoughts and feelings of others; (2) be self-aware; (3) apply the categories good and bad to events and to self; (4) reflect on past actions; and (5) know that a particular act could have been suppressed (Kagan, 1998; Wilson, 1993). This is attributed to the size and complexity of the brain, particularly the cerebral cortex, and its ability to integrate and coordinate stimuli from the environment and to allow the individual to abstract and be self-aware. None of these conditions, Ayala notes, could have taken place if humans had not crossed an evolutionary threshold, a feat that no other animal has done, allowing for intelligence and anticipation of the future. These adaptations have led some theorists to argue that, from the biological side of the equation, ethics is no more than raw inclusive fitness altruism, while others feel that when genuine ethics evolved it replaced inclusive fitness altruism (Mayr, 1988; see also Simpson, 1969).

Ehrlich writes that ethics and values evolved when humans were assembled in small-group communities as hunter-gatherers (Ehrlich, 2000). Individuals knew each other as well as the expectations inherent in these collective units. As such, there was no need for a complex set of codified rules to protect the values that held these groups together. There were rules and these were understood by all. Moral philosophy begins when people determine that their prevailing sets of rules are unsatisfactory. There is therefore an element of rational self-guidance inherent in the notion that morality is a social system. And there is a tension that exists between what society accepts as morality and what the individual agent accepts. This tension is tested because, in the words of Frankena (1963), we are more than just agents of morality – we are also spectators, advisers, instructors, judges and critics. Our decisions about what is right, and how we decide on this, under certain situations, form the primary question that is dealt with through ethics.

A significant challenge to the application of human ethics is in the extension of the human cultural group from an extended family to a society. This was necessary for purposes of territory, war and resources. For ethics to remain viable within larger groups, inclusive fitness altruism had to be extended to non-relatives. Mayr asserts that this is where reciprocal altruism may have played a role in developing cooperative relationships between individuals within the group (Mayr, 1988). Although the group selection theory has been shown not to be applicable in animal groups, it could have played a role in human groups because of the importance of culture. These groups, and the evolution of their ethics, were held together by leadership, communication, rituals, geography and so on, where some behaviours were preserved and others discarded because of their uselessness. It is no coincidence that the evolution of larger brains and larger social groups made ethical behaviour possible, leading to: (1) a selection reward for certain unselfish traits that had benefits for the group; and (2) ethical behaviour by choice and free will, rather than by instinct of inclusive fitness altruism. This has entailed the suppression of self-interest in favour of the welfare of the group. In order to adapt to changing circumstances within the group, ethical systems had to be continuously altered through the modifying influence of community leaders (Mayr, 1988).

Maynard Smith and Szathmary favour this interpretation of culture and behaviour in suggesting that rituals are culturally, not genetically, transmitted (Maynard Smith & Szathmary, 1995). They cite the work of Boyd and Richerson, who note that between-group selection may have favoured rituals that bind groups together culturally, not genetically (Boyd & Richerson, 1985). This led the authors to observe that 'there can be between-group selection for culturally inherited systems of belief that favour the success of the group, and there is individual selection for the genetically inherited ability to be influenced by the ritual' (Maynard Smith & Szathmary, 1995: 272–273).

But just because people have a moral sense does not mean that they are innately good. People must compete with other innate senses, such as power, sex, greed and so on, which, in turn, depend on our character, circumstances, and the cultural and political conditions of the day. For Midgley, morality evolved at the juncture of freedom and constraint:

> If freedom and morality are indeed closely linked . . . it is perhaps a rather paradoxical fact that the first effect of freedom should be to put us under these new constraints [impartiality, truthfulness, parsimony and so on]. Our freedom is exactly what gives us these headaches, what makes possible this moral thinking, this troublesome kind of search for priority among conflicting aims. By becoming aware of conflict – by ceasing to roll passively from one impulse to another, like floods of lava through a volcano – we certainly do acquire a load of trouble. But we also become capable of larger enterprises, of standing back and deciding to make lesser projects give way to more important ones. That, it seems, may be why moralities are needed. (Midgley, 1994: 9)

So, although the origins of ethics may be found in conflict in group situations, we have learned that people are just as adept at conflict resolution. After all, we must be if we have made it this far as a species. And it is just as easy to say that our conflict resolution strategies have evolved over countless generations to enable us to coexist. Along with the innate tendency for war, the Hobbesian way of thinking, we must also have evolved a kinder nature, including a sense of justice, morality and community, which allows us to select behaviour in keeping with the rules of society. It is as Pinker has noted in his work on the denial of human nature: 'If people are products of biology, life would have no higher meaning and purpose' (Pinker, 2002: 139). What this means is that, although human nature is dictated by biology by as much as 50%, a large amount of who we are is dictated by our peer groups, our experiences and even our parents.

The Naturalistic Fallacy

The philosopher David Hume (1739/1978) argued that we cannot use deductive logic or reason to prove the truth of moral beliefs. That is, there is a gap between that which is (the factual side of the equation) and that which ought to be (the moral side). In this regard, the assemblage of various facts does not necessarily allow one to reach a conclusion on whether something is moral or not. In considering the statement 'Murder is wrong', Hume suggested that we might only conclude that this is simply one reporting their subjective feelings towards murder, or their disapproval of it.

The central issue of the problem rests with the extent to which biology can explain the basis of human nature (Keith, 1947). Scientific knowledge (i.e. the fallacy of using natural science) is taken too far in considering what is right or wrong to the exclusion of other mechanisms of knowledge, such as intuition and experience. Biology, for example, provides no magic solutions in describing the behaviour of people, despite the importance of genes and natural selection as predictors of behaviour.

> The naturalistic fallacy suggests that we can never solely rely on what biology, or other sciences for that matter, has to say. There are broader moral questions that will continue to challenge the status quo.

Is and *ought*, therefore, belong to two different logical spheres, prompting Rolston (1986) to suggest that, while science is descriptive, describing what is the case, ethics is prescriptive and prescribes what ought to be. Biology should therefore be considered neutral when it comes to our understanding about what is right or wrong. Thus, while we are predisposed to acting in certain ways (the capacity) from a biological manner of thinking, it is humans who decide what is right or wrong. By contrast, the moralistic fallacy works in the opposite direction or is at least a variation of the naturalistic fallacy. That what *ought to be*, is in fact true, which might in fact ignore scientific truths. *Ought*, therefore, replaces *is*.

Conclusion

Bauman (1993) in his treatise on postmodern ethics argues that moral responsibility has no foundation or determining factor; no 'convincing case for the necessity of its presence' (p. 13). Morality, he says, is 'an act of creation *ex nihilo*, if there ever was one' (p. 13), or a mystery that defies reason. We should not be 'normally' moral, and it is only because of some special powerful cause that we are. The road to morality according to Bauman, it seems, takes place when we surrender our self-interest on the way to being social. The bulk of this chapter was meant to steer around what continues to amount to intractable incommensurabilities, or no common measure between forms and no possible way to imagine a fit as a result of disciplinary divergence. Biological and social science research like two (or multiple) ships passing in the night.

Evolutionary science argues that there are innate factors that drive us to be ethical. These have been described in the context of inclusive fitness, reciprocal altruism, commitment and emotion, and group dynamics. Think how frequently reciprocity enters into our lives in families and communities, and in the workplace: workers exchanging weekend shifts with the promise of making it up in the future; neighbours loaning garden tools, to be returned at a later date (hopefully in one piece); and a sister borrowing clothes for a special date. We know that the likelihood of entering into these bargains is diminished if prior experience shows a failure to return items, or if the deed is never repaid. It is in our nature to rely on one another because we are social beasts. But are we social beasts because of our innate natures (and there is mounting evidence to support this)? For many biologists, including Alexander (1987), the developments in evolutionary theory represent the greatest intellectual advance of the century. This perspective rests on the shoulders of the application of evolutionary theory to our understanding of human conduct and behaviour. At its root lies the paradoxical duality of selfishness and altruism in human nature, which has been at the heart of moral philosophy since the earliest times and that Alexander illustrates has not been resolved in any writings other than biology.

Not surprisingly, however, there is not universal acceptance that 'selfish genes' and 'inclusive fitness' are capable of explaining human nature. Kagan (1998), for example, dislikes these because they mislead us into thinking that proximal causes (an anatomical analogy) have less utility than evolutionary ones. He fails to understand why a Bangkok streetwalker sends her earnings to her parents (who sold her to prostitution) so that she can maximise the reproductive fitness of those who share her genes. But at the same time, Kagan offers no convincing evidence or no other alternative to the biological explanations above; no proof that there is a better foundation of knowledge that exists in explaining the way we are. But I am not sure the vast majority of evolutionary biologists would make this leap. What they may say is that we have identified certain innate traits, such as selfishness and altruism, and constituted systems of ethics that no doubt would be influenced by culture and circumstance.

Miller (1991) is one who believes that the moral qualities we have as a species have developed over time as a product of human wisdom through experience. At the same time, he supports that which has been put forward by Singer and Keiffer, respectively:

> When well grounded biological theories are relevant to an ethical decision, they should be taken into account. The particular moral judgements that we end up making may reflect these theories. For this reason, it is perfectly true that philosophers, along with everyone else, should know something about the current state of biological theories of human nature. To ignore biology is to ignore one possible source of knowledge relevant to ethical decisions. (Singer, 1981: 68)

> Systems of ethics are the product of human wisdom and the experience of human beings living together and not of the expression of human genes [alone?]. Though a rational system of ethics cannot be independent of evolution, neither can a system of ethics be derived directly from evolution. (Keiffer, 1979: 21)

So here we have come over a period of 50 years of research and millions of years of evolution to suggest that both culture and biology are essential in determining who we are and how we act. As such, in order to understand tourist behaviour we need to understand a bit about human nature, and to understand human nature we need to know a bit about genes and culture. Failing to do so diminishes our efforts to fully elaborate upon the motives and actions of tourists and tourist groups, and suggests that it is perhaps unwise to travel the highways of human behaviour (at least in the way we have done so from the tourism context) at full speed without first gathering momentum along the city or rural access routes.

3 The Basis of Ethical Discourse

Human social life cannot be understood apart from the deeply held beliefs and values that in the short run, at least, motivate and mobilize our transactions with each other and the world of nature

Harris, 1989

Introduction

In the previous chapter there was an emphasis on the genetic and cultural basis of human nature. This has provided the necessary backdrop from which to look at ethics more closely. We learned that morality is part of our innate makeup, in which each of us has a conception of what is right and wrong. This is also a function of the many influences that we have in our lives. Our behaviour in the context of right and wrong is both a product of the tension that exists between who we are as individuals and the norms established within society. This tension is not new. While Aristotle argued that we need to collectively derive a code of ethics within society for guidance, Socrates contended that it is our duty to continually challenge what is right within our communities. This chapter will examine ethics from the context of classical antiquity, and will continue with an illustration of many of the most important theories of ethics in providing a basis from which to examine more specific aspects of ethical thought. In particular, two main theoretical domains will be discussed in this chapter. The first, the absolutist theories, include teleological theories (utilitarianism, hedonism, egoism), deontological theories (theology and the Golden Rule, Kantianism, social contract ethics) and virtue (or character-based theories). The second group, the subjectivist theories, is represented only by existentialism.

Philosophical Terminology

Philosophy

The word philosophy stems from two Greek words, *philo* (love) and *sophia* (wisdom), and means love of wisdom. It is the purpose of philosophy and those who practice it to seek a more unified view of the universe and to do so through critical thinking. Perhaps one of the shortest and most concise definitions of philosophy is one posed by Honderich (1995), who says that it is 'thinking about thinking'. However, a definition that provides us with more information is a subsequent one offered by Honderich:

> philosophy is rationally critical thinking, of a more or less systematic kind about the general nature of the world (metaphysics or theory of existence), the justification of belief (epistemology or theory of knowledge), and the conduct of life (ethics or the theory of value). (Honderich, 1995: 666)

In analysing this definition further, philosophy is initiated by asking questions (thinking in a rational and systematic nature) about what the world is like (metaphysics is an historical term meaning that which cannot be explained by physics, or everything else beyond the physical world), about the types of knowledge that allow us to gain further insight into phenomena, and about how people ought to behave, and the values that they should subscribe to, in living a good life. In asking these questions about the world, we can begin to understand why the world is as it is, with the hope that we can make it a better place. This entails expressing a sense of wonder about any number of different phenomena, including tourism. This book is about the latter of these three aspects, i.e. ethics, and so will entail asking questions about what is right and what is good. In a general sense, i.e. without specific reference to tourism, questions of a moral nature have often surrounded issues such as: whether humans are inherently greedy or generous; the right to tell others what is good or bad; the types of acts that are always morally wrong; whether guidance about right or wrong comes from pre-established rules or consideration of the consequences of one's actions; or whether some people are 'better' at being moral than others.

Morality and ethics

As a branch of philosophy, ethics subsumes a body of knowledge that is based on some 2500 years of dialogue. It comes from the Greek word, *ethos*, meaning a habitual mode of conduct, and is concerned with answering the question: What should one do in order to be good? In judging what is good or ethical, we need to define what is meant by 'good'. Yaman suggests behaviour can be good in and of itself or because of some prescribed standard. In the case of the former, he says that, although there are many persuasive arguments:

> a final value that is somehow inherent in the 'act' is difficult to detect as a means of guidance: When we ask a child to be 'good and quiet' in a particular setting (in a library, for example) we do not necessarily mean

> that 'quietness' has some inherent goodness, or that it is good under all circumstances. We may prefer the same child to be 'good and play' in a playground, in the full knowledge that the child will be active and noisy. Consequently, whether or not certain human conducts have some inherent quality of 'goodness', or lack of it, should remain the domain of philosophical debate. (Yaman, 2003: 109)

Many theorists, including Guy (1990), note that ethics and morals are often confused. He believes that they are essentially synonymous, with ethics being derived from Greek, as above, and morality evolving from Latin. To some, however, there is a distinction. Ethics is usually viewed as the systematic general science of what is right or wrong conduct, while morality refers to the patterns of conduct and the rules of action. This can be seen in the work of Millar and Yoon (2000: 158), who write that 'Ethics are rules but morality is more than rules.' Other researchers take a similar tack in stressing that 'ethics is simply the philosophical study of morality, a study that enables us to arrive at a basis for conduct and action' (Miller, 1991: 12). In this way, ethics bases its existence on a reflective foundation for the comparison and contrast of the actions of humans, which are often in codes of ethics. In attempting to ascertain whether or not it is acceptable to take a piece of Uluru home at the conclusion of one's vacation, we can be aided by a systematic set of rules on what is morally right or wrong (a code of ethics) regarding such actions. By consulting these sets of rules, one may be able to proceed in a way that best protects the integrity of the environment in which he/she visits. But central to this is the notion of individual thoughtfulness, which must rest at the heart of ethical decision-making despite the atmosphere of moral codes. As such, we can say that excellence and integrity of the individual begins with ethics. Alternatives are selected on the basis of the values that lie at the foundation, within the individual and social norms, of sound ethical decision-making (Guy, 1990).

In general, and in reference to the discussion above, the key difference in distinguishing between ethics and morality might be application. For example, Velasquez (1992: 16) writes that 'Business ethics is a specialized study of moral right or wrong. It concentrates on how moral standards apply particularly to business policies, institutions, and behaviour.' Terms such as 'good' or 'wrong' are normative in that they depend on some implicit or explicit standards or criteria that are developed within society. Velasquez suggests that moral standards include both moral norms and moral principles. The former include standards that require, prohibit or allow certain types of behaviour (e.g. rules against lying or injuring). The latter are more general and are used for the purpose of evaluating individual behaviour, policies and institutions (e.g. principles of rights or justice as they are articulated in society).

Furthermore, Frankena (1963) writes that morality is social in its origin and function and may be viewed as an instrument of guidance at the level of the individual and that of the group. In such a way, we can identify it as a system or an institution that makes certain demands on an individual that are external to that agent. He notes that individuals may be the prime spokespersons for these demands and thus may internalise them (take them on as theirs for the purpose of regulating their own behaviour). It may be conceivable that the individual does not fully believe or agree with these tenets; however, these are adopted for the greater good of society and in an effort to apply conformity to what is expected of our actions.

Ethical systems are quite frequently developed for the purpose of delineating a shared set of values (e.g. a code of ethics for marine tourism operators). As such, the system in place must be a reflection of the values of not only the collective, marine operators in this case, but also of the public at large, as outlined above. This means the ethical system should be a reflection of the democracy in which it is entrenched, while serving the needs of an entity (marine life). What complicates our ability to know what is right or wrong is the fact that moral rules differ from one culture to another, or from one region/situation to another, because of the cultural influences directing stakeholders.

The sociologist Barbara Ann Strassberg (2003: 170) defines ethical systems as 'socially constructed complex systems characterised by paradoxes embedded in processuality, reflexivity intertwined with contingency, and plurality'. These systems have both social and cultural constituencies. The social system component is comprised of a number of different categories of people that determine the ethical system in time and space, as described by Strassberg. These include: (1) founders, including authors of sacred texts or secular codes of ethics; (2) interpreters, such as scientists, scholars or theologians, who process that which has been created by the founders; (3) teachers, such as parents, teachers, peers and the media, who further process the ethical precepts and thus contribute to a higher or lower level of moral literacy in society; (4) activists, represented by religious or professional organisations, who have a vested interest in a specific cause; and (5) consumers, which includes the people who live ethics in their daily lives. Important in her discussion is the notion that being good is a function of one's moral literacy and moral competence. The former, as a forerunner to the latter, includes the changing terrain of knowledge that comes from our understanding of culture (e.g. religion and codes of ethics), the social world, the environment and the universe. Moral competence is how we use this knowledge in the creation of sustainable societies that bring out the full range of human potential. Conversely, the cultural aspect of ethical systems is expressed through a cognitive dimension (e.g. the knowledge that we have from scientists); an emotional dimension (e.g. human response to values, norms and those who devise such societal beliefs); and the readiness for action, which includes the patterns of behaviour that permit us to repeatedly test and transmit ethical information from one generation to another.

A good summary of the aforementioned material can be found in the work of Ray (2000: 241), who writes that 'Ethics is the study of the rules, standards, and principles that dictate right conduct among members of a society. Such rules are based on moral values.' More specifically, Ray provides the following three definitions of related terms:

Ethics. The rules, standards and principles that dictate right conduct among members of a society or profession. Ethics are based on moral values.

Right conduct. Behaviour that is fitting, proper or conforms to legal or moral expectations.

Code of ethics. A systematised set of standards or principles that defines ethical behaviour appropriate for a profession. The standards and principles are determined by moral values.

Values

Adam Smith wrote that the word 'value' has two different meanings. The first, 'value in use', referred to those things in life that, although having great value (such as water), had virtually no exchange value (not so today, obviously). On the other hand, 'value in exchange' referred to items such as diamonds that have very little value in use, but have tangible value in that 'a very great quantity of other goods may frequently be had in exchange for [them]' (Smith, 1776/1964: 25). The word 'value' has surfaced often in the preceding discussion, with the insistence that at the root of any moral system is the importance of values. Rand (1964: 15) has defined value as 'that which one acts to gain and/or keep'. Values serve to act as the basic premise of ethics because they presuppose an entity that acts to achieve a goal in the face of alternatives. In cases where there are no alternatives, Rand notes, there are no goals and no values. The recognition of value – which can only be realised by living entities – provides the foundation for what we view as good or bad. 'What are the values his survival requires?' (Rand, 1964: 22) is the question to be answered by the science of ethics, and the necessity for a code of ethics. Hence, in order for humankind's survival, Rand feels that ethics is an objective, and metaphysical, necessity.

Values can also be substantive in reference to such things as liberty, justice and rights or they can be more instrumental and focus on economy, hierarchy, due process, impartiality and efficiency. The former relate to democratic ideals, while the latter are more bureaucratic. Some scholars argue that, although bureaucratic ideals can take on a moral character, they are not subsumed under a democratic ethos because they define more of the professional aspect of public service: 'they offer validity and legitimacy to public administrators in a way that is separate and distinct from democratic ideals' (Denhardt, 1989: 188). The bureaucratic ethos is built on the notion that administrators are bound, legally and morally, to enforce laws and implement policies that are the voice of the citizenry. The actions of the bureaucrat are thus designed to implement the public will in a morally neutral manner (Woller & Patterson, 1997). These are thought to be teleological (ends-based), to employ instrumental rationality, and to be based on capitalism and market values (Pugh, 1991). By contrast, the more recent democratic ethos exists today as a serious challenge to the 'overhead democracy' model of government that emerges in the bureaucratic approach, which is characterised by no direct line to the will of the people and, as such, decision-making is based on an alternative and instrumental value set that lies within the system itself. The democratic ethos is deontological (or means-based), is based on substantive rationality (holistic thinking within a system of values, according to Weber, 1927/2003, which may be interpreted as critical thinking over technocratic thinking), and stems from more classical values of law, duty and obligation to the state. This ethos emphasises citizenship, public interest and social equity, among other values.

Instrumental rationality (a topic to be expanded upon in Chapter 6) has a strong ideological overtone that substitutes for democratic principles, leading to a distortion of moral behaviour. Our lack of ethical behaviour within organisations and society stems from the overpowering influence of instrumental rationality (Pugh, 1991). And, where once we felt that our social institutions reflected our beliefs and values, there is now a resounding feeling that the opposite is true (Jos, 1988). The workforce, which could be so empowered as to provide a catalysing element to organisational tactics, is thus reduced to a mass that submits to whoever or whatever can assert its will. Increasingly we are chained to our institutions both economically and ideologically, with the need for a centralised authority that has calculable rules and directives and thrives on close supervision and impersonality. According to Jos (1988), this prevents us from passing morally autonomous judgements.

Values can also be referred to as instrumental (as above) and intrinsic, as is the case in the literature on the environment. Intrinsic values are those that are defined on the basis of value for its own sake, as opposed to that which is tied to instrumental ends on the basis of a cost-benefit calculation. So what is the value of a person, a community or a society? How about a forest or a stream? How do we compare one entity against another from the perspective of value? If we value something intrinsically, we avoid having to rationalise on the basis of a calculation, through acceptance of the entity on its own merit.

Researchers have argued that there are a number of core values that lie at the centre of the relationships between people, including caring, loyalty, honesty, fairness, accountability, integrity, promise-keeping, respect for others, pursuit of excellence and responsible citizenship (Beauchamp & Bowie, 1979; see Rokeach (1973) for a discussion of terminal and instrumental values). Abraham Lincoln expressed a core value when he said 'that government of the people, by the people, and for the people shall not perish from the earth' (Jacobs, 2004: 176). Writing in the context of the integrity and longevity of empires and culture, Jacobs observes that, as long as these core values, such as Lincoln's, are not lost to practice and memory, there is the real possibility that they can thrive for centuries. Other core values have been expressed by Kinnier *et al.* (2000) from their content analysis research of religious and secular text passages:

(1) Commitment to something greater than oneself (supreme being).
  (a) To seek truth.
  (b) To seek justice.
(2) Self-respect, but with humility, self-discipline and acceptance of personal responsibility.
  (a) To respect and care for oneself.
  (b) To not exalt oneself or overindulge – to show humility and avoid gluttony, greed or other forms of selfishness or self-centeredness.
(3) Respect and caring for others (i.e. the Golden Rule).
  (a) To recognise the connectedness between all people.
  (b) To serve humankind and to be helpful to individuals.
  (c) To be caring, respectful, compassionate, tolerant and forgiving of others.
  (d) To not hurt others (e.g. murder, abuse, steal from, cheat or lie to others).
(4) Caring for other living things and the environment.

Values also have an important role to play in the development of theory. Often, researchers want to commit to certain values such as equity, where administrators are worried less about the organisation and more about public responsibility (Garofalo & Geuras, 1999). However, theory is often not provided to support these values. How does equity differ from political stability or adherence to the public will? Without a clearly defined theoretical basis for comparison, conflict will occur in our understanding of each. This is why there needs to be a tight relationship between vision, logic and theory (Sowell, 1987). A vision is a subjective pre-analytic cognitive act, which we feel or sense before we try to attach some sort of reasoning. Vision provides the basis of causation, often in the form of a hunch or gut reaction of how the world works, and thus plays an important role in the development of theory. Because vision is not enough to provide the stuff necessary for the development of theory, it must be tempered by logic as the second key ingredient in turning the former into theory. In this way, logic, facts and other objective phenomena come later, but only after feeding on the raw material set forth by the vision (Sowell, 1987).

While visions in tourism development are typically market-driven, there are cases where the vision for national treasures has been determined by the values of many stakeholders. Such was the case for Banff National Park, Canada, where a Task Force determined that all stakeholders who maintained an interest in the environmental, social and economic aspects of the park would be included in a visioning process. This was less a consulting process, which is usually the norm, and more a process of asking these groups to take responsibility for the park rather than their differing positions. The values built into this process are articulated in the following passage from the final report:

> This vision is the culmination of a collaborative effort by a large number of Canadians to whom the Banff-Bow Valley is of great importance. The heart of the vision is not in the words but the spirit of cooperation and collaboration in which they were written. At a crucial time in the life of the Valley, it is an attempt to reflect on the past, understand the present, and imagine the future. (Ritchie, 1999: 281)

By including so many different stakeholders in the process, the actions of these groups have inspired government to join in the efforts to preserve this natural wonder. Indeed, it would be politically suicidal to do otherwise with such momentum. This prompted the Minister responsible for the Valley to announce a cap on future developments and a return of 17% of the Banff town site to parkland.

Norms

Although the discussion will not centre specifically around norms, it is important to make mention of these because of the discussion on both normative theories of ethics and business and organisational culture to follow. Norms are standards or sets of criteria that are held by groups. Research in outdoor recreation suggests that it is difficult to manage natural areas because of the number of users and the diversity of use within these areas. It is more practical, some contend, to look at the normative use of these areas (i.e. the use tendencies of specific groups, such as anglers, hunters, birdwatchers and so on) in attempting to develop management strategies that will prove useful both ecologically and socially. In business it is not unrealistic to find situations where employees see the norms of the organisation as different from the ones within society. Consequently, if they are asked to make decisions in the organisation that go against their personal beliefs, as nested within society, then they can justify these as unacceptable and unethical in the context of the normative approach in business.

Classical Antiquity

The origins of classical Western philosophy lie in ancient Greece, where a transformation occurred in how people began to think about the world and human agency around the time of *c.* 600 BC. Up until this period, knowledge and explanation were dictated by myth, using the stories of gods and heroes as a means by which to explain the world. Decisions on whether to go to war, support social programmes over economic ones, and so on, were largely dictated on the basis of these mythical narratives. But as Greece became embroiled in political and economic turmoil, society slowly began to change and new questions were asked demanding different answers. The transition came in the form of the *logos*, or the formation of strict argument as seen as a more philosophical, scientific and rational way of viewing humans and the world (*logos* quite literally denotes the word meaning), as noted by Honderich (1995).

The period from *c.* 650 BC to *c.* 500 BC, known as the Presocratic period, was characterised by thought that attempted to uncover the basic meaning of life and how the world was formed through natural and cosmological forces. This included the search for the essence of the basic elements (arche) of the earth and universe, including water and fire. For example, Xenophanes (*c.* 570–475 BC) speculated on the unity and eternity of the universe, as well as on the position of mankind relative to other animals, by suggesting that a higher understanding of the world needed to be sought outside myth and religion. During the same period, Pythagoreans were concerned with numbers and how these were the root of our understanding of the cosmos. Numbers provided the foundation for the development of the four Pythagorean sciences, including arithmetic, geometry, astronomy and acoustics (harmonies), which later formed the basis of the Seven Liberal Arts, and became the philosophical focus of the Sophists, Socrates and Plato. From the work of the Presocratics, there appeared in Greece a period of philosophical enlightenment that would spawn recognisably modern forms of thinking. This entailed a shift to a focus on human behaviour that, in part, resulted from disillusionment with Homeric explanations of morality. The questioning of Olympian gods was most likely due to expanded travel and trade (David Malloy, personal communication).

As philosophy became more entrenched, a number of differing perspectives developed in attempts to explain human nature. The Sophists, for example, were a band of teachers who had neither a shared organisation nor a professional respect for one another. These individuals travelled throughout the cities of Greece giving lectures and championing discourses, not on nature and the cosmology, but rather on subjects that had social and ethical significance. At the heart of Sophistry was an attempt to convince by rhetorical skill, rather than by philosophical argument, that there was no secure foundation for knowledge (e.g. no absolute standards for morality). It was the will of the people who were in power that determined right and wrong.

Socrates

This form of scepticism, above, was in marked contrast to the approach of Socrates (470–399 BC), who, instead of employing skills to win arguments, like the Sophists, formulated an approach, called *elenchus*, that was based on discovering the truth through debates which were grounded in philosophical dialogue (Honderich, 1995).

Socrates used a form of pretended ignorance that was supposed to remove one's over-confidence on knowledge, a tactic supported by Rawls (1971) in contemporary times, and thus clear the mind for the development of views particular to the discussion. Socrates was famous for saying that the only thing that he knew was that he knew nothing. He preoccupied himself with questions related to the foundations of virtue – his chief aim it seems was to learn how to live virtuously – specifically with a focus on justice and the nature of the good life. Although scholars can find no evidence that Socrates wrote anything, there is agreement that he believed virtue lay in having knowledge, particularly self-knowledge. In this capacity morality had a spiritual connotation, which is a state thought to be well beyond extrinsically based motivations to obey constructs such as laws. This self-knowledge allows us to know who we are and how we should act. Such knowledge could not be taught, but rather must be experienced by the individual in formulating his or her perspectives on acceptable behaviour. Socrates was imprisoned and later killed as a scapegoat for the declining state of the Athenian empire, where he was charged with corrupting the youth of Athens for instilling in their minds the notion that existence and being were not a function of gods, but rather that society was within the realm of human agency.

Plato

The aristocrat and student of Socrates, Plato (*c*. 428–354 BC), who wrote over 30 dialogues, gained recognition through his works, *The Apology*, in reference to the death of his teacher, Socrates, the *Gorgias* and the *Republic*. In these he wrote brilliantly on a wide range of topics, including ethics, epistemology, politics, metaphysics, psychology, mathematics, education and the arts. Much of Plato's work is hinged on the societal injustice that was built into a society and could destroy the virtuous Socrates. This is seen most explicitly in his *Republic*, which raises many challenging philosophical questions through views on the individual, the state and morality. Plato's ethics was tied to a society in which the idea of the 'good', standing above all other ideals, provided a foundation for individual and social action. All other ideals, including knowledge, laws, science, social practices and so on, were to be grounded in the good on its own. As such, being wealthy could not be equated with 'good' because this would lead to confusion about the qualities of each. 'Goodness' therefore must exist independently.

Plato's political philosophy led him to believe that the ideal state is one with three classes of individuals. The first or ruling class is one made up of philosopher guardians who were responsible for administering the state by virtue of their education and knowledge. The second class was comprised of police and the army, who were responsible for the protection of the state. The last class, the general population, contained the masses, who would produce goods for the state and in return be protected by the wisdom and strength of the other classes. In Plato's utopia, only the guardian class was able to understand eternal truths, so the organisation of a society that is to avoid deception was best accomplished through the rule of philosophers. Warriors were assigned qualities like courage and bravery, whereas the masses were characterised by their sensual appetites (Saul, 2001). Plato's absolutist approach to philosophy was practised through the Academy, which was dedicated to teaching the brightest scholars of Greece. But because he assumed that morality for the individual was the same as the morality of the state, he opened up the debate, taken up by Aristotle, of the closed society, where decisions by individuals were judged worthy only if they conformed to the wishes of the elites who ran these societies, despite the fact that many of these rulers could be corrupt. Plato's influence is woven into Christian theology, but also more generally throughout society in the way we think about culture and philosophy (Collinson, 1987).

Aristotle

Aristotle, student of Plato and founder of the Lyceum and of logic (the science of argument, which seeks to detect cause–effect relationships between things), moved away from Plato's utopian vision of society by focusing on the more pragmatic aspects of morality. While Plato believed that only a precious few could attain eternal truths, Aristotle believed that the common man could be moral through practice and through experience in much the same way that Socrates thought. Individuals were able to accomplish this by striving to be virtuous as the result of choosing between excess and deficit (the golden mean). For example, we can be brave by mediating between recklessness and cowardice. Aristotle's philosophy was decidedly ends-based, meaning that what qualifies as good was not an abstraction in the Platonic context, but rather one which was goal-oriented. Goals pursued for their own sake enabled individuals to flourish, which Aristotle called *eudaimonia*. Living a virtuous life, then, allowed the individual to flourish. These principles are laid out in his *Nicomachean Ethics*, which is organised into 10 books (Aristotle, 1998). Moral excellence is described explicitly in Books II to IV, with discussion of the general accounts of moral excellence and moral virtues in detail. For example, in Book II we see clearly the characteristics of Aristotle's ethical perspective:

'Moral virtue, how produced, in what medium and in what manner exhibited'

(1) It, like the arts, is acquired by repetition of the corresponding acts.
(2) These acts cannot be prescribed exactly, but must avoid excess and defect.
(3) Pleasure in doing virtuous acts is a sign that the virtuous disposition has been acquired.
(4) The actions that produce moral virtue are not good in the same sense as those that flow from it: the latter must fulfil certain conditions not necessary in the case of the arts.

In 'Definition of moral virtue'

(5) Its genus: it is a state of character, not a passion, nor a faculty.
(6) Its differentia: it is a disposition to choose the mean.
(7) This proposition is illustrated by reference to the particular virtues. 'Characteristics of the extreme and mean states: practical corollaries'.
(8) The extremes are opposed to each other and to the mean.
(9) The mean is hard to attain, and is grasped by perception, not by reasoning.

Virtue is thus a result of the agent's conscientious choice to select the golden mean – the middle course – between excess and deficiency. Supreme happiness can therefore ensue through a life of active contemplation.

Hellenistic ethics

Along with the physical expansion of the Greek empire came a cultural expansion through a mosaic of ethnicities and customs that were absorbed into Greek culture. New philosophical schools emerged during this time, including Cynicism, Stoicism and Scepticism. Cynicism emerged as a philosophy based on the outward denial of customs, institutions and opinions of the masses as the foundation for happiness. Those living by this school of thought felt that the individual was better off by retreating into his or her own world, away from the corruption of society, into a life of self-sufficiency. Stoics believed in a 'natural law', which viewed all life living in harmony, where virtue was the guide for all ethical action. Stoicism underscored the notion that we are all citizens of the cosmos, a unitary living organism as noted in the contemporary philosophy of Gaia, a perspective outlined by Diogenes. There is a turning away from the corruption of the city state towards a life of independence and self-determination. To be stoic suggests that the virtuous individual is able to remain calm in an otherwise chaotic and hostile environment. Scepticism was the school most hostile to ethics and more broadly to philosophy. Its proponents, referred to as philistines (self-interested individuals who felt that philosophy and matters of truth had no place in the world) advocated a complete move away from philosophy as it offered no answers to society's most meaningful questions. For example, if Hellenistic scholars argued that ethics was primarily about determining the *summum bonum*, or highest good (Seth, 1896), sceptics might argue that it is impossible to assemble all the knowledge required to know what the *summum bonum* is and how it is achieved (Barnes, 1990). Scepticism is often thought to be the precursor to nihilism, or the belief in nothing and the expression of little emotion, or in the words of Frankl (1985: 152), that 'being has no meaning'.

While it is not the purpose of this book to document the history of philosophy, and more particularly ethics, the foregoing discussion is essential in establishing a brief historical overview of ethics from the Western context. Roman society was influenced heavily by Greek philosophy through translations by Cicero, and applications by Seneca and Marcus Aurelius. By the Middle Ages (approximately 1000 years spanning from the mid-400s to the mid-1400s), the collapse of the Roman empire, countless wars and the influence of the Church all contributed to a stagnation of knowledge. Although scholasticism emerged between *c.* 800 and *c.* 1400, primarily as a means by which to merge Aristotelian philosophy with Christian teaching, philosophy was subsumed under theology. Important to philosophers of the time was the need to defend Christian doctrines through philosophy, with a focus on immortality, love and monotheism (one God). St Augustine, in the 400s, for example, wrote that the supreme goal of man is to strive for mythical union with God. In his work, the *City of God*, man emerges as the victor leading the way to secular philosophy and science, thus having direct implications for the Renaissance.

It was only during the Renaissance, alongside the Reformation, that there was renewed interest in art, history, philosophy and architecture, which quite literally transformed Europe into a centre of intellectual and technological progress. One of the main philosophical advancements to come about at this time was humanism (more on this later), which was premised on the union of theory and practice in the creation of men of liberal education, sound moral foundation and well-cultivated manners (Delius *et al.*, 2000). Human achievement was thus the cornerstone of humanism, with less of an emphasis on the role of God in human agency. The Renaissance catalysed new systems of thought that reached a zenith in the 17th and 18th centuries, through the works of John Locke, David Hume, René Descartes, Francis Bacon, Thomas Hobbes and Adam Smith. The influence of these individuals on ethical thought will be touched on briefly in the pages to come.

Teleology

G.E. Moore, in his *Principia Ethica*, thought that the primary foundation of ethics was to examine the meaning of good, which he viewed as a meta-ethical question, rather than the things that are good, which is the realm of normative ethics (Moore, 1903). He thought that good was a term that was indefinable, touching off a debate on questions related to whether good or right were the basic concepts of ethics. Rawls, for example, argued that good must be defined independent of right (Rawls, 1971). Teleological theories are said to define the good independently from the right, and the right is later defined as that which maximises the good. Conversely, deontological theories fail to specify the good independently from the right or do not interpret the right as maximising the good (Honderich, 1995).

Teleology is an ethical approach which suggests that an act is right or wrong depending on the consequences of its performance or non-performance (Brody, 1983). As teleology is 'ends'-based, it focuses on the achievement of the optimum outcome of a situation. Tourism researchers have noted that, because teleology is ends-based, it allows individuals to take a path to the future, with the opportunity of releasing themselves from past traditions and/or dogmatic views (Fennell & Malloy, 1995). This consideration is important in that it allows the decision maker to adapt to new and ever-changing social trends and circumstances. The basis for our ability to determine good or bad lies in our effectiveness at understanding the balance between these two forces. In the words of Frankena:

> an act is right if and only if it or the rule under which it falls produces, will probably produce, or is intended to produce *at least as great a balance of good over evil* as any available alternative; an act is *wrong* if and only if it does not do so. An act *ought to be done* if and only if it or the rule in which it falls produces, will probably produce, or is intended to produce a *greater balance of good over evil* than any available alternative. (Frankena, 1963: 13)

Teleology is dominated by three main theories: utilitarianism, hedonism and egoism, each of which is described in more detail below.

Utilitarianism

The theory behind utilitarianism, or 'the greatest good of the greatest number', was first proposed by Francis Hutcheson (1694–1746/7), a professor at Glasgow College (Boorstin, 1985). God, Hutcheson lectured, governed the world in accordance to this doctrine such that all would be able to share in his glory. This perspective was later adopted and refined by Joseph Priestley (1733–1804), a chemist and Unitarian minister, who argued for the 'greatest happiness for the greatest number' during the time of the Industrial Revolution. Priestley was brother-in-law to the ironmaster John Wilkinson, and so had the opportunity to significantly influence many industrialists of the day. Although the Industrial Revolution has traditionally been viewed as an era of hardened capitalists driven by self-interest, Priestley's vision was to use industrialism to create a transformation in the lives of the poor, through a higher standard of living and health. The new products of the Industrial Revolution – brought about through mass production in factories – were to provide these amenities not only for people in the cities, but also for those residing in the countryside.

The individual most often credited with the formulation of the greatest good for the greatest number principle, however, was Jeremy Bentham (1748–1832), who gained fame by formulating a calculus, in the figurative sense, based on the notion that humans were governed by the two sovereign masters of pleasure and pain. The action that provided the greatest amount of pleasure (the principle of utility), and therefore happiness, was the most morally correct (e.g. laws should be passed if they maximised pleasure for the masses and minimised pain). The type of calculus that Bentham devised is often manifested through practical examples of pleasure and pain units. For example, plus four units of happiness (+4H) would mean that someone would be classified as being overwhelmed with joy, while minus four units of unhappiness (–4H) could make someone close to suicidal, along the proposed scale. After summing the results of polls on, for example, the need for a roundabout in the centre of town, one would be able to collectively determine the units of happiness and unhappiness for the entire town. If the majority were in favour of the roundabout, on the basis of their collective happiness, then democratically the building of the infrastructure was in the best interests of the collective.

An example of the cost-benefit mentality of utilitarianism can be demonstrated through the work of McAvoy (1990: 70), who suggests that in the development of an appropriate environmental ethic for parks and recreation we should be able to use a 'moral goodness standard where good activities are those that add to the health and well-being of the individual, the society, and the environment'. Therefore, an activity like off-road trail-riding might be lower on the moral goodness scale than walking on an interpretive trail. The calculation needed to determine what is good and bad about these activities, for whom, and in what environments, would be a difficult task. There would also need to be a consideration of people's rights and freedoms to pursue activities within the precepts of the law. Prioritising activities is a very different task from legislating them, and, in the former, we run the risk of violating people's rights.

One of the most noteworthy advocates of utilitarianism was John Stuart Mill (1806–1873), who stated that actions are right to the extent that they promote happiness and wrong whenever they produce unhappiness (Mill, 1861/1979). He wrote that the progressive change that accompanies each successive age in society creates changes in our character as the result of outward circumstances. We have more knowledge, but are we more knowledgeable? Do our memories allow us to recite, recall or recount? The innumerable stimuli that we encounter on a daily basis may play with our ability to contain information. Mill was therefore a hardened blank slater, and this was reflected in his utilitarianism, especially regarding differences in races and individuals, which he felt was subject to circumstance. Where Mill deviated from Bentham was on the basis of values. While Bentham chose to measure happiness according to calculus (a decidedly quantitative view of happiness), Mill did so in consideration of social goods and qualitative values. That which is good can only be determined if it has broader social value. For example, Mill might suggest that, while one may view the Taj Mahal as an attraction that generated the most amount of pleasure, another may find a smaller attraction more pleasurable. The only way Mill could discern happiness in such circumstances was to leave the choice to someone who was able to appreciate the value of both attractions.

Utilitarianism is further subdivided into act- and rule-utilitarianism. In act-utilitarianism 'an act is considered morally obligatory if and only if it produces a greater balance of pleasure over pain, or of desire satisfaction, than any alternative action available to the agent' (Honderich, 1995: 890). For example, if a tour guide has two options: one to take tourists on a two-hour hike and the other to take them on a one-hour hike with the chance to sit by the waterfall for the second hour, the best act or option would be the one that produces the most amount of happiness for the group at that time. It may also mean that an act is superior if it produces greater overall well-being. For example, if we can only save one of two people on a desert island, one being a surgeon and the other an unskilled labourer, it could be argued that we should save the former because he or she has the ability to bring about more happiness in the future through saving lives than the unskilled labourer. Act-utilitarians therefore can break moral rules, and justify this, if they can create greater happiness by doing so. Actions are thus evaluated in terms of their own consequences, and not a predetermined set of rules or social norms. By contrast, rule-utilitarianism maintains that good ends need to be achieved through some adherence to a set of rules. As such, the rightness of the action becomes a function of these rules, and not necessarily the act itself. We thus need to obey rules that have been shown to produce the greatest happiness for the greatest number. So, an act-utilitarian could justify visiting a marine park that holds captive orcas because he or she is writing a newspaper article on the pros and cons of these parks (an individual act under exceptional circumstances and in light of the consequences of this act), whereas a rule-utilitarian would view travel to such parks as completely unethical – as a general rule – if the money generated from tourism went towards the purchase of more orcas, especially given the short lifespan of these animals while in captivity.

Although utilitarianism is considered by many to be a strong ethical viewpoint, it has been criticised for its inability to consider matters of social justice (see the example of the moral goodness scale in parks, above). The argument is that, if happiness is maximised for the greatest good, it must be achieved to the exclusion of minorities in order to secure greater benefits (e.g.

the displacement of a minority group to make room for the development of a hotel complex and associated infrastructures). Therefore, the mild happiness of the majority may severely outweigh the misery of a few. When we focus too much on the ends themselves, instead of the means, we have a tendency to treat people wrongly. What is good for the collective is not necessarily good for other groups that may be further marginalised through utilitarian decision-making. There is also the problem of time. We just do not know when or for how long to calculate the consequences of our actions. There may be immediate benefits to local people from the development of a hotel complex, but these may wane as time goes on, especially given the complexity of the social and economic system in which the hotel is enmeshed and the powers that act to control these.

Hedonism

In reference to the quality of pleasures discussed by philosophers, the utilitarianism of Bentham and Mill is said to be hedonic on the basis of the greater commitment to more pleasure over pain. This hedonistic motive clashes with the older Aristotelian (ancient) perspective, which maintains that flourishing, or *eudaimonia*, is not the result of inner sensations (e.g. believing one is admired by others, but in reality being ridiculed behind one's back), but rather in having a relationship with the world – the life worth living and the conditions of its achievement. In the end (*telos*), what really matters is the reasoned and purposeful engagement with others that can lead to a good, healthy life. This is what is meant by the 'natural' or 'intended' life, and must be separated from the hedonistic context brought forward by Bentham.

Hedonism regards pleasure as the goal that renders participation in an activity worthwhile. For centuries people have debated as to what constitutes the good life (e.g. relaxation), and how it should be attained (e.g. travel to warm climates). Hedonism surrounds issues related to the greatest good and least amount of pain for the individual. It is egoistic in nature, and has traditionally treated pleasure and pain respectively as the sole good and bad in the life of a human. This stance, however, has been broadened to include approaches that seek to discover whatever makes people happy. Variations on this theory consider the forces that shape short-term or immediate pleasure and long-term pleasure (Brandt, 1959; Gosling, 1969; Sprigge, 1987). Pain includes all sensations that are aversive or that are disliked by the agent. Similarly, pleasure is a sensation that is difficult to quantify. It therefore becomes a matter of allowing the individual to determine what is pleasurable. It could be argued that an overnight flight to Europe is more likely viewed as a painful experience to a frequent business flyer than to someone who is a first-time pleasure traveller and who has never experienced such a flight before.

Egoism

The main distinguishing feature of egoism is that the only ultimate goals that a person has are ones that are self-directed as an end in and of themselves. Caring for the welfare of another only takes place if there is an expectation of a benefit to the agent. In this regard, egoism is characterised as follows: (1) one has as a basic obligation the promotion of the greatest possible balance between good and evil; (2) in making second and third person moral judgements an individual should go by what is to his or her own advantage; and (3) that in making such judgements an individual should go based on what is to the advantage of the person he is talking to or about (Frankena, 1963: 16–17). Selfishness has been discussed by Blackburn (2001), who says that self-interest is one of seven threats to ethics within society. Although we can say or pretend that we are anything but selfish, we appear to be very good at deceiving others and ourselves. People may ask us about our feelings on things like staff management, and we may tell them that we take good care of the people who work for us. But our actions may speak a different language. What we do for our staff may be good in the context of our own world view, but then again there is a universe of other examples of staff management, compared to which we fall short. This aspect of selfishness is tied to the concept of egoism.

Psychologists have suggested, not unlike the literature on self-interest and altruism, that ego-satisfaction is human nature and self-love is the only basic motivation of individuals. Each of us strives to do what is best for oneself as regards knowledge, sex, happiness, power and social status (Butler, 1950). Contained within this is the notion that, while we are driven to satisfy our own needs, the drive to help others may be strong enough to overrule our own selfish pleasures. This is why altruism, from the biological standpoint, is defined as it is – where there is a net reduction in the benefits accrued to the individual through his or her altruism (Chapter 2).

In steering our way through the channels of hedonism, egoism and altruism, researchers have suggested that egoism differs from altruism in that the latter refers to the fact that agents sometimes care for the well-being of others as an end in and of itself. More specifically: (1) altruism is incompatible with egoism; (2) hedonism is a type of egoism (all hedonists are egoists, but not all egoists are hedonistic; an egoist may be motivated to make money as an end in itself, but a hedonist may not be motivated in such a way at all); and (3) an egoistic ultimate desire is self-directed, whereas an altruistic desire is other-directed (Sober & Wilson, 1998).

Following from the previous discussion, Frankena (1963) notes that an ethical egoist need not be called an egoist because his actions are determined by honesty and modesty and might benefit both himself and those around him. Egoism is also differentiated on the basis of differences between being selfish and self-directed. Being selfish carries with it negative or disapproving characteristics. One cannot state that Julie is being selfish by leaving a party if she is experiencing a severe headache. She is being more self-directed in her actions, not selfish. However, she may be considered selfish if she chooses to leave a weekend family reunion if she does so in order to increase her quota at work. In this way, egoism should be viewed as self-directed behaviour and not necessarily selfish.

Deontology

In general, deontology, or non-consequentialist theories of ethics, can be grouped into three categories: theology and the Golden Rule, Kantian ethics and social contract. To these we could add theories of justice, which are essentially deontological, but provide somewhat more guidance for decision makers than other deontological theories. Justice therefore has an objectifying aspect, in the sense of deontology, but also a fairness element in regards to social justice (justice will be examined in greater detail in Chapter 4).

Deontologists contend that an action (or inaction) is right or wrong even if it does not promote the greatest good for the greatest number. Unlike the teleological perspective, there is less of an overriding comparison between the moral value and the non-moral value. It is the moral principle itself or the set of moral rules, and one's duty to uphold these, that is most important to the deontologist. Ergo, the duty towards a moral principle such as 'never lie' is the most important factor in determining whether something is morally right or wrong. Most important in deontology, therefore, are the normative rules, guidelines, duties and/or principles that have been established by society. The many codes of ethics in use today are deontological (in fact, the word 'deontology' means 'codes of ethics' in Greek), as they only tell actors what to do and what not to do without discussing the implications of their actions (although it could be argued that codes of ethics are rule-utilitarian as well; see Malloy & Fennell (1998a)). Using the example above, a deontologist code of ethics would prescribe that tourists and operators should 'stay at least 100 m away from killer whales', even though venturing any closer may or may not have any effect on the whales. Regardless, the deontologist tourism operator or tourist follows the letter of the law without compromise. Since the actors are not told why they should not venture any closer to the whales, the consequences of their behaviour are not the focus of the message (see Malloy & Fennell, 1998b).

Deontological ethics can be further examined through the use of a situation developed by Garofalo and Geuras (1999) on ranking subordinates on an evaluation scale. It is a difficult thing to objectively rank subordinates because favouritism and the desire for approval often creep into the process. Because of these factors we often find that subordinates are ranked higher than they might otherwise be from a strictly objective standpoint. Those who are 'good' are ranked as 'very good' and those who are 'very good' are ranked as 'excellent'. However, if all of us were to follow the universal rule of 'rank subordinates one cell higher above their deserved rank' we would encounter problems because the rule would be self-defeating. Taken to its logical end, perhaps questions as to why 'x' was rated higher than 'y', and so on, we may find that all are rated at the highest level creating a useless system, because the rule that was developed to advantage someone, when universalised, would render the system useless and eliminate any advantage. The Kantian deontologist would suggest, in an effort to avoid absurdity, that superiors should rank subordinates according to their true merits. Unfortunately, lying, stealing, deceit and over-evaluation cannot be made consistent universal rules. In the above example, no one would really know their true level of merit (more on Kantian ethics to follow).

Deontology, like teleology, can be split into rule and act domains. Rule-deontology focuses on judgements that are particular, over ones that are more general. The values upheld in the rule-deontology perspective are still independent of whether or not they promote the greatest good for the greatest number, thus steering clear of the utilitarian position. They differ from the act-deontology view by suggesting that acts need not be hard and fast rules either through induction or deduction. In this latter perspective, there are no firm standards for determining what is right or wrong. Judgements are thus made and rules based on these judgements more from an intuitionist perspective – with no guiding principles. The standard that upholds something as right or wrong is subject to rules that are either concrete or abstract. In the former context we could argue that rules can be loosely derived, such as 'We should always stay on the path', or, in the case of the latter, rules can be personal viewpoints on the relationships between one agent and another: 'I will treat all people the same because I can no more determine differences in my clients than rearranging the stars in the sky'.

The act-deontologist would suggest that 'we should always stay at least 100 m from marine wildlife'. There is some basis from which to make an assertion based on research or other forms of knowledge. We always need to adhere to this rule because it is based on knowledge or fact. The only manner in which to overrule this guideline is through more accurate knowledge. We cannot dispute a rule, that is, but we can replace the rule in a form of hierarchy such that there is an evolution of rule-making that is built into, for example, the protection of heritage sites. This may be the case at Stonehenge where, over the past many years, there has been an evolution in the management of the resource for its overall protection.

As deontological ethics ordains how persons should act, there is the recognition that judgements must be based on some higher authority. An interesting element of deontology is the universality of rules and guidelines. Kant argued, via the categorical imperative, that ethical rules could be established on the basis of universal applicability (Kant, 1788/1977). Accordingly, moral or ethical laws are maxims that can be applied universally (e.g. why put off until tomorrow what one can do today?). There is the argument that, because the world is a much smaller place, at least from the perspective of communications, such universalities are much more applicable today than ever before. The principal concern over the deontological perspective is in understanding where individual duties originate? Who lays down the law? Do some people have special rules and guidelines over others? If so, why? (Broome, 1991; Brody, 1983; Gowans, 1987; Kohlberg, 1981; Raphael, 1989).

Theology and the Golden Rule

In the Western world, most people either implicitly or explicitly live by a code of conduct that is tied to a religion and expounded by, for example, the Bible, the Koran or the Bhagavad Gita. The Ten Commandments (do not murder, commit adultery and so on) that Moses supposedly received from God and brought down Mt Sinai provided a template from which to organise the ethical beliefs of society. It is the sacred duty of those who subscribe to these religious doctrines to follow and abide by the rules that define the religion, and that makes them deontological.

Religion (defined as the worship of supernatural forces) is not the basis of morals, but rather an aid to them (Durant, 1935/1963). Morals exist without religion and must therefore be given higher priority. Religion does not sanction an absolute good, for there is none, but rather the norms of society that emerge from our social systems. This is substantiated by Blackburn (2001), who notes that although religion is often seen as being synonymous with ethics, it is not the foundation of ethics but rather a showcase or a symbolic expression of ethics. So it is not a source of the standards that we call ethics, but rather a projection of our standards dressed up through an absolute authority. This dichotomy may be examined in the following thought experiment:

> What would be the right thing to do if God had commanded people to be selfish and cruel rather than generous and kind? Those who root their values in religion would have to say that we ought to be selfish and cruel. Those who appeal to a moral sense would say that we ought to reject God's command. This shows – I hope – that it is our moral sense that deserves priority. (Rachels, 1990, cited in Pinker, 2002: 189)

Religion is premised on the belief of a known God, with the corresponding development of rituals and moral codes that stem from this belief. Philosophy holds no such belief, but rather examines the logic behind religion and uses certain tools or methods of investigation in seeking truths. Religion, as such, is but one part of philosophy.

A universal commandment, which can be found in most societies and doctrines, both secular and religious (Donovan, 1986), is the Golden Rule, which is written as: *Do unto others as you would have them do unto you.* What is suggested by the Golden Rule is that each and every person should be treated as an end rather than just a means. We are all, therefore, no less or no more than one, equally.

In light of the differentiation between deontological norms that exist across societies, some suggest that if we were to live by one commandment only it would be the Golden Rule that would be the one rudimentary guiding principle (Heintzman, 1995). This is reinforced in a wide range of different religious scriptures, including those of Judaism, Christianity, Islam, Hinduism, Jainism, Buddhism, Confucianism, Taoism and Zoroastrianism (Hick, 1992). It exists as a testament to the fact that moral systems exist not only in the Western world, where we find a great deal of the scholarship on morality, but also in other cultures. It says that we have the ability or willingness to put ourselves in the shoes of others (there is thus the element of reciprocal altruism built into the Golden Rule).

Unhappily, the Golden Rule has been used to condone certain forms of depreciative behaviour, including the practice of inflicting pain on others. But we should be clear that these masochistic tendencies and interpretations of the Golden Rule are the practical choices of specific individuals. This means that although individuals would not like to experience the same level of discomfort (i.e. do unto others as you would have them do unto you), the Golden Rule was never intended to operate outside other principles of good conduct.

Kantian ethics

The philosopher Immanuel Kant felt that, in consideration of what is moral and non-moral, the former is premised on the sense of duty rather than doing what we think is right or focusing on the consequences of our actions. To Kant, by following well-founded rules or 'imperatives', good consequences would naturally unfold. In following these rules there can exist a state of cognitive dissonance because the duty to act from the societal context and that of the personal or existential context may be at odds. This often creates a state of unhappiness for the agent who must conform to the rules of society rather than their own personal motives.

Kant's philosophy was grounded in the notion that humans are rational beings with the ability to reason and perform complex intellectual tasks. But this theoretical reason also plays against the backdrop of practical reason, or the need to serve their own goodwill. Goodwill, Kant notes, is the underlying motive that propels us to do good for the benefit of society, and our practical and theoretical reason helps us to accomplish this. He notes this in his first categorical imperative (absolute command), which is very close to the Golden Rule, stated as follows: *'Act according to a maxim that you can will to be a universal law'*. In more closely examining this rule we find exceptions. Society cannot have universal laws if negative virtues like lying and stealing are morally acceptable (the case in some circles) (Garofalo & Gueras, 1999). Furthermore, as we saw above, one cannot make laws universal if exceptions are made for personal advantage; the exception renders the law *not* universal. Also, things must have value as ends in and of themselves, not simply as means to an end. Art has meaning not simply as a decorative piece on the wall or because of its value on the open market, but because it is a thing of beauty in and of itself – it has intrinsic value.

If an individual is afforded special value, we must also extend this to all members of humanity. Kant expressed this in his second formulation of the categorical imperative: *'So act as to treat humanity, whether in thine own person or in that of any other, in every case as an end withal, and never as means only'* (Garofalo & Gueras, 1999: 70). Universality of the first iteration gives way to what is to be valued. Kant claims that only those who are capable of rational thought are able to make moral decisions; a status generally reserved for humans. Other animals or the lower animals are driven by innate instincts – inclinations in Kant's words – which do not allow for rational thought (Fennell, 2012a). More specific to human agency, even acts such as giving money to the poor are deemed unworthy of moral status because they are designed to promote one's reputation or gratify one's ego (see Chapter 11 in reference to volunteerism), and must be classified as morally neutral actions. As such, acts that are truly moral are those that are performed dispassionately and out of one's duty as a human being.

As such, we may choose not to hire people as waiters for fear that we may be using them as a means to an end. In considering this, Kant notes that the interpretation should be that a person is never always used as a means to an end (they may at times be used as a means), in efforts to secure his or her own status as an end. The waiter can thus be employed if the superior maintains the proper attitude towards the employee. Paying these employees fair and equitable wages is a way to secure the belief that these individuals are ends in and of themselves; conversely, appropriate work ethic by these employees is also a way of treating the employer as an end, rather than simply a means. Kant, in his practical imperative, however, also felt that one should never use another 'simply' as a means to an end. Thus we can hire and use a waiter, but the employer and customer must treat the waiter with dignity (and all that it involves) as we use them to accomplish organisational goals.

As Garofalo and Gueras note, the value of rational beings as ends in and of themselves provides the foundation for Kant's notion that autonomy comes only through the freedom to exercise reason: freedom from emotions, desires, wants and the control of others (Garofalo & Gueras, 1999). Treatment of people as ends leads to freedom, which might also lead to the ability to make solid moral decisions. But as the authors note above, all people cannot be free at all times. The freedom to walk about Stonehenge, for example, conflicts with other people's notions of protection or conservation. This conflict led Kant to a third iteration of the categorical imperative, which suggests that a 'truly consistent world is one in which each person acts in accord with the "idea of the *will of every rational being as a universally legislative will*"' (Garofalo & Gueras, 1999: 72). In an effort to move away from the dislocations and inequalities of society, therefore, we ought to strive for the development of communities where all people are treated as equal and as ends in themselves. Consequently, we all have each others' interests in mind in our actions as well as our own, in the development of the ideal state. People are thus able to live in a society where all are 'free and equal participants in a conflict-free pursuit of their own ends' (Garofalo & Gueras, 1999: 73).

While this is a noteworthy goal it still does not mean that the possession of freedom will be used properly in economic, political or ecological circles such that conflict will be minimised or eliminated. Although laws and codes of ethics are important in facilitating the ideals of a conflict-free society, society is still reliant on the moral behaviour of its citizenry at the very core, as Kant suggests.

The categorical imperative is an absolute command, which means that all moral laws under the Kantian perspective would have no exceptions – for example, no exceptions on the basis of situation, setting or time. Such rigidity becomes problematic in a dynamic world. Garofalo and Gueras note that exceptions could only be made on the basis of higher considerations, but they ask what can be a higher consideration than morality? Making exceptions therefore creates hypothetical commands that have no absolute power (Garofalo & Gueras, 1999). For example, suppose we have two rules. The first one says 'Do not kill endangered species'; while the other says 'Do not let your family starve'. To be good providers parents must be able to ensure the survival and good health of their children. But at what cost? Some would say any cost. But this puts the agent in the position of violating other rules, such as the first rule, above, in order to secure the second. Citizens of the developed world might often wonder why those in the lesser developed countries behave as they do (e.g. slash and burn agriculture) and vice versa, perhaps because we cannot grasp the magnitude of the problems that catalyse certain behaviours. Citing the work of Singer, Garofalo and Gueras write that we can deal with the problem of the generalisation of rules by noting the importance of situation or circumstance: '*What is right (or wrong) for one person might be right (or wrong) for any similar person in similar circumstances*' (Singer, 1961: 5, as cited in Garofalo & Gueras, 1999).

Kant explains that good deontological acts are done dispassionately without any satisfaction derived from the acts. This means that feelings of benevolence and altruism are morally insignificant. But we are not unfeeling, dispassionate machines. We go back to our earlier discussion of the donor who wishes to garner public approval through gifts of large sums of money.

Is this for that person's own societal benefit (self-interest), or is it simply a matter of the benevolence virtue coming to the fore? Kant would no doubt reject the action, in either regard, because of the underlying motives and the associated consequences of the action.

Social contract theory

Jean Jacques Rousseau's philosophy of the social contract, first published in 1762, analysed political authority with reference to the terms of a contract and how subjects reacted to this authority. Rousseau's work is very much a political commentary, as the terms of these contracts are a manifestation of the legitimate political authority. The term 'contract' has a legal connotation inferring that there is an analogy between political and legal obligation. This is an analogy and not necessarily a reality because the political contract is a much broader force than the legal contract, which is merely one aspect of the former. As such, the social (political) contract must rest on processes and systems that are developed prior to civil law (Lessnoff, 1986).

What comes before civil law is natural law, as we have seen in previous chapters. In his book *Leviathan*, Hobbes (1651/1957) underscores the importance of natural laws as guiding human agency. Even before the development of social institutions, such as politics, men were seen to have a natural right of liberty. But this belief in the natural state of man works as a foundation, not as a factor that guarantees human stability. As Hobbes notes, there must be general agreement among people that a covenant (contract) is required, through legitimate political authority, so that the individual can obey the laws of nature – what he referred to as the true moral law. Consequently, it is only through legitimate political authority (Hobbes used the social contract to defend authority) that we can manage to survive. He was clear in illustrating that the political state was not natural to individuals, but that it was essential to their renouncing their natural liberties and unrestrained desires. Furthermore, this would suggest that the state of nature was naturally a state of war.

The philosopher John Locke felt that men had the right to accumulate whatever possessions they could, thereby supporting the inequality of private possessions, through work and by abiding by the rules of the social contract. In this he championed the cause of a justification of the right to resist political authoritarianism in the name of individual rights to life, property and liberty (Lessnoff, 1986). Locke acknowledged that natural laws and rights, if disrespected and left unchecked, would dislocate the systems within society. Government and the political institutions for a civil society are therefore required to ensure that people respect these natural tendencies, and, further, there is the need for the individuals to submit or consent by their own free will to be members of a civil society. Consequently the aspect of will is critical in the development of a state governed by the individual into one that is governed by the collective. Locke also realised that while submission to a political institution was important, such institutions and rulers could not enforce such submission under conditions of absolute rule. In such cases there remains only one ruler and one judge of this rule. Impartiality, a common judge of individuals and rulers, is the only way to distinguish between a civil society and an uncivil one and therefore one that goes beyond the natural state of humankind.

The significance of social contract theory lies in four features. First, it is voluntaristic in the sense that it is dependent upon acts of human will. People (rulers and the rest of society defined as individuals or a collective) adhere to it because they want to be subject to it. Second, this voluntarism is consensual in the way that members of society submit to the contract. Third, that even though the social contract has collective implications, it is individualistic in how it must be embraced by individuals within society. Finally, the social contract must be tempered by rationalism (Lessnoff, 1986). That is, individual will must be rational in order to reach consensus. As such, there is a tension between freedom of choice (the individual's free will) and this rationalism, the latter of which suggests that we need to follow rather definite lines in our choices.

In this there is agreement to forego some of the independence of individuals for security from the state. It became the foundation for the economic man theories of the 20th century, which were based on cost-benefit calculations. The modern theory of evolution also relates to the social contract concept in that complex adaptations evolved to benefit the individual, not the community or the state. Social organisation evolved when the long-term benefits to the individual outweighed the costs. Many scholars argue that the theory of reciprocal altruism is the social contract but repackaged in biological terms. Thus, costs to the group must be outweighed by benefits to these groups, with individuals and groups struggling to mediate between self-interest and cooperation. The implications of this perspective are that we can no longer sequester the social sciences and humanities on the basis of unique and separate realities.

Social contract theory has been criticised as one of a few voluntaristic theories (i.e. people willingly, spontaneously, rationally and voluntarily give up their individual sovereignties and unite with other villages to become a state), which are never fully subscribed to by human groups. Instead, Carneiro (1970) advocates the theory of circumscription, which is an ecological hypothesis whereby geographical, resource and social factors emerged to be more important in forming the state. States were thus shaped not by social contracts but by: (1) geographic factors, such as mountains that restricted the movement of peoples; (2) the quality of the resource base, such as rich soils and sources of water, which were important in influencing people's movement; and (3) social circumscription, which was said to be a function of the proximity of villagers to the central and peripheral areas of the overall community (those in the more central hub easily comes under the influence of higher-ranking officials). The result is the development of higher-order political units, which are derived from nested hierarchies which were arranged according to family, tribe, chiefdoms and finally to the state.

Discourse ethics

The German pragmatist philosopher Jürgen Habermas has gained widespread recognition for his work on communicative rationality, particularly how agents pursue rational interests (Habermas, 1979, 1984, 1993). A major focus for Habermas is how the use of reason, or the potential for the use of reason, may lead to a more just society. He argues that as a species we have evolved and been endowed with an enhanced competency in communication, but this ability has been hampered in major institutions like the market, organisational contexts, and the state because of the pursuit of instrumental and strategic rationality over communicative reason. Whereas instrumental reasoning is said to be a means to an end, communicative reason is practiced in a community, as opposed to individually, and it is an end in itself. Its value lies in the ability to attain intersubjective understanding as stakeholders learn the rules of the game and how to appreciate the views of all involved. The concern is that the dominant entity, say a large business organisation, may use its power to achieve its own self-interest in the face of so many perceived recessive interests, as a reflection of the power imbalances that exist between participating parties. As explained by Noland and Phillips (2010: 43):

> the firm's decision to engage must be in no way strategic or instrumental, and conditions must be established for the engagement which ensure that power imbalances are removed from the interaction because... if engagement is to count as moral, the parties must communicate solely for the sake of reaching agreement rather than in order to pursue any particular interests. Moral engagement of stakeholders must be distinguished from strategic engagement.

The deontological connection here is the universal acceptance of rules and procedure, in the tradition of Kant, and the focus on democracy. As Reed (1999) affirms, organisations must be disciplined enough to rein-in their own interests by imposing voluntary restrictions on their power. What is problematic in this process, therefore, is balancing moral and strategic concerns (Gioia, 1999), especially when organisations are wired towards profit as a first priority (Noland & Phillips, 2010).

Virtue Ethics

A third normative theory of relevance to our discussion is virtue ethics, which frames the question: what sort of person should I be? Also referred to as objective teleology, this theory has an ends-orientation through its focus on *eudaimonia*. Aristotle remains one of the chief sources of knowledge on virtue ethics. He believed not only that one who lacks virtue lacks happiness, but also that moral virtue was a state of character that could only be obtained by shaping our desires in the correct fashion. Such qualities cannot be encompassed by rules, but, rather, a virtuous person should be able to perceive and act upon situationally unique moral requirements (Honderich, 1995). Important in this view on virtues was the belief that they were deeply ingrained and honed through training and habituation.

Virtue ethics has benefited from a resurgence of interest in the past few decades. It focuses on the traits of a morally good person rather than on the surrounding social environment. Examples of virtues include moderation, order, resolution, industriousness, sincerity, humility, justice, integrity, honesty, wisdom, courage and temperance (Rand, 1964: 25). What is deemed moral is the particular character trait that the act exhibits (Garofalo & Gueras, 1999). As such, the focus of virtue ethics is on the person, rather than the action itself. These character traits are part of human nature according to some theorists, who see virtues such as courage and honesty as part and parcel of the human condition in the same way that aspects of the phenotype (e.g. a nose, arms and feet) are (Foot, 1959). This is different from Cooper's perspective – that virtues are 'traits of character, more or less reliable tendencies to conduct oneself in a generally consistent fashion under similar conditions. Furthermore, virtues are not innate and, therefore, must be cultivated' (Cooper, 1987: 323). Accordingly, virtuous conduct is not merely conditioned reflex in an 'unthinking' manner, but rather a function of reason through pre-established attitudes and will. This contrast implies that virtues are difficult to understand, like other social science factors, because they are outcomes of how we use our human qualities. But it is important to keep in mind that, although virtues exist outside the realm of moral principles, they may be seen to help shape such principles (Benn, 1998).

Some scholars believe that there are two cardinal virtues held by people that, although separate, work in tandem and spawn a number of other virtues (Hart, 1984). The first, *eudaimonism* (defined as a system of ethics basing moral obligation on the likelihood of actions to produce happiness (*Concise Oxford Dictionary*, 1976)), focuses on the individual and the ability of the agent to be true to itself. The second, benevolence, focuses not on oneself but rather on others. From this perspective, as noted above, we are born with these virtues and these remain innate guides enabling moral behaviour.

Where virtue ethics has more of a social utility is in the understanding that no two persons, groups or situations are alike (Soifer, 1997). Virtue ethics is concerned with the ability of an agent to overcome self-interest in the care of others, one's community or even the natural world. The notion of seeking the golden mean (see Aristotle, above) in the context of virtue has salience to sustainable development, which purportedly represents the middle path of extremes in regard to community and the natural world. Individuals are also known to approach situations from different perspectives depending on a number of historical antecedents: a one-parent family history, an economic interest in tourism, a disdain for tourism, political motivations and so on. Consistent with this approach is the need to understand whether it is more virtuous to overcome strong temptations, or to lack such temptations altogether. In reference to tourism, some individuals might never think of developing a tourism resort in a sensitive region. For others, however, it would be more commendable for a developer to overcome the temptation to exploit an area. The argument against virtue ethics is that it is too vague in relation to which character traits should be seen as virtues and how they should be considered in different situations. In addition, there is the typical deontological and utilitarian response of knowing which actions are right, over wrong, without the need to tease out selected traits as being virtuous (Foot, 1978; Slote, 1992; Soifer, 1997).

Perhaps the most practical application of virtues is manifested in the leadership, character and judgement that people demonstrate in their positions of authority. Those who are successful leaders can balance virtues such as benevolence and *eudaimonia*. Although benevolence may make implicit the intrinsic nature of these actions, it is worth noting that benevolence should include actions for the sake of others and not necessary for others with other extrinsic motivations.

Existentialism

The subjectivist theory of existentialism is said to be based in the experience of what it is like to exist as a human being (Collinson, 1987). Its proponents are associated with two dichotomously opposing schools. The theistic school has its roots in the work of Sören Kierkegaard (1813–1855), Paul Tillich (1886–1965) and Martin Buber (1878–1965). The atheistic school evolved from the work of Friedrich Nietzsche (1844–1900) and includes theorists such as Martin Heidegger (1889–1976), Jean-Paul Sartre (1905–1980) and Albert Camus (1913–1960). Despite their fundamental metaphysical differences, both schools share the belief in radical human freedom. With a primary focus on responsibility for one's free will, the existentialist makes ethical decisions not on the basis of pre-existing ethical theories or moral authorities, but rather on the basis of their own subjective value set. A separate strand of commonality among some existential and phenomenological theorists is their focus on care for the other (Shuster, 2015).

Kierkegaard's philosophy of existentialism is based on an attack on rational humanism – the importance of reason and intellect – as the foundation of knowledge. He felt that we needed to take a 'leap of faith' in which passion and feeling of the individual were the sources of meaning and value

(Collinson, 1987). Although a passionate Christian, he was critical of organised institutions like religion that removed the individual's ability to think as an autonomous unit. He believed that life is comprised of a number of choices that the individual must make free of tradition, religion or reason. This duty to love the other is grounded in a state of indebtedness. As such, we must remain in love's debt; we are never finished or done. In taking care of the hearts of others we set ourselves free. Fennell (2008) argued that Kierkegaard's philosophy based on love and ethic of care for others should be the basis for responsible tourism; a contemporary type of tourism that is in dire need of philosophical grounding.

Heidegger's concept of Being-in-the-world pertains to the deep understanding of what it means for something to be rooted in the world, or as Steiner (1978) argues to be fixed, embedded and immersed in the physical, literal, and tangible facets of our day-to-day existance. Care (Sorge) according to Heidegger is the primordial state of Being, and is said to be the proper tool for rooting oneself in everydayness, and in efforts to remain authentic in the face of angst. Care, Heidegger argues, can be for many things including others, a care for the ready-to-hand, and for Being itself (Hornsby, nd). This concept of rootedness was used liberally by Heidegger. In *Discourse on Thinking* (1966) Heidegger contrasts two dichotomous spheres of thinking. Calculative thinking is based on utilitarian and instrumental patterns of thought, where productivity, empiricism, efficiency and productivity reign supreme. By contrast, meditative thinking compels us to open ourselves up to the broader horizon of interests and values. We may therefore consider the interests of the impoverished, marginalised cultures, or the natural world, for example, in formulating a more holistic and less rational approach to decision-making (Fennell & Malloy, 2007).

The concept of care is also a central theme in the work of Emmanuel Levinas, particularly in his magnum opus, *Totality and Infinity: An Essay on Exteriority*. A point of departure from Heidegger is Levinas' belief that dwelling or home is an essential phenomenological category worth investigating. For Levinas, it is the sensuous engagement with the other (Face-to-face) that is the benchmark for intentionality and human consciousness, which he recognises as the primordial production of Being (Mensch, 2015). The notion of freedom, so important to Kierkegaard, reemerges in Levinas:

> For Levinas, by contrast, self-separation and hence, freedom occur through the Other. They arise when I respond to the Other's alternative perspective. The distinction of my will from inclination or appetite occurs when I have to explain myself to the Other, that is, present him with my reasons for what I say and do. As for reason, its origin is precisely this need of explaining oneself. The question of reason, the question 'Why¿' originates in the encounter with the Other's alternative perspective. (Munsch, 2015: 130)

It is for this reason that Levinas is known as the philosopher of the other human being (Bernasconi, 1999). But this is surely based on Levinas' terminology surrounding 'other', because, as above, philosophers such as Kierkegaard and Heidegger, and others, built their theses on responsibility for the other. And like these others, Levinas did not focus only on the two but rather the third party, whether that be three individuals or the collective as a the third party. As illustrated in Bernasconi (1999: 79), 'The face of the Other does not ask only for him- or her-self, as if there were only two of us in the world. My responsibility to the Other does not allow me to put aside my responsibility to the others of the Other'. This perspective resonates in work by Grimwood (2013), who used Levinas' (1985) concept of responsibility to better contextualise how interactions between inhabitants of the Thelon River, Canada and tourists, might lead to more culturally sensitive forms of responsible tourism. Responsibility, Grimwood contends, is fundamental to who we are at the very core.

The atheist existentialist Jean-Paul Sartre explored at length the nature of what it is to be human, and in particular, that existence takes priority over essence, i.e. existence, freedom of choice, and responsibility all are antecedents to essence. In *Being and Nothingness*, Sartre (2003) argues that agents have unlimited freedom despite the many social and physical limitations and constraints that they are confronted with in everyday life. Sartre believed that every one of us is unique, so efforts to derive generalisations about human nature are fruitless. This individuality provides us with the ability to be free and so traditional definitions of human nature do not hold to be true. Those who choose to deny this freedom are said to be inauthentic and cowardly. Sartre therefore renounces the normative rules of society because of the limits that these impose on the freedom of individuals (Honderich, 1995). As such, our human natures cannot be determined by anything, including society. However, although Sartre argued that each person has the capacity to be free and to choose to be and do whatever they wish, they must choose their values for all of humanity in the most unselfish of ways. This means that one's private values may not necessarily be in opposition to the collective. For example, existentialism has social implications that are embodied in liberal democracy through the Declaration of Independence (Stevenson & Haberman, 1998). This doctrine underscores the importance of the right of each person to freely pursue his or her own idea of happiness, and that such pursuits ought not be constrained by the institutions in place within society. Essential in this configuration is that although we may choose to embrace the same values as others, it is the burden of making up one's mind for oneself that is most challenging.

Existentialism has been criticised because it is rather like a *manner* in which to identify a form of moral freedom to which people subscribe, rather than a *means–end* ethical process. What this means is that existentialism underscores a process of moral thought but without a link to any firm and meaningful conclusion. In addition, existentialism may be criticised on the basis of this radical freedom. It means that because the agent has the authority to make his or her own decisions, free of any binding moral normative dogmas, there is less knowledge or guidance at hand that may be activated to make choices. For example, an existential tour operator may decide to reject any pre-existing socioecological rules in freely choosing to programme plan for his participants in a way that underscores his own private values rather than those of the collective. Notwithstanding, we can note that his behaviour may be above and beyond the expectations of appropriate conduct, but conversely the opposite is true as well. Local people and local ecological systems may be of little concern to the operator. With freedom comes the responsibility of living a life that has been chosen. This means that, if the agent chooses wrongly, then he or she must shoulder the responsibility for this decision and live with the anguish that results.

The meaning *in* life

Perhaps one of the clearest examples of how existentialism and responsibility apply in our lives is found in the work of Frankl (1985: 123). He thought the term existential had three main connotations: (1) *existence* itself, i.e. the human mode of being; (2) the *meaning* of existence; and (3) the striving to find a concrete meaning in personal existence, i.e. the *will* to meaning. Frankl's life in a Nazi prison camp enabled him to develop more fully his work behind the concept of logotherapy – *logos* being the Greek word for meaning, as noted previously. Despite the conditions of the concentration camps – quite possibly the worst that anyone can imagine – Frankl realised that 'love is the ultimate and the highest goal to which man can aspire' (Frankl, 1985: 57). Even in the midst of all the misery, one could still find meaning in life through those people and things (e.g. nature) that one loved. For some, however, anything that was not connected with keeping oneself alive in the camps lost value. This loss of meaning was manifested by behaviours that seemed to discard the humanitarian values that make us social beings, leading to a loss of dignity and will.

It is his third aspect of existentialism, dealing with the will, that has the most significance to our discussion here. Even though we may have all of our worldly possessions taken away from us, including our families, we still have the ability to choose our destiny in life through our attitude and the spiritual freedom that emerges from our decisions. So a prisoner's fate was not so much tied exclusively to external circumstances, but rather came about as a result of his inner decisions. This freedom to make decisions is what makes life meaningful and purposeful. Under the conditions of the camp, Frankl noted that only a small percentage of prisoners were able to reach a high moral standard through the preservation of their inner liberty and values. Suffering and adversity could thus be said to bring out the best and the worst in human nature. In this he quotes Nietzsche, who wrote: 'he who has a *why* to live can bear with almost any *how*' (Frankl, 1985: 126).

Perhaps one of the most moving statements made in this wonderful book is in reference to one's perspective on the meaning of life: that 'it did not really matter what we expected from life, but rather what life expected from us' (Frankl, 1985: 98). By this Frankl asserts that life is about taking responsibility through right action and conduct in all that we do. From this perspective, the 'why' for his existence, in the Nietzschian context, is delivered through one's responsibility to others. This led Frankl to develop a personal maxim: 'Live as if you were living for the second time and had acted as wrongly the first time as you are about to act now!' (Frankl, 1985: 175). This is the undercurrent that stimulates responsibility.

One of the other main messages in Frankl's work is the folly behind striving towards a state of homeostasis. Frankl says that homeostasis is in fact dangerous because it reflects a tensionless state between what one is and what one should become. This dynamic field of tension that we all must maintain is the tension between the meaning that is to be fulfilled, at one pole, and at the other pole the person who is to fulfil it. In reference to the existential vacuum, the ultimate meaninglessness in one's life is developed when the individual has no instinct that tells him what to do. The equilibrium has been established, with no further action towards a more actualised end. And because people are unable to fill their leisure time with meaningful pursuits, there appear to be more ailments to solve from boredom than there are from distress. As such, efforts to reach a state of self-actualisation are not possible because the more one strives to attain it, an egotistical aim, the more one is bound to miss it because of the inherent self-centredness in the act itself. The same can be said about happiness and pleasure. These are merely by-products, because they are destroyed if made a central goal in life – they must ensue, they cannot be pursued. So although travel is perhaps the best exemplar of our search for happiness, as De Botton (2002) notes, does it really make us happy? Referencing the travels of Des Esseintes, De Botton marks the words of Huysman, who wrote that 'the imagination could provide a more-than-adequate substitute for the vulgar reality of actual experience' (De Botton, 2002: 27; see Brown, 2013, and Steiner & Reisinger, 2006, for discussions on tourism as a catalyst to express one's authentic self).

Care Ethics

One of the criticisms of normative ethics is the over-reliance on the application of abstract principles in decision-making. Utilitarianism and Kantianism, for example, provide different objectivist ways in which to understand the moral world, but their guidance may not always suffice in understanding people's lived experiences. Stemming from the seminal work of Carol Gilligan (1982, explained in more detail in Chapter 10), and bridging from the perspective of care as a serious philosophical topic from the existentialists, there is recognition that women demonstrate more of an ethic of care towards others. Care ethics is rooted in understanding people's – particularly women's – lived moral experiences. Important in any differentiation between how women moralise, however, is that we ought not to reduce such moralisation according to just proximal or intimate causes or relations. The moral outlook of women is not simply tied to notions of nature, reproduction and instinct, or the biological domain, but rather that care-giving and relationship-building characteristics of women might also be nested within or alongside other normative ways of seeing the world in the creation of an extremely complex moral landscape. This unique complexity provides an opportunity to question the more rigid objectivist way of interpreting the moral world. Care ethics has thus advanced around sets of critiques and assumptions that provide the scaffolding necessary for a fully functioning, coherent philosophy in its own right. Held (2004) and Robinson (1997) have written liberally on the topic of care ethics.

Held (2004) argues that our institutional systems are too heavily wedded to a masculine rhetoric in relation to universal law, rights, and equality. The failure to wrestle ourselves out of this mindset, or perhaps better stated as predicament, has prevented us from embarking on a broader horizon of thought and action, following Heidegger's meditative thinking (Heidegger, 1966). Viewed 'as a new and distinct kind of moral theory' (Held, 2004: 143), Held argues that care ethics is different than virtue ethics in that the latter focuses on individuals and their qualities, while the former is viewed as relational and interdependent: the ethics of care is marked by the intermingling of self with others based on mutual interest, attentiveness, trust, responding to others' needs, to build strong, cooperative social bonds (core elements of care ethics to Robinson, 1997, include attentiveness, responsibility, and responsiveness). For Held, the proper type or style of moralising may not demand the implementation of highly abstract rules and principles, but rather empathy, sensitivity and responsiveness. More along the lines of a bond of attachment rather than a contract of agreement (Robinson, 1997). The ethics of care is not only a value, i.e. we value caring for others, but also a practice and, as such, it has direct or immediate application. Care is therefore not a set of principles and rules, which makes it rather more ambiguous (Robinson, 1997). Held (2004) is also clear in illustrating that although care ethics stems from feminist thought, it should not be considered a form of feminist ethics because it perpetuates stereotypes of women as selfless nurturers who must stay home and leave the public realm to men.

Held (2004) also discusses the applied nature of the ethics of care outside the family or close relations (the private sphere). She argues that care at a broader level, i.e. the public sphere of life, might liberate cultures from excessive commercial domination. Decisions about what is best for a community, perhaps a community in a lesser-developed context caught in a web of the commodification of people and artefacts, may come about through open dialogue and discourse. The same holds true for the natural world, as values and priorities switch towards a more holistic view of our relationship with local places and the planet in general. Such a broader conception of care ethics has been discussed by Robinson (1997), who argues that there is application of care ethics to the global scale 'provided that it takes account of the social relations, institutional arrangements, norms, and structures through which perceptions of difference and moral exclusion are created in the global system' (1997: 114). The global 'culture of neglect', according to Robinson, would instead be replaced by caring virtues hinged upon a new culture of attentiveness and responsibility.

Conclusion

The rich history of debate from ancient Greece provided fertile ground for the development of philosophy and ethics, which prevails in the 21st century. We know the impact that Socrates had on philosophy through the writings of Plato, and how the utopian view of Plato's ethics differed from the virtue ethics of Aristotle through the enduring history of their scholarship. Philosophy, or the love of wisdom, allows us to seek a more harmonious view of the universe, human nature or, indeed, tourism, by providing the context from which to think more critically. The branch of philosophy known as ethics is thus important to tourism because it allows us to know better what is right or good vs. wrong or bad. But it is important to understand that we in tourism have only just started to ask questions about good and bad from a truly philosophical standpoint. This is a trend that must not continue if we are to appreciate more fully how ethics and philosophy can come to the aid of tourism. This chapter distinguished many different theoretical types of ethics, as well as their pros and cons, including the absolutist theories of intuitionism, teleology, deontology and virtue ethics, in addition to the subjectivist theory of existentialism. These will be further examined in later chapters of the book.

4 Applications of Ethics

There is no greater miracle than man
Sophocles

Introduction

This chapter is an extension of the previous one, with a focus on the application of ethics in a general context. The chapter begins with a discussion on the polarised positions between ethical relativism and ethical universalism. This is important because it sets the stage for the dichotomy of thought that exists in ethics between a site-specific perspective and more cosmopolitan views. This is followed by a discussion on the basis of justice, including rules and laws, and later on rights, with a focus on marginalised populations in general as these relate to tourism. Responsibility and responsible tourism are analysed, along with free will and determinism, for the purpose of further contextualising catalysts and constraints to ethics. The chapter concludes with an examination of Singer's circle of morality and its implications for tourism ethics.

Relativism vs. Universalism

Relativism

Popularised through the work of the anthropologist Ruth Benedict (1934), cultural relativism is based on the belief that different cultures have unique and acceptable ways of life. Benedict widely discussed cannibalism without explicitly condemning it, with the inference that anything in culture could be right or wrong. This theory posits that there are no universal moral codes, but, rather, unique practices and mores for different cultures. The assumption is that what these various groups believe to be right, is morally correct. Different cultures have different customs (e.g. tipping waiters, treating women as inferior) and all can theoretically be right and there is no absolute criterion from which to generalise. The attraction of relativism, both ethical and cultural, lies in the belief that no one system of morality is adhered to in space and time; that is, one system cannot explain the diversity of thought and application because people and cultures are different, bringing us to the conclusion that all truth is relative (Benn, 1998). Consequently, certain motives can only be developed under certain cultural conditions. Right or wrong is thus a function of the customs of the society in which one resides. But along with this is the notion that we cannot judge one society's codes as more worthy over another's (that would be objective), nor can we hold the moral code of one society over another (Rachels, 1989). Those who make judgements of right or wrong on the basis of objectivity are considered naive by relativists.

This has led proponents of the theory to suggest that there are some basic strands that have enabled us to articulate an ethics based on relativity (Edel, 1964). These are: (1) the discovery that ethics is a human product; (2) there is an expectation of change that takes place over space and time; (3) we as individuals and groups have a tendency to act and think according to our own needs, and that this forms the basis of determining what is good, bad, right or wrong; (4) we are subject to the struggle for power that makes societies different in their structure, hierarchically speaking; (5) psychological and educational theory has told us that we can be fashioned in unlimited ways (the behavioural school); (6) cultural differences imply that what is good or right is a function of the mores that exist within the community; and (7) linguistics and semantics have enabled us to examine the meaning of judgements regarding what is good, bad, right or wrong. To this we can add that cultural relativism allows us to be tolerant of one another (Blackburn, 2001).

In support of relativism, theorists see no hard evidence that allows us to definitely conclude that ethics is ingrained in the human mind as a universal (Slobodkin, 1993). This is a perspective adopted by Tierney (1994), who illustrates that moral responsibility hangs precariously in the balance of authority, discernment and circumstance. In the case of authority, there is really no universal authority to appeal to in determining what is right or wrong. As regards discernment, it is difficult to judge between ethical alternatives if these come from polemic visions of what is good. Finally, in reference to circumstance, and perhaps in reference to the aforementioned two, our responsibility to be moral is based on who, what, where and why. As Tierney notes, doctors have the moral responsibility to help those in need, while children are divested of this concern by virtue of their station in life, just as a healthy male is more ethically obligated to intervene in a mugging than an elderly female. As such, cultural relativism is the only legitimate alternative because it is rooted in the cultural norms of the society. He suggests that objectivist theories such as deontology provide little practical guidance that may be particular to the setting and situation:

> the more precise and comprehensive our theory of the principles involved in concrete situations of choice and decision, and the more thorough our characterization of the reasons why these principles should be followed, the less relevant and applicable our theory becomes, both to the exigencies of the situation and to the actual motivations of persons involved. (Tierney, 1994: 10)

Tierney further notes that ethics needs to be about people as subjects rather than just principles and persons as objects. This does not mean that we should abandon the notion of acting on principle, but merely that abstract principles are incomplete if they do not allow us to understand what is right for the people within the circumstance.

On the other hand, the arguments against relativism are daunting. Freeman and Gilbert (1988) write that there are three compelling reasons to reject cultural and ethical relativism. The first is that there is a problem of conflict between the norms of different societies. The advice offered by the cultural relativists is 'Do whatever you like, because you violate a cultural norm (the law) whatever you do.' In this way, no matter what choice you make it will be both right (to one group) and wrong (to the other). The result is that we often give up reasoning through tough situations at all because of this ambiguity. Second, there is the problem of tie-breakers. It is often uncertain which norms are to count as the prevailing ones, given the heterogeneous nature of society. Third, they ask the question, 'What norms are to be justified in a society¿' (Freeman & Gilbert, 1988: 37–38). This is absolutely essential, because, as stated previously, it made no ethical sense to accept the norms of Nazism, but for a time Nazis ruled the day. The prevailing norms of a group simply take over society. We must continue to question new or prevailing norms against a benchmark. Ethical relativism becomes a challenge for business in our globalised economies because of the insistence that the people of one country follow the rules of another. Also, it may be that disingenuous principles or norms within a society are not necessarily the prevailing morals, but rather the motivations of a group in power to persecute or suppress the less empowered.

According to Broom (1987: 38), cultural relativism 'destroys one's own and good', which means that where openness once used to be a virtue in allowing us to use reason in an effective manner, it now means that we have a tendency to accept just about anything we read or see, to the denial of the power of reason. But by attaching reason to cultural relativism we have implied that reason in this way establishes values. In this regard, relativism has the potential to deform ethics because relativism eliminates choices and supports only processes and interest. Situational ethics, which has been the rallying call of the technocrats, is a 'denial of reality and therefore of ethics' (Saul, 2001: 95).

Universality

Critics of relativism argue that there are certain universal values that cultures maintain (e.g. life, knowledge, reasonableness) and that provide guidance to rule-making (Finnis, 1980). This universal 'common ground' provides the needed impetus to structure objective criticisms about the practices of one culture over another. Furthermore, the distinct patterns and practices of cultures have slowly eroded over time. What this means to cultural relativism is that there may be more global homogeneity on the basis of shared global values. There is also the question of whether individuals sharing a certain culture should be allowed to harbour beliefs contrary to the practices of the region in which they live and, if so, whether they should be taken seriously (Finnis, 1980; McNaughton, 1988; Williams, 1985).

The universality of ethics has much to do with an evolutionary ethics that has developed for all of humanity over evolutionary time. Just as 99.9% of our genetic base is shared by all human beings, there is the notion that we have evolved a set of universal traits that are common across humanity. This is the argument put forth by Tooby and Cosmides (1990), who contend that, although a human universal nature is legitimate, it does not mean that people are not genetically or ethically unique. These authors support the notion that our universal human nature allows for the variability of traits and behaviours among individuals and different cultures on the basis of variability across different settings and circumstances. This perspective is adopted by Richards who wrote:

> Evolution has equipped human beings with a number of social instincts, such as the need to protect offspring, to provide for the general well-being of the community (including oneself), to defend the helpless against aggression, and other dispositions that constitute a moral creature. (Richards, 1987, cited in Petrinovich *et al.*, 1993: 477; see also Ruse & Wilson, 1986)

The salience of the universality claim has been supported by the research of Petrinovich *et al.* (1993), who sought to better understand the universality of moral intuitions in their study of US university students on a number of different ethical dimensions. In general, these researchers discovered that there is evidence to support the universality hypothesis, and this was corroborated in a companion study of Taiwanese university students who adopted the Eastern religions, but who showed a similar pattern of results to the US study. It may very well be that, just like the nature via nurture discussion of Chapter 2, we are endowed with an innate set of moral traits (e.g. do not cause harm or injustice, or violate the rights of others, or indulge in incest or cannibalism), however limited or expansive, that provide the foundation for the development of culturally derived ethics, which are thus switched on by the environments and circumstances in which we exist. This would account for the similarities that have been found by Petrinovich *et al.* as well as the differences between cultures found by other researchers. We can only hope that a new applied science of ethics can combine the theoretical aspects of biology, psychology and cultural studies to develop interesting new approaches to understanding the cultural relativism of ethics as well as its universality. Tourists could be an important population of subjects to include in these studies, because of the cross-cultural interactions that take place as a result of travel over borders, but with careful consideration of the ethics behind multiple stakeholders (e.g. government, community members) with competing interests (Moscardo, 2010).

The Basis of Justice

Rules

In his doctoral research in southeast Asia, Brown (1991) recounts a story of sitting with three Brunei Malays in front of a house. Two were seated with him on the bench and one was seated on the rung of a ladder. Tired of sitting on the bench, Brown slipped down to sit on the ground. No sooner had he done this than all three of the other men immediately moved down with him. Insisting that they move up to their more comfortable original position, Brown later realised that they did so because in Brunei society it was impolite to sit higher than anyone else, as a higher order of seating meant one was of a higher status. Despite Brown's insistence that the men sit comfortably, they were unwilling to let anyone see them break this most important rule in the etiquette of rank within society. Formal or informal in nature, rules do not encompass the whole of morality but surface at the point where clashes between right and wrong are most obvious. In determining why rules for society matter, we merely have to look at how society would be without them, and if there was ever a time in the distant past when society did not need them (Midgley, 1994).

Moral systems are societies with rules, which include systems of ethics and normative conduct. Rules are based on an understanding of what is wrong and how they translate into agreements about what is permitted within society and what is not, along with appropriate rewards and punishment (Alexander, 1987). To behave in a moral way entails following the rules that have been put in place. But rules often change and are put into place by tyrants, such that what surfaces as a rule is not always morally right.

As a species that endures pain, death, disease, war, oppressive governments and morality, it is inevitable that these would bring us to that point where clashes demanded the development of rules. Morality causes pain as much as any of the others, as it allows us to understand how we have behaved, according to our own standards and the standards of others. The pain simply shows that we have the capacity to know what would have been the right course of action. But out of this pain emerges the basic core of the person we want to be or strive to be. This reminds us how much a husband loves a wife, or how important their children are to them – strong deterrents, or rather strong symbols, of the capacity of the individual to love another.

The sociologist Emile Durkheim felt that it was not governments but rather social conventions that would provide the needed power to instil order within society. Because different societies maintained different social conditions, each would develop its own moral system to invoke solidarity. Interestingly, Durkheim felt that the normative structure of society was like a collective consciousness that allowed people to understand and appreciate the relationships that were needed in securing long-term relationships and societal order (Pinker, 2002). In reality, governments fail to have the authority to institute order. A show of force cannot compel all individuals within society to obey rules and regulations. This means that social contracts fail to work if there is not the *willingness* to obey contracts (Smith, 1993). Order is upheld not in such forceful ways (as we see in many of the dictatorial regimes in Africa), but rather by the natural sentiments and emotions that we feel towards others (Smith, 1759/1966). Happily, most of what we refer to as universal behaviour does not come in the form of rules at all, but rather impulses, such as the impulse to care for one's own children (Smith, 1993). Rules are not being enforced (they do not have to be), but rather impulses are being obeyed.

Alexander (1987) provides a biological explanation for the importance of rules, by illustrating that the interruption of another's expectations (e.g. getting into line before others in securing a seat in a theatre) is costly to the 'interuptee' if that individual has invested certain costs into securing the experience. The higher the costs in preparing for the event (purchase of petrol, asking someone out for a date, paying for a meal before the show and so on), the more the expectations will be dashed. Alexander notes that the expense of investing in expectations is the reason for the development of rules. He defines rules, then, as:

> aspects of indirect reciprocity beneficial to those who propose and perpetuate them, not only because they force others to behave in ways explicitly beneficial to the proposers and perpetrators, but because they also make the future more predictable so that plans can be carried out. (Alexander, 1987: 96)

Implicit in Alexander's perspective on rules is the importance of intent. Due to the fact that reciprocity involves promises and punishment as well as reward, such systems lead to avoidance of selfishness as well as positive acts of altruism. This is why we have a propensity to examine other individuals as either being moral or immoral, rather than being good or bad at one time or another. Accordingly, we use motivation and honesty in one circumstance to predict the actions of individuals in others. Seeing morality as self-interest through reciprocity allows us to consider one of two outcomes:

(1) If I do an immoral action to someone I am apt to suffer costs greater than the benefits. These costs are imposed on me by members of my own group.
(2) If I do an immoral action to someone I will foster reactions within my group that will eventually impose a cost on me greater than the benefits.

Such costs can take the form of a change of rules that go contrary to my long-term interests or the disintegration of group cohesion, which in turn will prompt my group to be more watchful of my behaviour and increasing the costs of immoral behaviour. From Alexander's perspective, the consequences of indirect reciprocity involve the spread of altruism (rewards), the spread of rules (punishment), and the spread of cheating (deception). These are illustrated in Box 4.1.

The cost-benefit calculus that we do in protecting ourselves or benefitting from these situations has also been outlined by Alexander. He notes that we either consciously or unconsciously make these calculations, and we have evolved to be extremely accurate in making them. Table 4.1 provides an elaborate breakdown of many different motivations for social acts and the various outcomes of these acts. The motivations fall into nine specific categories that are determined on the basis of: (1) not thinking or knowing about what one is doing; and (2) doing things deliberately. The outcomes range along a continuum of self-interest, inclusive fitness, reciprocal altruism and, finally, what might be viewed as pure altruism. The outcomes are further characterised as moral, neutral, immoral, unlikely and problematic (Prob.). The contents of this table will be more fully discussed in Chapter 11.

Laws

Durant observed that the Polynesians gave the word *taboo* to mean different prohibitions sanctioned by religion (Durant, 1935/1963). In essence taboo meant that something was untouchable, whether it be food, certain words, times of the week or year to hunt or avoid work, or women. Traditional

Box 4.1 The consequences of indirect reciprocity

Indirect reciprocity (rewards)
Rewards (why altruism spreads)

(1) A helps B
(2) B helps A
(3) C, observing, helps B, expecting that
(4) B will also help C.

Or

(1) A helps B
(2) B does not help A
(3) C, observing, does not help B expecting that, if he does
(4) B will not return the help.

Indirect reciprocity (punishment)
Punishment (why rules spread)

(1) A hurts B
(2) C, observing, punishes A expecting that, if he does not
(3) A will also hurt C.

Or

(4) Someone else, also observing, will hurt C, expecting no cost.

Indirect reciprocity (deception)
Deception (why cheating spreads)

(1) A_1 makes it look as though he helps B
(2) C_1 helps A_1, expecting that A_1 will also help him
(3) C_2 observes more keenly and detects A_1's cheating and does not help him (avoids or punishes him)
(4) A_2, better at cheating, fools C_2
(5) C_3 detects A_2's cheating (etc.)

$C_1 \rightarrow C_2 \rightarrow C_3$ = Either learning or evolution
$A_1 \rightarrow A_2 \rightarrow A_3$ = (Or both)

Source: Adapted from Alexander (1987)

Table 4.1 Interactions of motivations and outcomes in determining morality and immorality of social acts

	Motivations								
	Doesn't know or think about what he is doing		*Does things deliberately*						
Outcomes of social acts	**A** Considered to be insane or incompetent (i.e. *cannot* know – includes non humans)	**B** Considered to be lazy or thoughtless – does these these things without thinking about them (could know but doesn't)	Believes he is selfish and expects to win because of selfishness		**E** Believes he is altruistic and expects to win *because* of altruism	Believes he is altruistic but expects too win *despite* altruism			
			C Sees his way of life as satisfying; acts this way because he enjoys it	**D** Sees his life as a burden (as compared to other lives possible)	Sees his way of life as either satisfying or as a burden	**F** Expects reward on earth		**G** Expects reward in heaven	
						Satisfying	Burden	Satisfying	Burden
Helps only self	1 Neutral (e.g. baby)	2 Immoral	3 Immoral	4 Unlikely	5 Immoral	6 Immoral	7 Unlikely	8 Unlikely	9 Unlikely
Helps only self and relatives	10 Neutral	11 Immoral	12 Immoral	13 Unlikely	14 Prob.	15 Prob.	16 Prob.	17 Prob.	18 Prob.
Helps self, relatives, and friends who are likely to reciprocate with interest	19 Neutral	20 Immoral?	21 Immoral	22 Unlikely	23 Prob.	24 Prob.	25 Prob.	26 Prob.	27 Prob.

Helps self, relatives reciprocating friends, and others in the presence of potential recriprocators	28 Neutral	29 Moral?	30 Immoral	31 Unlikely	32 Prob.	33 Prob.	34 Prob.	35 Prob.	36 Prob.
Helps all of the above, and also helps strangers when it is not too costly even when not in the presence of reciprocators	37 Neutral	38 Moral	39 Unlikely?	40 Unlikely	41 Moral	42 Moral?	43 Moral?	44 Moral	45 Moral
Helps anyone who needs it even if the immediate cost is great	46 Neutral	47 Moral	48 Unlikely	49 Unlikely	50 Moral	51 Moral?	52 Moral?	53 Moral	54 Moral
Helps others indiscriminately while maintaining self at approximately the lowest level consistent with doing this effectively	55 Neutral	56 Moral	57 Unlikely	58 Unlikely	59 Moral	60 Moral?	61 Moral?	62 Moral	63 Moral

Notes: I have speculated as to how each category of act is likely to be judged. Problematic cases illustrate the difficulty of deciding questions of morality when self-interest is broadened to include reproductive (genetic) interests, and when motivation comes to include realisations of the nature of such interests. Squares 25–32 and 41–48 would probably be marked 'moral' by those unaware of biological considerations because they seem to involve self-sacrifice. An evolutionary biologist might regard all square as representing possible behaviours, and as all possibly representing self-interested behaviours, but he might also regard squares 55–63 as less likely than would non-biologists. Biologists would also be more likely to search for ways in which squares 55–63 could represent behaviours that serve the actor's interests.

[b]Behavious that will probably be seen by most as immoral because of outcomes and despite motivations.

[c]Behaviours that will probably be seen by most as moral because of outcome and motivation combined.

[d]Desirable behaviour even if morally neutral.

Source: Adapted from Alexander (1987).

societies constructed countless taboos surrounding pregnant women, childbirth and menstruation. In the more advanced traditional societies taboos took on the role of laws by forbidding members of the unit to behave in certain unacceptable ways.

Laws codify the various customs, ideals, morals, beliefs and ideals of a society (Guy, 1990). We have problems in society when values are seen to be contradictory. This means that, while some values are maximised, they are done so by lessening others. A judge basing action on an ethical code without any legal standing would result in chaos. As such, laws are different from codes of ethics because they are more formalised (Slobodkin, 1993).

An example of the rule of law in tourism is the implementation of an eco-tax in the Spanish Balearic Islands, where tourism accounts for 84% of the economy of the archipelago. As a result of increasing levels of visitation, and associated impacts on the fragile natural environment, the Balearic Parliament approved Law 7 of 23 April, 2001, levying a direct tax on tourists staying in five-star hotels or one- to four-star aparthotels, for the purpose of developing a fund for environmental conservation. Article 19 states that the revenues generated by the fund will be used for:

(a) Redesign and renovation of tourists' areas to improve their quality, including the installation of systems for the efficient use and saving of water and techniques for the saving, efficiency, and development of recyclable energies.
(b) Acquisition, recuperation, protection and sustainable management of resources and protected natural areas.
(c) Defence and recuperation of assets belonging to the historic and cultural heritage in areas of tourist incidence.
(d) Revitalisation of agriculture as a competitive activity, giving special attention to the use of purified water from plants.
(e) Sustainable management of natural spaces to allow the lasting conservation of their bio-diversity. In particular, the development of the Biosphere Reserve, Nature Reserves and others included in current legislation (Bouazza Ariño, 2002: 170).

Even though local people and tourists were in favour of the eco-tax, the Central Government of Spain and the accommodation sector in the Balearics both opposed the tax for fear that it would have a negative backlash on the industry in the region, leading to a reduction in tourist visitation (Cantallops, 2004; Gago *et al.*, 2009). These short-term economic concerns mirror a vast number of other actions in other destinations that are based on instrumental reason: that the best and perhaps only solution is one that is geared towards maximising revenues, with little concern for the overall quality of the environment, which could sustain the industry in the future.

One of the most applicable areas of law to tourism includes the realm of torts, which has been defined as 'Breach of duty (other than under contract) leading to liability for damages', with tort-feasor defined as the 'person guilty of tort' (*Concise Oxford Dictionary*, 1976: 1224). It is thus the area of law that examines wrongful acts that are non-criminal, but that result in damage to something or someone. Tort law governs the relationships that exist between people, organisations and institutions like government, and includes the following main categories (Peterson & Hronek, 1992):

Negligence. Premises; programme supervision; and facilities supervision.
Strict Liability. Animals; product liability; food service; and drinking water quality.
Nuisance. User injuries; land use; and controls.
Intentional Torts. Personal (assault and battery, defamation); property (trespass to land); and constitutional torts (invasion of privacy, due process, right to freedom of movement, right to own property, speech, religion, equal protection and civil rights).

Negligence is an area of tort law that is directly applicable to tourism and recreation. It is 'the unintentional violation of a duty to take into consideration the interests of others that results in injury . . . [or] to act as a reasonable and prudent person would have acted under the same circumstances' (Russell, 1982: 230). Certain elements must be proven in order to take a case of negligence forward in the court. These include: (1) it must be proved that the defendant has a legal duty of care, i.e. is legally responsible to the plaintiff; (2) the plaintiff must prove that there was either a failure to perform a required task or a breach of duty; (3) there must be some direct connection between the damages and the actions or lack of actions by the defendant; and (4) a plaintiff must prove that he or she suffered damages, for example, physical anguish, mental anguish or financial loss (Peterson & Hronek, 1992: 9).

At the heart of the matter is the duty or standard of care that should be extended to the participant while participating in tourism and recreation activities. In light of the risks, many companies and organisations have taken it upon themselves to develop extensive standards for the purpose of protecting participants and providers. Examples include standards established by the American Camping Association, the Association for Experiential Education, and the various ecotourism accreditation programmes that are beginning to surface in different regions of the world (Fennell, 2002). The following few examples serve to identify the issues and responsibility for different populations and under different conditions (Peterson & Hronek, 1992):

Example 1. If people choose to engage in white water canoeing, they assume a risk that the canoe may turn over with resulting injuries or even death.

However, if the outfitter that furnishes the canoeing equipment provides canoes that are known to be dangerous, then the outfitter may be negligent. If the agency fails to warn the canoeist of dangerous high water, then the assumption of risk defence may not be applicable.

Example 2. A plaintiff was sailing on an inland lake when a wind tipped the hired boat, causing injury to the boat and the participant. The courts would not allow legal action because the wind was considered an act of God. The management agency and the boat rental facility are not able to warn for such acts of God, which they cannot predict.

Example 3. A participant is injured in an off-road vehicle accident on a mountain trail. He says that the management agency failed to inform him of the uncertain conditions of the trail. After an investigation it was found that in fact the participant had gone off the trail at excessive speeds, which contributed to his injuries. If a person contributes to his own accident, they are barred from suing.

Example 4. A woman who was cross-country skiing came upon a sign that read 'Danger, do not enter, avalanche area'. Although she saw the sign she decided to ski around it and ended up becoming entrapped and died under an avalanche. The woman's family decides to sue the management agency for loss of life because of the failure to stop her from entering the area. The courts would find this inadmissible because the deceased had the last clear chance to avoid the accident.

Example 5. A park visitor is upset that the government management agency responsible for the area failed to provide enough money to properly maintain the site. The visitor however would not be able to sue because policy and planning decisions are not subject to suit because of government immunity.

In their efforts to avoid costs to themselves, tourism service providers should consider the development of risk management plans. Such costs include damage to equipment, to facilities and to people. These may escalate in programmes that involve greater levels of risk, such as mountaineering or moving-water trips. Risk management has been defined as the 'formal process of assessing exposure to risk and taking whatever action is necessary to minimize its impact' (National Association of Independent Schools, as cited in Ammon, 1997: 174). In general, the risk management plan will involve a series of steps that will enable the operator to identify risks, classify these, develop risk control measures, implement these measures, and monitor and modify (Watson, 1996). The identification and classification of risks may be accomplished by outlining a number of steps, including (Ford & Blanchard, 1993):

(1) General description: name of programme, type of activity, level.
(2) Dates and times: as determined.

(3) Goals and objectives: organisational, activity.
(4) Location: site/area, weather, routes/accommodation, facilities.
(5) Transportation: mode, routes/destination.
(6) Participants: number, skill level, characteristics.
(7) Leaders: number and roles, qualifications.
(8) Equipment: type and amount, control.
(9) Conducting the activity: pre-activity preparation, group control.
(10) Emergency preparedness: policies, health forms, telephone numbers.

Risk management is an ethical issue in tourism because those who fail to or choose not to have a risk management strategy put themselves and participants at risk. As service providers, we should do everything possible to ensure not only that participants are having fun in our programmes, but that they are safe too.

Justice

The most comprehensive treatise of justice to date is that put forward by Rawls (1971) in his examination of social justice as the basic structure of society. Rawls felt that justice is the 'first virtue of social institutions, as truth is of systems of thought' (Rawls, 1971: 3), and, as such, stands as a social contract theory where principles of justice are based on fairness within society. His theory of justice was constructed in contrast to utilitarianism, on the basis of the fair and equitable distribution of goods in society as well as individual freedom (see the discussion in Chapter 2 on social exchange theory). This generated a renewal in normative ethical theories that spilled over into the 1970s and 1980s, corresponding with the movements in both environmental justice and business ethics. Justice emerges in his theory as providing a standard where the distributive aspects of a society at its most basic level can be assessed, and where principles (if just) are important in dictating social cooperation, regulating agreements, and providing a basis from which to establish government.

Justice is something that we find within all cultures. It is hinged on what is right for individuals in association with what is right for the group. For example, Rawls writes that the sufficient condition for equal justice among people is a moral personality; there is no race or recognised cohort within society that lacks this attribute. Moral persons are distinguishable by two main features. First, they are found to have a conception of their good; and, second, they have a sense of justice through their desire to use and act upon principles of justice, defined in societal terms.

When they fail to have a conception of good, we may rely on our punitive systems in reinforcing the principles of justice which encompass our societies. So the impetus behind keeping promises for persecution makes justice a tangible concept. This was nicely summarised by Oliver Wendell Homes, who said that:

> If I were having a philosophical talk with a man I was going to have hanged (or electrocuted) I should say, 'I don't doubt that your act was inevitable for you but to make it more avoidable by others we propose to sacrifice you to the common good. You may regard yourself as a soldier dying for your country if you like. But the law must keep its promises.' (Holmes, as cited in Kaplan, 1973: 16)

This means that there must be policies that are absolutely and unequivocally upheld in keeping order in society. These policies can be hardened in the case of laws, or less stringent in the form of local norms or guidelines. For example, if an interpreter allows tourists to cut into a queue ahead of others at an attraction, this quite clearly constitutes a violation of our sense of justice, especially if we have been waiting in line for a long period of time. We may even be moved to retaliate at a personal cost if we perceive an act to be an extreme example of an injustice (Daly & Wilson, 1988), demonstrating an unwillingness to be exploited. This was the case with the tourist who was ripped off by the rickshaw drivers in Toronto (Chapter 1). Associated with this is the willingness for people to rally around an injustice and support the victim in a variety of ways. The most adept at this are the media, using the same rickshaw example, who often force decision makers to acknowledge the situation (city councillors) and revisit policies for the protection of all stakeholders. When such matters affect the social and economic fabric of a community, action is likely to be more swift and decisive, suggesting a degree of severity of the transgression, than if such matters are more trivial, and where accountability – personal or financial – is less of a concern.

Justice has been criticised in its avoidance of sentiment (at the expense of logic), which was so central to the work of Adam Smith and David Hume. Even in the more contemporary work of Rawls, sentiment is missing, where people are asked to decide upon rules of social justice from the perspective of lacking knowledge about themselves and their own special talents (Frank, 1988). Social contractors are therefore not asked to examine their own sentiments, such as sympathy and envy in decision-making. Rawls argued that the best way of making socially just decisions was to set aside one's distinguishing social characteristics, such as race, gender, social class, marital status, talents, wealth and so on, a position he termed the 'veil of ignorance', in order that we try to be as fair to everyone in our decision-making as possible. Otherwise, people will strive to choose what they perceive to be in their own best interest. The manner of justice can also be different according to different interest groups. Egalitarians, for example, push for the fair distribution of society's wealth through equity and social equality. They favour welfare programmes and taxation schemes that are designed to redistribute wealth. Libertarians, who support the notion that there needs only to be a limited influence of the state (e.g. protection and enforcement of contracts) in the affairs of the citizenry, feel that left-wing measures go too far towards constraining individual liberty through a levelling of the playing field. The egalitarian would support Rawls' difference principle, which says that resources within a society can be allocated differently only if this allocation benefits those who are disadvantaged (Rawls, 1971).

Tourism is very much a justice issue (D'Sa, 1999). It is fraught with disparities, racism and corporate power, which might in practice be the antithesis of what Rawls would advocate in his theory of justice. It is also a place where essential principles are shelved and community rights ignored. The lack of respect for the rights and autonomy of local people is often secondary to the concern over the preservation of wildlife. While the priorities of rights of humans and other beings is rather contentious, as we shall see in the next section, decisions to marginalise and disrespect certain cohorts within society is a measure of priority, which in turn is symptomatic of a psychology of profit that can diminish human welfare to the most sub-standard levels imaginable (Hall & Brown, 2006). More recent research in tourism studies has endeavoured to more firmly situate justice issues into a theoretical framework. For example, Higgins-Desboilles (2008) argues that the development of justice tourism, as a tourism type, should be radically different than other forms of alternative tourism through a transformative process involving humanistic globalisation as radically different than capitalist globalisation. Lee and Jamal (2008) construct a tourism justice framework, premised on environmental justice and equity as necessary components for sustainable tourism and ecotourism. Their paper provides important inroads into several environmental factors (e.g. water and air quality, water consumption, forest resources, public health, accessibility to natural goods and so on), predicated on distributive and procedural justice grounds. Following from the latter work, Jamal and Alejandra Camargo (2013) contend that if we are to achieve sustainable tourism, we will need to tackle the social and cultural well-being of those populations that have been marginalised and disadvantaged. The authors use Rawls' theory of justice and Fainstein's (2010) concept of the Just City to develop a joint ethic of justice and care. A paper that is built in the same spirit as Jamal and Camargo's, is work by Mihalic and Fennell (2014) on trading tourism rights in an effort to develop a more just international tourism situation. The authors also draw on Rawls (1971), as well as the work of Nozick (1974), in suggesting that deprived citizens of the world should be able to trade away their right to travel (in the form of tradable tourism certificates) to citizens in the developed world and, in so doing, realise certain economic benefits.

Rights

The extension of rights to marginalised populations has a long history as a function of the widening of the circle of morality. Proof of this lies in the considerable changes that have taken place in the US over the last 200 years plus, including the Declaration of Independence (1776); the Emancipation Proclamation, freeing slaves (1863); the Nineteenth Amendment, allowing women to vote; the Indian Citizenship Act (1924); the Fair Labor Standards Act (1938); the Civil Rights Act, acknowledging the rights of American blacks (1957); and later the Endangered Species Act of 1973 (Nash, 1989). This latter act, which gave rights to animals where before they had none, represents one of the most dramatic expansions of morality in the course of humanity. This is substantiated by Stone (1979), who argues that a natural progression in our evolution as a just and benevolent species is the conferring of rights to non-human entities, such as forests, oceans and other natural objects – indeed, to the natural world as a whole. In defence of his stance, Stone argues that it is no good to say that a forest cannot have rights because it cannot speak. For that matter, neither can corporations, states, estates, infants, universities or incompetents. From the legal standpoint he argues that we ought to treat the natural world as we do incompetents, where someone such as a guardian or conservator steps in to manage affairs. Courts, he says, do the same for corporations when they are unfit to operate on their own. (More on this in Chapter 7.)

Becker (1977) illustrates that the word 'right' is often equated with ideals, social goods or virtues. From the perspective of the latter, he discusses the notion of a person who is 'property-worthy', suggesting that we are not all equal in our ability to be virtuous towards a land holding (where 'virtuous' means to use property to good effect). He uses the example of art in explaining that, while some can ascribe *intrinsic worth* to art, others merely view it as having *extrinsic value*. Becker discusses two cases of property ownership and the underlying moral discussion that follows: (1) people who will use property to good effect, and (2) people who will manage property well. Both can be used to examine the basis of some ownership (through purchase) or management of large expanses of land, such as a tropical rainforest, in lesser developed countries. Efficient management of these lands would entail the following:

(1) The general justification for acquiring property rights in a thing is at least in part to secure the success of the carrying out of purposes for or with the thing.
(2) It is wasteful for things not to be well used – that is, for them not to be employed to the greatest advantage.
(3) Waste is morally objectionable in the case of scarce goods.
(4) Some people are better able to manage certain goods to the greatest advantage than others, and where good management is necessary to good use, it follows that:
(5) if goods can justifiably be owned by individuals at all, those goods (at least if they are scarce) ought to be owned by the people who can and will manage them to greatest advantage.

We can say that certain people deserve land holdings by virtue of their ability to manage the resource to its greatest advantage. However, as Becker notes, this may have nothing to do with the overall judgement of a person's moral fabric. A wicked person can be the recipient of good fortune, not because they deserve the resource, but because there may be other legitimate reasons for his or her claim to the property. This means that people or groups own things by chance, and not because of their intelligence or moral character. But, more typically, we feel that good people receive things because they are good, while those who are bad go without. When we read that an NGO has taken hold of lands for perpetuity we tend to feel more at ease with this than if such lands are left to the destructive nature of developers and other landholders. In 2003, southern Ontario, Canada, was embroiled in a debate over the fate of Marcy's Woods, a prime ecological tract of land with species of plants and animals that are more typical of the US Carolinas. When it was proposed that this land be sold to a business, instead of an NGO, people did not feel comfortable with the decision because of the perceived negative implications. It took many successive rounds of public relations on the part of the business to ease the minds of the citizenry through the intent of leaving the land in an untrammelled state. The same applies when we read about the development of a new hotel complex in Costa Rica that we feel will be unacceptable because of the potential impacts on the ecology of that region, even though we may travel to another mass tourism destination at a later date and be either unaware or uncaring of the impacts that have taken place there.

The sheer magnitude of rights issues has rather recently compelled tourism researchers to appreciate more fully the extent of such issues in the travel industry. For example, in 1995 a bomb exploded in the offices of a Greek environmental organisation that had been peacefully lobbying for more restrictions on tourism hotel development and for the creation of a marine park to protect an endangered sea turtle. There is also the case of village fishermen in Malawi where, in 1991, the owner of a hotel on Lake Malawi was given permission by government authorities to evict an entire community in order to develop more tourism infrastructure. Over 70 homes were flattened by bulldozers with no plan for relocation (Sachs, 1995). The unethical nature of this example is obvious and the response is also not hard to envision. But there are many considerations that should be taken into full view in creating clear direction for the use of such lands. First, prior informed consent needs to be considered in any attempts to use or develop lands for tourism purposes, along with a clearly delineated list of pros and cons attached to these developments. In addition, aboriginal people need to have the right to say no to tourism development; to be able to access all information on a tourism development project; and to have access to and participate in policy-making. There is also a need to support models and case studies developed by aboriginal people; to support economic diversity within communities; and to support the development of indigenous community programmes (Pera & McLaren, 1999).

The extent of the human rights problem in tourism has also been captured by Drummond (1998), who reported on the Paduang 'human zoo', close to the border between Thailand and Burma, where 33 people – an entire tribe from Burma – had been kidnapped by a Thai businessman (Thana Nakluang) and put on display for international tourists. As one of the world's rarest indigenous tribes, the 'giraffe women' of this tribe wear several rings around their necks that elongate the muscles. The group was lured from their home in the south of Burma, under army control, through the promise of being reunited with their relatives over the border in Thailand. Instead they were taken to a jungle area and presented to the authorities under the guise of helping the group develop a model tourist village. Soon thereafter, tourist agencies found out about the village and began promoting it in Bangkok and Chiang Mai. The villagers were paid in rice and oil and the equivalent of £42 (GBP) pounds per month, depending on their behaviour. Those who attempted to escape were beaten by guards who patrolled the perimeter of the village. Having learned about the village through a report in the London *Times*, Prime Minister Chuan Leekpai ordered the closure of the camp and had formal charges laid against Nakluang. With no rights to speak of, this group's future is said by Drummond to hang in the balance. If they are deported back to Burma, they face a similar fate, or a risk of being treated as rebels by a brutal military regime. Recent evidence illustrates that the women are starting to take their rings off, and this is a practice supported by the Burmese government (Morgan, nd; ThaiMed, 2009).

In response to these and other rights issues, tourism organisations have become more vigilant. The NGO Steering Committee Tourism Caucus (1999) prepared a paper on tourism and ethics that nicely summarises important concepts of justice and rights in tourism. In their section on Areas of Concern, they emphasise the following on ethics and tourism (Committee Tourism Caucus, 1999: 1):

> In the same way ethical principles apply to all individuals, communities, and societies, they also apply to all sectors in tourism in their respective specific roles. This entails both rights and responsibilities. ... This means, for example:
>
> - democracy and peace in the management and resolving of conflicts connected with tourism, which includes the openness and preparedness for a reciprocal understanding and the observance of the general principle of justice;
> - solidarity with those who are directly and strongly affected by tourism and who suffer from unjust structures connected with tourism,

and solidarity with those who need material or political and philosophical support in defending their interests and rights which are threatened by tourism development or which are ignored in decision-making process;
- justice in a world tourism order, an aim which intends to change all structures of injustice that exists in the fields of economics, politics, social, and cultural life.

The group further illustrates that human rights transgressions are caused by tourism developments that are inconsistent with the needs of communities:

> In many Southern destinations, international human rights standards are daily violated within the tourism sector, including racist and sexist practices. Indigenous peoples are particularly vulnerable to market-driven tourism, losing their customary lands and resources, religious freedom, and ultimately their cultures and capacity for self-sufficiency. Women and children are also at high risk, where tourism economies are built upon exploitative labour practices, and where sex tourism occurs. (The NGO Steering Committee Tourism Caucus, 1999)

Perhaps the most comprehensive treatment of rights in the context of tourism has been undertaken by the UK-based Tourism Concern, whose *Tourism and Human Rights* document makes several links to The Universal Declaration of Human Rights (1948) in illustrating that tourism is far from a positive representative of the spirit of these rights (Keefe & Wheat, 1998). The authors note that, despite the magnitude of the tourism industry, there has been little attention paid to the many human transgressions that take place in large volume. And, although tourists and tourist developers are usually treated with respect, it is usually the hosts who are not afforded the same basic rights as these other players. Although it is beyond the scope of this book to fully examine The Universal Declaration of Human Rights, it is worthwhile to illustrate at length many of the Declaration's Articles used by the authors in their application to the tourism industry:

Article 13. 'Everyone has the right to freedom of movement and residence within the borders of each state; and everyone has the right to leave any country, including his own, and to return to his country.'

The rights inherent in this Article do not always apply to members of the Lesser Developed Countries (LDCs), where, as noted by Keefe and Wheat, tourism often restricts the movement of people in their own country. The authors cite a case in Kenya where private beaches restrict women from collecting crabs, which remain an important nutritional supplement for their families.

Article 17. 'Everyone has the right to own property alone as well as in association with others; no one shall be arbitrarily deprived of that property.'

But, through tourism, people continue to be displaced from their homes to make way for the construction of hotels, golf courses and other leisure services. Such was the case in Malaysia in 1989, where several shop owners had their shops destroyed by the Tourism Development Committee because they were seen to be an eyesore. Other shops were later built in the same space by the development group.

Article 12. '(Social progress and development shall aim at) the elimination of all forms of economic exploitation, particularly that practiced by international monopolies.'

The very nature of tourism is defined by transnational corporations that operate from the world's most developed countries. They have both the resources and power to radically transform the markets of any region. This was the case in Sri Lanka, as suggested by Keefe and Wheat, where the local soft drink market was taken over by Coca Cola, and the sweet potato market was abandoned in favour of chips (forcing the import of potatoes that cannot be grown there).

Article 25. 'Everyone has the right to a standard of living adequate for the health and well-being of himself and his family.'

The construction of hotels and leisure facilities is an example of situations where people have been endangered because of the products used in such development. The health of people living adjacent to golf courses has deteriorated through consumption of water and fish that have been contaminated by herbicides and pesticides.

Article 22. 'Everyone, as a member of society, has the right to social security and is entitled to realisation . . . of the economic, social and cultural rights indispensable for his dignity.'

As an agent of change, tourism has had a substantial impact on the authenticity of culture. Dress, dance and food, for example, have been subject to the standards of Western countries. In Peru, during the 1980s, whole families and communities were forced to move to more accessible places in order that tourists would have better opportunities to photograph 'natives' performing their rituals.

Principle 9. 'The child shall be protected against all forms of exploitation. He shall not be the subject of traffic, in any form . . . and shall not be admitted to employment before an appropriate age.'

Keefe and Wheat illustrate that the 13 to 19 million children working in the tourism industry in the late 1990s represent between 10 and 15% of the global workforce in tourism. As a result of the earning power in tourism, many of these children are recruited to work in tourism and deprived of an education. In Sri Lanka, children who beg often make more money in a day than their parents do in a month.

Article 24. 'Everyone has the right to rest and leisure including reasonable limitation of working hours and periodic holidays with pay.'

The authors note that it is one of the great paradoxes in tourism that those who work in the tourism industry in the LDCs are those who are most unlikely to be able to take a vacation themselves because of low pay, seasonality of the industry, long hours and poor working conditions leading to poor health.

The authors recommend that the following actions be taken by relevant UK governments in addressing the significant level of human rights dysfunctions that are rampant in the tourism industry globally: (1) develop policies that take direct account of ethical and human rights issues based on the Manila Declaration on the Social Aspects of Tourism (see http://www.univeur.org/cuebc/downloads/PDF%20carte/70%20 Manila.PDF, for the full text on this document); (2) create clear ministerial responsibility for out-going tourism, incorporating ethical tourism issues into its work; (3) become a member of the WTO, and adopt the Manila Declaration; and (4) form a cross-departmental group in government on ethical tourism policy, incorporating human rights issues, and consulting with relevant NGOs and the British out-going tourism industry (Keefe & Wheat, 1998).

Responsibility, Free Will and Determinism

> **responsible** adj. 1599, corresponding or answering to something, in Ben Johnson's Every Man Out of His Humour; borrowed from obsolete French *responsible*. From Latin *respōnsus*, past participle of *respondēre* to RESPOND. The meaning of answerable or accountable is first recorded in English in 1643, and that of trustworthy or reliable in 1691, in Locke's writings. The meaning of involving obligation or duties (as in a *responsible position*) is first recorded in 1855. (Barnhart, 1988: 918)

As one of the newer tourism concepts in vogue, responsibility relates to our efforts to be as accountable and ethical as possible in our touristic behaviour (cf. Chapter 3 and the discussion on existentialism). In this regard, Frankl's logotherapy, or healing through meaning, is really meant to be an education towards responsibility. Each and every one of us needs to be responsible for something or for someone, which translates into meaning in our lives. This something or someone could be the natural world, members of our community, or even tourists, through the ability to draw the line between self-interest and the interests of others (Guy, 1990).

But is responsibility synonymous with ethics as implied? The thinking behind this is reflected in the work of Velasquez (1992), who illustrates that moral reasoning is often directed at those who are responsible for *wrongful* acts (as different than being morally responsible for good acts). The moral judgement is used to determine whether or not someone should be blamed or punished for their actions or inactions. If a manager willingly exposes an employee to dangerous conditions, then he or she may be found to be morally responsible for any injuries that may result. This is explained at length in the following text:

> A person is morally responsible only for those acts and their foreseen injurious effects: (1) which the person knowingly and freely performed or brought about and which it was morally wrong for the person to perform or bring about, or (2) which the person knowingly or freely failed to perform or prevent and which it was morally wrong for the person to fail to perform or prevent. (Velasquez, 1992: 40)

There is also an expectation that individuals will take responsibility for their inappropriate behaviour by punishing themselves, for example, expressing remorse or compensating the victim, or being open to punishment by others. However, unless the person is willing to take ownership of the situation and accept the consequences of his or her actions, claims of responsibility ring hollow. This was the case for Richard Nixon, who was ridiculed for not initially accepting responsibility for the Watergate scandal. Nixon later 'accepted responsibility', but did not incur any costs for his actions by stepping down as president, apologising or firing aides (Pinker, 2002).

In the case of corporate moral responsibility, Velasquez (1992: 42) notes that those who 'knowingly and freely did what was necessary to produce the corporate act are each morally responsible'. This may include individual actions or also actions or omissions by several people cooperating together in jointly producing the act. Interestingly, the traditional view of moral responsibility in the case of a jointly produced transgression is that each individual is held accountable for the collective action or inaction. Critics, however, charge that it was the corporation that acted, and that was responsible, and not the individuals who may not have been privy to all the various permutations and combinations of the decision that led to the action or inaction (see Chapter 6).

Can we say that those who build large-scale hotels are morally responsible for their acts if there are adverse ecological consequences as well as displacement and marginalisation of the local population? Yes and no. No, because morality can also be argued on the basis of law, which can open the door to development and divest developers of any responsibility of these impacts, rightfully or wrongfully (remembering that laws are not always moral). This elimination of a person's or group's moral responsibility lies outside the two widely regarded conditions that free an individual of his or her responsibility: ignorance and inability. In reference to ignorance, people cannot control matters where they are ignorant of facts, so they cannot have a moral obligation with respect to these matters. Inability is a result of internal and external factors that render a person unable to complete certain tasks. This may include having insufficient skills (mental or physical) or resources to carry out the task. A manager who places a new staff member in the position of operating a craft, for example, a jet boat, without sufficient skill or knowledge, may cause a catastrophe. In this case, it is not the staff member at fault, but rather the manager, who may be simply motivated by the need to save money.

Notions of responsibility as good and bad also come from the work of Kagan (1971, 1984) on the development of children. Kagan sees responsibility as one of only a few fundamental, uniform emotional capacities that underscore unpleasant emotion. These include: (1) anxiety, over physical harm or social disapproval; (2) empathy, including with those who are at risk or in need; (3) responsibility, which relates to those who would cause harm to others; (4) fatigue/ennui, which follows repeated gratification of a desire; and (5) uncertainty, which involves inconsistency of beliefs and poorly understanding events. What Kagan infers is that individuals are driven towards the avoidance of unpleasant emotional states as a motivating force behind moral behaviour (see also Kohlberg, 1984). Kagan's research, like Velasquez's, allows us to see responsibility in contrasting ways. The first way is in the positive context, which views benevolent and virtuous actions as being worthy of praise. The second implies that there is an understanding that one is responsible for violating some normative standard or practice. Again, what this means for tourism is that we can uphold the responsible tourism platform on one side of the coin, but there is also the other side, which may examine those who are responsible for a social, ecological and/or economic impact (e.g. 'I'm the one who is responsible for taking photographs of people without their permission'). Responsibility is thus a double-edged sword.

What individuals do for themselves and their families in the name of responsibility may not be the same as being responsible according to other criteria. Clearing the rainforest of more than one's fair share is a responsible thing to do if it equates to feeding and clothing one's children. However, it may not be the most responsible decision in light of the tenets of sustainable development. And if the manager of a cruise line feels that allowing more tourists on large ocean-going vessels is good because it creates more jobs – the responsible thing to do for these employees and their families – this might not be the most ethical course of action for those at the destination (although one may justify it on the grounds that these tourists spend more money as excursionists) or for the local ecology. The point is that, if responsible tourism wants to be both legitimate and useful, it will need to take into consideration not only the virtuous side of behaviour, but also the not so virtuous side too. Alongside this it is important to realise that there are preconditions for being responsible, which lie in having (1) a conscious, and (2) a moral framework (Kalisch, 2000). Responsibility may thus be viewed as being one step removed from ethics, the latter of which is central to the former. We have yet to fully grasp this fundamental understanding in tourism.

Linked to the concept of responsibility are the dichotomous terms determinism and free will. Determinism refers to the sense that there is a force that takes away from our ability to control our own choices. Determinism is said to be biological (e.g. the control the brain has on our actions), environmental (e.g. the control the environment has on our actions), cultural (e.g. that our behaviour is a function only of the external cultural forces around an agent) or religious. But, Pinker (2002) notes that there cannot really be any biological determinism simply because the brain responds to senses, which are cued to the environment, with the purpose of selecting an action that has foreseeable consequences. He writes that:

> If the most ironclad form of determinism is real, you could not do anything about it anyway, because your anxiety about determinism, and how you would deal with it, would also be determined. It is the existential fear of determinism that is the real waste of time. (Pinker, 2002: 175)

In the past, philosophers feared that the human nature side of the nature/nurture debate will eat away at personal responsibility (cf. existentialism in Chapter 3). That is, if we are biologically predestined to act in certain ways, this becomes an excuse to avoid taking responsibility for our actions (i.e. biology makes us all blameless). Those who commit heinous crimes like murder would be blameless (Pinker, 2002). But people are not blamed if they unaware of the consequences of their actions, as noted above; children are not prosecuted because they do not know the broader social implications of their actions; and people are not incarcerated if they unintentionally (and are not under the influence of illegal substances) hit another in a traffic accident. However, we are blamed if it can be proven that there was motive for murder.

The argument follows that, while determinism removes the ability of the human agent to make ethical decisions, free will guarantees a sense of control. But this element of responsibility sneaks back into the equation when discussing free will. Not unlike the human nature debate, free will also does not allow us to hold people responsible for their actions (Dennett, 1984). Offenders could not really be deterred by forms of punishment or feelings of guilt because the agent, operating on his or her own existential level, would be unable to understand cause and effect and therefore the underlying basis for regulations, codes of ethics and the like. Free will then presupposes behaviour that, if punished, would fail to yield any certain effect on the future actions of the offending agent. The point is that, if the agent was not truly under the influence of a free will, and therefore influenced by punishment, he or she is no longer operating under a free will.

The Circle of Morality

Singer (1981) illustrates that humanity has continued to expand the mental line that demarcates what we view as morally considerable, where once it only included men (see the previous section on rights). The circle extends outwards from family to community to society and to humanity, and includes women, children and the foetus. Environmental ethics has expanded the circle further outwards in the consideration of plants and animals. Proof of the expansion of the moral circle does not mean that there exists some magical formula that compels us all to be better than we were. As we saw in previous chapters, the drive can still be a selfish one through our interactions with others in a very complex world. We cooperate together because both parties may benefit from a relationship. This is the notion of the zero-sum game, which finds that two can benefit, through cooperation, for the purpose not of being overly civic-minded, but rather in an effort to avoid punishment or conflict.

Although Singer's work has been widely accepted and cited, and is most applicable to our discussion, the widening of the circle of morality is one that has been considered for decades. For example, Durant observed, although he may have got it only half right, that 'Moral progress in history lies not so much in the improvement of the moral code as in the enlargement of the area within which it is applied' (Durant, 1935/1963: 55). What this means is that the number of people contained within the expansion of morality has increased, but not necessarily the type of morality under question. So, Durant would have us believe that the moral code over the ages has changed very little and is not so superior to that which was expressed by primitive man. This is obviously contrary to what is discussed above, and we thus need to be clear in our distinction of what constitutes the moral code.

More contemporarily, the expanding circle of morality has meant that the standards that we use to judge morality are much tougher (Wright, 2000). In this regard, it is not only appropriate that people are not enslaved, but also that they get paid a decent wage for their efforts and work under sanitary conditions. As such, morality can be said to have a relative context. In this regard, Wilkinson (2000) has observed that people today in the lesser developed societies are a great deal better off than they were many years ago. They have televisions, access to food and water, telephones, refrigerators, sources of heat and so on. But equality exists on a sliding scale and is subject not to real gains in prosperity, but rather to relative gains in prosperity.

Well-being then becomes not necessarily a function of taking care of one's physiological needs, but rather an assessment of one's social status. Those at the lower end of the scale may feel defeated despite the fact that they may be better off than many others. Wilkinson says that these feelings of defeat have a biological implication as they contribute to poorer health and lower life expectancies, because of an evolved stress that is triggered under these conditions. Critical energy to sustain this stress reaction takes away from the ability of the immune system to repair other physiological dysfunctions in the body. The structural inequalities in tourism would therefore exacerbate this situation in the lesser developed countries, providing a new area of research that could corroborate the knowledge that we presently have on the effects of tourism on such areas.

Moralisation

Moralisation is a process whereby 'the changes in which an activity that was previously outside the moral domain enters into it' (Rosin, 1997: 379). This means that objects or activities that were previously morally neutral later take on moral qualities. Associated with this concept is the internalisation of values. Theorists argue that an action that is performed in the service of a value is oftentimes more resilient and durable. As such, behaviour can be judged according to preferences and values, with the latter holding more weight. Good examples of moralisation include concerns over smoking, slavery, children and marginalised groups (moral condemnation). In this latter example, moralisation can be more easily activated if the target group has been stigmatised or marginalised, is a minority group or is sick.

While in general the circle of morality has increased in size, it can also shrink through our efforts to dehumanise people. A representative example of this is sex tourism, where an individual is reduced to something less than a person. We can also involve ourselves in desensitisation, by thinking in a vacuum that is turned on and off by society, by technology, by politics or by a combination of these. When we submit to this (the analogy is stepping on the moving walkway) we give away our will or free agency to the walkway, which controls our actions. It means that we have given away the faculties that we need to guide our actions; given away our education and knowledge; and given away the diversity of outside influences that we need constantly to affect our own behaviour – rather like a system of checks and balances. But when we fall victim to this vacuum we limit our ability to consider other viewpoints; to see beyond ourselves. For the profit-minded, this means the economic bottom line supersedes all else.

The predominance of what has been referred to as a mass neurosis of the modern age has resulted from too many people unable to find meaning in their lives (Frankl, 1985). During the 1920s, and in the affluent countries, happiness became the endeavour of individuals and, because of this, people turned inwards (Klingberg, 2001). This has held true to the present day, in the home, in schools and through our religions. The fall-out of this egocentric perspective on life is the loss of community, the breakdown of marriage as an institution, boredom, greed, substance abuse, violence, loneliness and promiscuity.

The push for humanity to step even further away from morality, to immoralise or get rid of morality altogether, in an effort to make us all conflict-free, was proposed by Nietzsche. This he articulated through moral nihilism, which 'holds that there are in fact no moral rights, no moral obligations, and that nothing is morally better or worse than anything else' (Benn, 1998: 32). While this stance is thought to represent a complete removal away from morality, nihilism usually surfaces as the need for people to reject one set of moral priorities and to replace these with another set of priorities. In this they do not reject morality completely, but merely have different preferences (Benn, 1998).

As the most demanding and least romantic of our qualities, Saul observes that ethics has the potential to slip into extremism, and this occurs when ethical principles are misplaced into certainty, which 'convinces the holder of the *truth* that he has the right to harm others' (Saul, 2001: 86). Ethical principles can be taken to the extreme (e.g. Hitler) and used to manipulate the group, the citizenry or whomever, according to one's own agenda. Saul continues by suggesting that ethics should not be romantic, moralistic or wishful thinking, because this is when it gets taken out of its proper context. And to keep ethical principles in perspective they must be rooted in everyday, normal life, where they are subject to common sense, memory and imagination, and where we can see the implications of our actions. We should exercise our ethical principles in the same way we exercise daily to keep well, eat healthy food for longevity and so on.

In tourism, the recent discussion of the new moral tourist, although largely untheoretical from both sides, has shown a tendency to judge people not only on what they *do* but on who they *are*. We are so preoccupied with wanting to classify types of tourism that we lose sight of the people who are actually participating in these activities, and eliminate the opportunity to appreciate the individual alone. Our tendency to group or to divide and conquer is an outward projection of instrumental reasoning.

Conclusion

The focus of this chapter was on the examination of certain applications of ethics in society. The dichotomy between relativism and universalism was included here instead of Chapter 3 because it represents two vastly different ways of envisioning how ethics and society mesh. It was also decided to include justice in this chapter, including rules, laws and rights, because of the critical role that these play in formulating a just society through the acknowledgment of the intrinsic right of other humans and other beings. A discussion of tourism and human rights provided the grounds from which to examine some of the most pressing issues in tourism. The chapter also illustrated the importance of responsibility, free will and determinism in human nature. In regard to the former, it was argued that ethics lies deeper and more fundamental than responsibility, and that we have yet to grasp this reality in tourism scholarship. Further, it was important in this chapter to outline the importance of the widening circle of morality. Over time, it is said, we have evolved to become more ethical. Similarly, the concepts of moralisation and dehumanisation were seen as important in the context of this work. Our propensity to allow ethics to slip into extremism is an outward growth of the often negative forces that work in society.

5 The Nature of Politics and Economics

Trade was the great disturber of the primitive world, for until it came, bringing money and profit in its wake, there was no property, and therefore little government
Durant, 1963

Introduction

Profit appears to be a central part of the human condition. We see it in all cultures and societies. This is as true in the 21st century as it was 800 years ago, tracing back to the early years of commerce and banking. St Francis, who founded the Franciscan order, was a victim of the psychology of profit and possession during the 1100s and early 1200s. Francis felt that wealth corrupts and that in order to be truly free one needed only to cast off earthly possessions. He also felt strongly that it was discourteous to be in the company of anyone poorer than oneself. For this he lost the favour of his family (his father was a wealthy businessman in Italy) and the ability to influence society beyond the bounds of a relatively few dedicated followers. In fact his cult of poverty was rather short-lived, as it failed even to be sustained beyond his lifetime. Those who chose to follow his doctrine were burned at the stake as heretics and although the Church might have helped, it had already become intricately tied to the rapidly developing international banking system that was originating in Italy at the time.

Obviously, men like Francis are in a minority, but they stand out because they have the fortitude to make us pause and think about the things in life that are truly most meaningful. This chapter is designed to contextualise the tendency for humans to be involved in politics and trade as a function of our human nature. That is, we seem compelled as a species to tie ourselves to these institutions, and tourism is nothing more than an extension of these needs.

The Evolution of Trade and Cooperation

In considering the importance of politics to society, a leading scholar on the subject defines it as social manipulation to secure and keep influential positions (de Waal, 1989a, 1989b). He notes that we are all, at some level, involved in politics through family, work, schools and how we cultivate connections at these places. We do not typically distinguish between behaviours and motives because we are well versed at camouflaging our true intentions. In further elaborating on the topic, he offers the following:

> And yet . . . power politics are not merely 'bad' or 'dirty'. They give to the life of the community its logical coherence and even a democratic structure. All parties search for social significance and continue to do so until a temporary balance is achieved. This balance determines the new hierarchical positions. Changing relationships reach a point where they become 'frozen' more or less in fixed ranks. When we see how this formalization takes place during reconciliations, we understand that the hierarchy is a *cohesive* factor, which puts limits on competition and conflict. (de Waal, 1989b: 208–209)

In your mind, as the reader, is this a fair assessment of politics within society as you know it? This leading scholar is Frans de Waal, who was not writing on human politics, but rather the politics of chimpanzees. De Waal, in his observations of chimpanzees at the Arnhem Zoo in the Netherlands, became convinced that the roots of politics are much older than humanity. He bases this on the fact that chimps formalised ranks within the community, they showed the capacity to have influence over the group, they established coalitions and stability within relationships, they used manipulation as a social instrument, and they developed exchanges. In this latter regard, the chimps were found to exchange social favours rather than goods, and they also used a central individual, a broker, who had and used prestige to provide group security.

When presented with bundles of food, chimps responded by kissing, embracing and group celebration. What was most striking, however, was the willingness of individuals to share the feast outside dominance hierarchies. Added to this was the fact that the most dominant members of the troop were often found to be the ones most likely to give food to any other member of the group. Doing so would evidently secure the individual's virtuous status within the group. However, sharing was not completely altruistic because it often came packaged with the assumption of a favour to be returned sometime in the future (reciprocal altruism). This concept of trade for benefit was observed time and time again, leading de Waal to suggest that although chimpanzees have demonstrated a degree of selflessness – more so than other simians – their level of sharing and cooperation does not rival the altruistic tendencies of humans, based on his extensive observations.

We get to know a bit more about the seeds of political and economic cooperation in humans through research in anthropology, and in particular the work of Hill and Kaplan (1985) and Hawkes (1993). Conventional wisdom in anthropology has been that food-sharing in societies was a function of egalitarianism. It was in the best interests of society if food was shared equally regardless of who caught it or grew it. This had benefits for the commons because there was little incentive for those to kill or grow more for their own benefit, if it was to be shared communally. However, Hill and Kaplan addressed the problem from a uniquely different perspective in their studies of the Ache, a hunter-gatherer society in Paraguay: from the perspective of the individual. Understanding the motivations behind the sharing of food necessitated an analysis of individual needs as different from social functioning. Societies are the sum total of individuals, and it is only from this standpoint that a true understanding of society could take place.

What is important in the altruistic hunting tendencies of the Ache, and the work by Hawkes on the Hadza of Tanzania, is that large game – a public good – is shared on overnight hunting parties with all of the hunters involved. This, however, is not the case when the hunting party returns, where game is given only to the immediate family of the one who killed it. The hunter who has been unsuccessful in the hunt still gets to return with food for the family (plants, grubs and other smaller food sources, which are not normally shared outside the family unit). But why is large game shared? There are at least two reasons for this. One is that hunting is a cooperative activity. Cooperation in the hunt therefore demands cooperation for the spoils. Another is success, or lack thereof. If I am unsuccessful at the hunt for a week straight, my family can still be assured of a good meal. When it is my turn to be successful I am merely returning the favour – reciprocal altruism at its finest. Still another is what has been referred to as tolerated theft:

> if resources are large and asynchronously acquired, then one forager's successful capture will be of potential consumption value to many. If the acquirer tries to consume it all, the consumption payoff gained from each additional unit consumed will be less than the consumption payoff hungry others would get from those same units. (Hawkes, 1993: 346)

Simply stated, the individual who kills a large animal will not be able to consume the kill before it goes bad, especially when others could have more use of it. Lions and other beasts practice tolerated theft, when they can no longer eat a carcass, and abandon it to other animals.

But there is also another edge to this scenario. In the debate with Hawkes, Hill and Kaplan argue that the reason why food is shared (big food like giraffes in the case of the Hadza) is because it can be exchanged for another commodity or currency in the future, such as a resource or service, which is worth more fitness to the acquirer than the food given up. And that currency usually comes in the form of envy from other men and extramarital rewards with the women of the group. Sex is one of the commodities, because the hunter can bribe women with the best, tastiest parts of the giraffe (remember the selfish gene and the importance of the immortality of one's genes). Much like chimpanzees and vampire bats, there is the expectation that at some later date there will be a tangible payback. Hawkes, on the other hand, believes that the payoff is one of social recognition only. In her words, 'the incentive for providing widely shared goods is favourable attention from other groups members . . . [such that] they have a larger, readier pool of potential allies and mates' (Hawkes, 1993: 341). Success at the hunt of big, not small, animals translates into prestige, sex and reciprocity – commodities that have an immediate and very important role within the community. Furthermore, these scholars found that, just as easily, if the game is not treated as a common pool resource, the hunter is subject to gossip and ostracism if he does not share it, losing a measure of status within the community.

While this research explains how and why members of a single group are found to cooperate, how does cooperation factor into the equation when members of other groups (inter-group dynamics) come into contact? One would assume that the more these individuals interact, the more they would be prone to conflict. But this does not necessarily happen, and it is because of trade. An oft-cited study that legitimises the importance of trade as a fundamental condition of human social existence is the work of Sharp (1952), on the Yir Yoront of north-eastern Australia. At the time of colonisation, the technology of the Yir Yoront was stone-aged, with the most prized possession among men being the stone axe. But these axes were made of stones from quarries over 400 miles to the south of the Yir Yoront. Sharp (1952) notes that the only manner in which to procure an axe was through frequent trade with regular trading partners:

> Almost every older adult male had one or more regular trading partners, some to the north and some to the south. He provided his partner or partners in the south with surplus spears, particularly fighting spears tipped with the barbed spines of sting ray which snap into vicious fragments when they penetrate human flesh. For a dozen such spears, some of which he may have obtained from a partner to the north, he would receive one stone axe head. (Sharp, 1952: 19)

Oftentimes, middlemen who made neither spearheads nor axe handles would receive a certain number of these two items in making a profit. Furthermore, the axe was so important to the culture of the Yir Yoront that a man would periodically prostitute his wife to a stranger in procuring further status (when the steel axe finally arrived in the colonial days). The two most important themes underscored in the work of Sharp are as follows. First, trade itself developed as a cultural form that helped to strengthen traditional kinship relations and social status within the group. Second, these groups used a middleman, who played the same role as the distributor in today's economy: they brought people together through the distribution of needed materials for exchange, when these parties could not otherwise make such a connection.

What this research shows is that we are continually drawn into reciprocal relationships for mutual benefit with others, which we would not otherwise need if we were not compelled to trade. This goes in stark contrast to what theorists thought many years ago regarding trade (Heaton, 1939). They thought that primitive man did not engage in trade, but rather obtained all of what he needed by catching it or collecting it. But what of the tribe or society that has seemingly everything it needs for its own survival? Ridley (1998) suggests that people have gone so far as to create exchangeable commodities and therefore a division of labour not necessarily for technological reasons (i.e. to fill a technological void within the village or society), but rather to stimulate further trade and consolidate alliances. These frequent encounters then provide the means by which to reinforce relationships, which might otherwise be unstable and subject to belligerence (see also Chagnon, 1983).

But Ridley makes two other valid points that help to explain why commerce appears to be so fully embedded within our human natures. Trade, he says, is not the consequence of politics, law and justice, but rather the precursor. Modern commercial law is therefore a consequence not of governmental policy development, but rather of merchants themselves. He reinforces his point by noting the development of *lex mercatoria*, which was a voluntarily produced, adjudicated and enforced set of rules that provided trade protection to agents within and between countries and that, incidentally, had to be independent from the policies and procedures of individual political states (see also Boorstin, 1985).

What these scholars observed about trade is substantiated by other researchers. For example, the Leakeys found evidence of a site established for the sole purpose of making axes and another for sharpening axes in their excavations of the Olorgesailie site in Nairobi. These spots were about 10 km away from their communities, which seems to point to the notion that the region was a 'kind of factory', which stayed in business for about a million years (Bryson, 2003). Also, because human skills and natural resources were unequally distributed, communities could produce certain articles more cheaply than neighbours (Durant, 1935/1963). The surplus of articles in these traditional societies was often sold to other groups in exchange for articles that they lacked. This was the case with the Chibcha Indians of

Colombia, who, according to Durant, exported rock salt to their neighbours in return for cereals. It seems after all that:

> human males associate less by desire than habit, imitation, and the compulsion of circumstance; he does not love society so much as he fears solitude. He combines with other men because isolation endangers him, and because there are many things that can be done better together than alone. (Durant, 1935/1963: 21)

Durant believes that this form of interaction led to specialisation throughout the world on the basis of these resources and skills and led to trade relationships on every continent, of any number of items, including skins, ornaments, weapons, knives, stockings, blankets, horses and wives.

Specialisation in the division of labour is a concept that has done us well as a species. This is chiefly for two reasons. One is that it has enabled people to avoid competition, which would create a significant degree of havoc in our communities, and the other is that, it has perpetuated what Adam Smith observed during the 1700s: that social benefits are derived from individual vices. It has allowed us to flourish through a self-interested reliance on the goods and services made by others, as counter-intuitive as this may seem. While self-interest led to unexpected benefits, government tinkering generally provided pitfalls. I say self-interest because, as noted by Ridley (1998), progress is achievable not by benevolence, but rather by selfish ambition. As observed by Adam Smith:

> In almost every other race of animals each individual, when it is grown up to maturity, is entirely independent, and in its natural state has occasion for the assistance of no other living creature. But man has almost constant occasion for the help of his brethren, and it is in vain for him to expect it from their benevolence only. He will be more likely to prevail if he can interest their self-love in his favour, and show them that it is for their own advantage to do for him what he requires of them.... It is not from the benevolence of the butcher, the brewer, or the baker, that we expect our dinner, but from their regard to their own interest. We address ourselves not to their humanity but to their self-love, and never talk to them of our own necessities but of their advantages. Nobody but a beggar chooses to depend chiefly on the benevolence of his fellow-citizens. Even a beggar does not depend upon it entirely. (Smith, 1776/1964: 13)

Through the concept of the division of labour, Smith helped to free the Western world of the doctrine that all wealth was of a limited capacity (i.e. that one state could only increase its level of wealth by taking from another). He argued that nations could accumulate wealth by enlarging markets on the world stage, and through the division of labour. This latter aspect was a function of education and training for the workforce, and also an antecedent to improvements within society. The system worked best not because of government intervention, as noted above, but rather on the basis of people's moral sentiments. This included self-interest, benevolence and sympathy, which were essential in helping them as part of society. And as Wright (2000) contends, human density played a large part in human destiny. Higher densities of people contributed to a flurry of cultural and technological advancements that literally catalysed these civilisations in a way never seen before – creating the impetus for specialisation through the division of labour.

Of particular interest in this discussion is the work of the famed economist Friedrich Hayek (1962), whose most controversial book, *The Road to Serfdom,* took Adam Smith's work further by recognising that Darwin had the strongest hand in allowing us to understand Smith's concept of the Invisible Hand (Smith obviously did not have the benefit of Darwin's theory). Success and the opposition to capitalism could thus only be found in the moral rules that were determined on the basis of trust, honesty and reciprocal altruism that emerged through evolution, rather than any social contract. We fail to understand capitalism, Hayek notes, because we live in two distinct worlds. The first of these is the smaller hunter-gatherer world of altruism and solidarity; while the second is the world of modern economic systems and huge political units. Hayek foreshadowed current evolutionary psychology by suggesting that our brains, through evolution, are better designed to cope with the first of these two worlds, and why we continue to be unable to understand the magnitude of the second world. We are just not programmed evolutionarily to understand current revolutions like globalisation because of the magnitude of the economic and political systems that they are founded upon. This 'fatal conceit' (or the belief that we can plan society through economics and politics), as termed by Hayek, may also explain why we are finding it difficult to implement sustainable development. Sustainable development is based on principles, economic and otherwise, that simply do not fit within the broad economic system in which we currently exist or, worse yet, do not subscribe to the fundamental aspects of Hayek's first world, above, and Darwinism (more on this in Chapter 7).

Politics, Power and Capitalism

Trade was also one of many by-products of the Crusades (Boorstin, 1985). Crusaders returned home from the East with silks, spices and perfumes, among other items, catalysing the Italian banks to invest in the Crusades in their efforts to finance kings and popes, as well as other travellers to these regions. In this, Marco Polo helped to solidify Venice as the leading centre for trade in the West during the 1200s due to his efforts to establish trading houses in Constantinople and Soldaia. The Italian focus on trade and banking with the East became more formalised in the 14th century by the efforts of individuals such as Francesco Balducci Pegolotti, an agent of a Florentine banking family, who wrote a guide for merchant travellers that ventured into the East. The guide made reference to distances, local hazards, weights and measures, prices and exchange rates, customs and other regulations, food and where to sleep:

> Whatever silver the merchants may carry with them as far as Cathay the lord of Cathay will take from them and put into his treasury. And to merchants who thus bring silver they give that paper of theirs in exchange. This is of yellow paper, stamped with the seal of the lord aforesaid. And this money is called; and with this money you can readily buy silk and other merchandize that you have a desire to buy. And all the people of the country are bound to receive it. And yet you shall not pay a higher price for your goods because your money is on paper. . . .
>
> (And don't forget that if you treat the custom-house officers with respect, and make them something of a present in goods or money, as well as their clerks and dragomen, they will behave with great civility, and always be ready to appraise your wares below their real value.) (Boorstin, 1985: 140)

This system of banking and commercial equity in Italy, which was based on the dominance by a few, is not at all unlike the economic climate we find ourselves in now. Power was held in the hands of a few wealthy merchants, within an emerging guild system, where workers had no say whatsoever (Clark, 1969). (Interestingly, Dowd (2000) contends that, although capitalism first emerged in Italy during medieval times, and later Holland, Britain should command our attention because capitalism there included both economic and social processes along with relationships that went beyond the basics of production and trade for profit.)

During the 1400s, Florence and much of Italy was embroiled in sweeping changes that were reflected in architecture, art and governance. These reinforced the belief that learning was essential in achieving a happy life, that free intelligence was important in guiding public affairs, and that the belief in community – the dignity of man – above all, was essential. In contrast to the humanism brought about by the Renaissance was the society that Niccolò Machiavelli (1469–1527) lived in and that compelled him to write his book, *The Prince*. His book provides inroads into the political and economic manoeuvring of the Italian elite, who had amassed great fortunes through a cloth (wool and then silk) industry, which allowed for unprecedented wealth, both for the merchants and the bankers who wielded control over the economy. Machiavelli served in government from 1498 to 1512, and was dismissed from his position due to a change in power. During the ensuing year he wrote *The Prince*, to show his prowess as a political analyst and win back a position in the new leadership (Grafton, 1999). (Note: *The Prince* teaches tactics and characteristics for persons in positions of power who wish to hold on to this power.) What was most intriguing about his book was that it examined virtue not from ethical and religious standpoints, but rather from the perspective of power and manipulation. A prince, therefore, could not be constrained by conventional morality if he hoped to do his job effectively and secure his status and holdings (Colish, 1978).

What is clear from Machiavelli is that: (1) politics and commerce were often diametrically opposed to morality; and (2) the ruler could not follow hard and fast rules in his daily deliberations, but rather needed to examine each issue independently. This might mean a liberal attitude in some cases or ruthlessness in others:

> when he has the chance an able prince should cunningly foster some opposition to himself so that by overcoming it he can enhance his own stature. (Machiavelli, 1513/1999: 69)
>
> I shall remind princes who have recently seized a state for themselves through support given from within that they should carefully reflect on the motives of those who helped them. If these were not based on a natural affection for the new prince, but rather on discontent with the existing government, he will retain their friendship only with considerable difficulty and exertion, because it will be impossible for him in his turn to satisfy them. (pp. 69–70)
>
> A prudent ruler cannot, and must not, honour his word when it places him at a disadvantage and when the reasons for which he made his promise no longer exist. If all men were good, this precept would not be good; but because men are wretched creatures who would not keep their word to you, you need not keep your word to them. (p. 57)
>
> It is far better to be feared than loved if you cannot be both. One can make this generalization about men: they are ungrateful, fickle, liars, and deceivers, they shun danger and are greedy for profit; while you treat them well, they are yours. They would shed their blood for you, risk their property, their lives, their sons, so long . . . as danger is remote; but when you are in danger they turn away. (p. 54)

In a climate of commerce, power and control, the West on the whole grew rich because of an improvement in medieval markets, where prices initially were set on the basis of justice (Rosenberg & Birdzell, 1986). Improvement was later determined by efficiency, where pricing started to reflect supply and demand and where growth was intertwined with expansion and trade. This massive change initiated a recurring ethical dilemma between justice (ignoring the worth of human beings) and economic efficiency. During the 1400s, the West began the process of replacing medieval institutions with modern ones. Central monarchies were developing in France, England, Portugal and Spain, which laid the foundations of the modern nation state containing completely different values (Rosenberg & Birdzell, 1986). These included economic motivation, private productive property, free enterprise, free markets, competition and limited government, as outlined in Adam Smith's *The Wealth of Nations* (Behrman, 1981).

Protestantism, too, was said to hasten the movement into what we view as Western capitalism, because of a move away from ethical ideals by the Church, which viewed business as inconsistent with the ideals of Christianity. The removal of these restrictions justified certain practices, such as interest-taking as a legitimate practice of those who lent money (McHugh, 1988). This opened the door to the view by some, both religious and secular, that business success was the fruit of a rational way of life. To others who continued to espouse a more traditional approach to religion, ethics and commerce, a tension between ethics (very much tied to religion at the time) and business emerged. For those in business, the other camp, a belief emerged that ethics had little to offer in the complexities of the corporate world, but had the consequence of leaving business without a systematic ethic of its own. This void was conveniently filled by Thomas Malthus and other theorists during the late 1700s to mid-1800s, whose ideologies reflected a political economy that stressed the inevitability of poverty, inequality, competition, private property and wage labour. These aspects of society were argued to be beneficial through the goodness and wisdom of God. So self-interest, avarice, pride, luxury and competition were seen as cornerstones for progress and the advancement of civilisation (McHugh, 1988).

These early years of capitalism were marked by growth in trade in association with these changing values, which over time ultimately amounted to a 'head start' in economic dominance. This is in essence what Heaton was suggesting when he said: 'The history of commerce is the story of selling more kinds of goods to more people in more places' (1939: 20). The momentum generated from this head start helped to define the three cornerstones of capitalist societies: expansion, exploitation and oligarchic rule:

> Throughout its history, capitalist profitability has required, and capitalist rule has provided, ever-changing means and area of exploitation (where 'areas' signify both geographical and social 'space', as will be seen). The central relationship making this possible is the ownership and control of productive property: a small group that owns and controls, and a great majority that does not, and whose resulting powerlessness requires them to work for wages simply to survive. Those social relations between these two classes are the basis vital for capitalist development. (Dowd, 2000: 5)

The preceding quote makes reference to two tribes: one that is small and profitable and one that is large and very poor. Capitalism ensures that the gap between these two groups is ever widening. The gap will only start to close if there is a willingness to shift power from the few to the many. As noted by McChesney: 'Capitalism benefits from having a formally democratic system, but capitalism works best when elites make most fundamental decisions and the bulk of the population is depoliticized' (1999: 3).

Tourism and Development

The concept of two tribes, as noted above, is alive and well in tourism, as we have seen in the work of Britton (1982), Hills and Lundgren (1977) and Jenkins (1982), regarding peripheral nations that have been dominated by transnational corporations for years. Although acquisition is not necessarily the correct terminology to use in these cases, as in the Machiavellian context, which obviously deals with a different subject matter, there is an element of domination and control in Third World economies that we cannot ignore.

What we view historically as dependent countries were not considered economic entities until the late 1890s, when a new 'lofty' responsibility regarding dependent peoples emerged through the encouragement of British Africa to participate in the cash economy (Brookfield, 1975). This lofty purpose was coupled with a crude hypocritical view of natives (the noble savage) and administration, where the distance between ruled and ruler increased. A dual mandate emerged, where external power had a responsibility to develop the resources of dependent countries for the benefit of the whole world. Some saw the acquisition of overseas dominions as motivated by strategic reasons. Others sought only economic incentives. The origins of development via capitalist countries broadening or diversifying production into new sectors was to be achieved internationally through a widening of resource exploitation of many of these developing nations. The most aggressive industrial and trading nations, such as Britain, which was successful at applying pressure on colonies to open trade, benefitted most through a laissez-faire philosophy marked by tremendous growth and optimal production (Brookfield, 1975).

For this relationship to continue to work in favour of developed countries, a form of development based on dependency has resulted and perpetuated core–periphery and a dualistic economy on local, national and international scales. The benefits to Britain, following the example above, include a dynamic, growing central region marked by high growth potential. For states that are dependent, economies grow more slowly or stagnate and are marked by declining rural economies with low agricultural production (loss of primary resource), with low rates of innovation, productivity and an inability to adapt to change (Freidmann & Alonso, 1974). A dislocating result of the efforts of the dependent economy to comply with the demands of the core is often a tiering or dualism of the social and economic conditions within the region. Dualism is a term that came into being with the colonial status of many underdeveloped nations (e.g. two Jamaicas, two cultures, two ways of life, including the African Jamaica and the European Jamaica), and is a condition in which a modern commercialised industrial sector has developed alongside a traditional subsistence agricultural sector. The condition creates an increasing divergence between rich and poor nations, and rich and poor peoples at various levels, which persists over long periods of time. Dualism, therefore, occurs both nationally (local and regional) and internationally, and is marked by the following four characteristics (Todaro, 1983):

(1) Different sets of conditions of which some are inferior and others are superior can coexist in a given space (e.g. wealthy highly educated elites with masses of illiterate poor people).
(2) This coexistence is chronic and not merely transitional; it is not a temporary phenomenon.
(3) Not only do the degrees of superiority or inferiority fail to show any signs of diminishing, they have an inherent tendency to increase.
(4) The interrelations between the superior and inferior elements are such that the existence of the superior elements does little or nothing to pull up the inferior element, and may push it down.

While some are able to benefit from the development of the new sector within the economy of the peripheral region, the vast majority of others are not. Such is the case in tourism, where the lesser developed countries have suffered from a history of colonial domination and from what appears to be an established pattern of multinational interests. Dependency, therefore, can be conceptualised as a process of historical conditioning, which alters the internal functioning of economic and social subsystems within an underdeveloped country (Britton, 1982). This does not occur from processes within that economy, but rather from demand from overseas tourists and new foreign company investment. The manner in which this demand and investment surfaces is through the efforts of multinational enterprises. After a potential tourist destination is identified (based on unique biophysical or cultural conditions), the involvement of the multinational increases. Foreign companies greatly influence the image of a destination country through promotion. Such advertisement leads tourists to perceive the host country in terms of this image and the nature of hotel accommodation, attractions and other tourist services as publicised.

Tourism researchers have demonstrated that metropolitan airline and hotel chains have the greatest influence on tourist movements, and undertake the most extensive advertisement campaigns (Britton, 1982; Jenkins,

1982). Sophisticated marketing programmes, computerised reservation facilities and established trade links are powerful advantages. The vertical integration between airlines, tour companies and hotels has further strengthened the power of the foreign tour generators vis-à-vis the host country (Jenkins, 1982: 236).

The domination, however, does not stop with airlines and hotels. Because of the inability of agricultural and manufacturing producers in most underdeveloped economies to guarantee a quality supply of goods and services for international luxury standard tourist facilities, there is a strong reliance on imported supplies for both the construction and operation of tourist facilities. Middle and senior management of tourism development in underdeveloped countries is occupied by expatriates (Winpenny, 1982). With their command over resources and location within tourist markets, metropolitan companies provide the most vital services (e.g. package tours, transport and marketing). Local counterparts – elites – may also predominate in the accommodation, tour and travel agency sectors. As mechanisms of control and dominance, local elites ensure that the interests of foreign firms are promoted within the hierarchy. Enterprises of a third, lower tier are involved outside the realm of dominant sector firms. Their options seem limited to small-scale inputs related to handicrafts, taxis and services related to budget accommodation and the retailing of souvenirs. The ultimate consequence of this system is that, while all participants in the industry hierarchy profit to a degree, the overall direction of capital accumulation is up the hierarchy.

There can be no doubt that an industry with such a marked influence in a region might have a negative effect on hosts simply by the nature of the tourism experience. The oft-quoted claim of Evan Hyde, a Black Power leader in Belize in the early 1970s, that 'Tourism is Whorism' (Erisman, 1983), reflects the frequent claims that tourism leads to conflict between locals and hosts. Locals also get a misguided interpretation of tourists. Tourists take on a personality that is often not representative of their values displayed at home (e.g. spending patterns). There is also the impact of the demonstration effect, where local patterns of consumption change to imitate those of the tourists (Britton, 1977; Hope, 1980; Mathieson & Wall, 1982; Rivers, 1973). Alien commodities are rarely desired prior to their introduction into host communities and, for most residents of destination areas in the developing world, such commodities remain tantalisingly beyond reach. Hosts are the direct consumers of this licentious behaviour. It is the process of commercialisation and an evolving wage society that can erode local goodwill and integrity:

> Cultural expressions are bastardized in order to be more comprehensible and therefore saleable to mass tourism. As folk art becomes dilute, local interest in it declines. Tourists' preconceptions are satisfied when steel

bands obligingly perform Tony Orlando tunes (and every other day the folklore show is narrated in German). (Britton, 1977: 272)

Traditionally, dependency is viewed from an economic or political context only. Erisman provides a comprehensive view of cultural dependency in the Caribbean, and tourism as a possible agent for facilitating its development. The following four theories delineate a form of external domination from which the region has long suffered (Erisman, 1983):

The trickle down theory. Controlling entities from the North 'Americanise' the Caribbean upper class, culminating in the diffusion of metropolitan values into the general population by this co-opted leadership. Central to this thesis is the local elite subculture, which is dependent and dominant. Dependent in the sense that its social orientations are determined by exogenous influences, dominant as it represents a hierarchy to which those in the lower echelons of Caribbean status aspire.

The commoditisation theory. These same controlling entities are solely concerned in maximising their profits by providing a desired product to as many consumers as possible while keeping overhead costs to a minimum. The islands are looked upon merely as commodities to be sold or leased to vacationers. This ideology affects Caribbean social interactions and cultural traditions, and precludes the emergence of a national identity.

The mass seduction theory. Although similar to the trickle-down theory, this theory rejects the idea that local elites are necessary as intermediaries. Instead it states that the industry's demonstration effect on the general population operates on the basis of direct contacts between hosts and guests: vacationers overrunning the region. The confrontation generates cultural impacts, exposing the archipelago's inhabitants to a lifestyle based on affluence, leisure and luxury consumption.

The black servility theory. Tourism pressures those involved to adopt an accommodating manner towards the customer. While practically all visitors are white, the locals who serve their needs – in low-paying/prestige positions – are mostly black. Mass tourism gradually imbues Caribbeans with the metropole's norms and values concerning race relations, creating a climate of cultural dependency (more on this in Chapter 10).

While cultural dependency theories have not been studied extensively, there is at least consensus among researchers that a latent relationship – that mass tourism has the potential to produce negative cultural effects – exists. These effects are aggravated if the afflicted country takes issue with the dependent relationship, by several tactics which may be employed by the multinational, especially if they work out of two or more countries. First, the multinational has the ability to shift its operations out of the country if the LDC becomes inhospitable. The multinational may thus choose another country which offers cheaper labour, less stringent laws, and more of an accommodating political regime. The other advantage, which presents itself as an ethical dilemma, for the multinational, is the ability to transfer raw materials and goods between countries for the purpose of escaping taxes and other obligations that firms who are tied to just one country must bear (Velasquez, 1992).

In his work on forms of sociality, Fiske (1992) writes that human life may be based on four different psychological models of social relations. The first of these is *communal sharing*, where people treat each other as equivalent and undifferentiated. They share a communal substance, like blood, and are altruistic to their own kind. Rituals are important in solidifying group membership. The second, *authority ranking*, is marked by asymmetry among the group, where people are ordered according to a hierarchical social dimension (people above or below others). Those in higher positions have more prestige and privileges, and control and protect those of lower orders. *Equality matching*, the most popular, and a form of reciprocal altruism involving more distant non-relatives, is based on a model of balance, justice, in-kind reciprocity, tit-for-tat retaliation and eye-for-an-eye revenge. Primary concern is for keeping track of how far out of balance relationships are, with the aim of entitling people to their own fair share (e.g. carpooling). People feel that such exchanges are important by serving the purpose of long-term cooperation. Aggressive retaliation results when cheating is detected. *Market pricing*, the final social relation, is based on a model of proportionality whereby components are fixed values for the purpose of comparison of qualitatively and quantitatively diverse factors. People are thus able to make comparisons of proportion to others, including one's share in a business venture or factors related to rents or commissions. Fiske notes that cultures use different rules to implement these four models. The foregoing discussion on development and dependency reaffirms the fact that tourism is very much about authority ranking, where there is great asymmetry among those who travel and those who cater to tourists.

It would seem plausible that policies could be enacted to offset many of the problems and issues that are created through these lop-sided relationships. In reality, though, tourism policy is both a relatively new phenomenon and based principally on two goals: maximising visitation and improving the balance of payments (Edgell, 1999). This has been reinforced by other theorists who say that national tourism policies or strategies are often geared specifically towards economic development, including the generation of employment and regional development (Hall, 1994). Even in countries like Australia, where policy seems to be better balanced with ecological and social elements, the economic goals are given much higher priority. Like it or not, governments are often drawn into the debate on tourism development with uneven results. Depending on the government and the situation, different decisions can be made for different reasons, as suggested by Ryan (2002). The most draconian of these often take place without due regard to all of the affected stakeholders, even though power 'over', should mean 'responsibility for' (Spratlen, 1973, as cited in Ryan, 2002).

A compelling addition to the literature on tourism, politics and ethics comes from Baptista (2012), who argues that both NGOs and tourists gain legitimacy through the performance of selfless, virtuous acts – a process by which to 'affirm their meaningful selves' (Baptista, 2012: 649). What has taken place in many developing country contexts (Baptista's work was done in Canhane, Mozambique), is the de-governmentalisation of governance. African nations are moving away from a traditional governance model towards one driven by neo-liberalisation processes.

The relevance of neoliberalism to the tourism industry demands further analysis. David Harvey (2007) provides one of the most comprehensive treatises on neoliberalism, arguing that it is a set of beliefs that has emerged as an ethic in and of itself, i.e. in matters of the market, privatisation, and finance, neoliberalism has been acting as a blueprint for all human action for decades. Market forces take priority, while state intervention, often in the form of regulation (Duval, 2008) becomes more recessive while at the same time becoming more responsive to these market forces, amounting to what has been referred to as 'growth fetishism' (Hamilton, 2003). Several years of sustained economic growth, Hamilton adds, theorised to make us all better off, has led to the corruption of social priorities and political institutions, and affluent individuals who are no happier than when they had less. By inviting lesser developed countries to the 'party', as illustrated by Baptista (2012) above, distributive justice erodes because of the corruption of institutions: values change in reference to employment practices and standards, wage disparities become more uneven, environmental regulations are compromised, and there is more inequality and poverty (Stilwell, 2002).

Wonders and Michalowski (2001) argue that neoliberal strategies are at the root of why prostitution has heightened in many cities. Prostitutes do not cause prostitution in the same way that the impoverished do not cause poverty. Global forces such as tourism, commodification, and migration have altered key institutions in these cities, stimulating the production of sex workers and the commodification of their bodies in creating demand. Similar processes exist in reference to medical tourism. Meghani (2011) contends that neoliberal strategies play a role in changing the medical landscape in places like India. Wealthy Westerners access the medical system, which in turn exacerbates pre-existing health care systems with inequalities created along gender and caste lines.

Governance is also being defined by the actions and interests of NGOs who place the interests of nature (and themselves) over people in the development process (Butcher, 2002). Dubois (2001) calls this 'assumed legitimacy' because NGOs attempt to regulate the actions of business and state sectors without actually questioning their own values. The prevalence of this mindset is rampant in tourism as NGOs attempt to co-opt emerging sun-sectors like responsible tourism and ecotourism to suit their own particular agendas (Fennell, 2015b). Volunteer tourism, too, has been criticised as a site for the ethical consumption of tourism experiences. Mostafanezhad (2013), in speaking on the geography of compassion, uses volunteer tourism as an example of the further expansion of neoliberal moral economies. Africa is seen as an 'authentic' volunteer site, while the child is said to be the ascendant signifier of humanitarian efforts usually made legitimate by the efforts of female celebrities and their adopted children. The notion of ethical consumption is no better articulated than in the edited compendium by Weeden and Boluk (2014). In their introduction, Weeden and Boluk discuss the ethical turn in society as new ethical practices move in the direction of global solidarity. Taken from a tourism context, the mobilisation of interested stakeholders has catalysed the boycott of several destinations where social and environmental injustices have taken place.

Efforts to side-step the neoliberal, according to Higgins-Desboilles (2006), will necessitate going back to earlier conceptualisations of tourism as a force of social change: to move away from profit and emphasise justice. Social tourism has been vitally important in setting new values and priorities in this regard. Minnaert *et al.* (2006) argue that social tourism has both demand and supply facets to it in reference to families that are impoverished, accessibility in hotels, and ecological holidays. Christian ethics, Marxist ethics, Kantianism, and utilitarianism all have relevance in building a more just tourism industry on so many different levels and scales. By contrast, Wearing and Wearing (2014) argue that fixing North-South relations in ecotourism demands a strong counter-discourse away from neoliberalism and towards the decommodification of local tourist markets to a new environmental morality based on justice. A required ethical framework for this modification emphasises the decolonisation of moral encounters, and full engagement and participation of local people in policy and planning. From a practical sense, this would include principled ethical standards such as human rights and fair trade, and from an applied perspective, this would include the gamut of practice-based ethics for all the various stakeholders in tourism to facilitate compassion for the oppressed 'other as host' (2014: 127). Compassion, empathy, inclusiveness, cooperation, equality, human flourishing, and justice and rights are all calling cards for the new morality in tourism. For Caton (2014), this amounts to a complete re-shifting of tourism from the little imaginative trappings of capitalism (tourism as a bastion of instrumental reason and activity), towards a humanistically-inclined tourism that emphasises the qualities noted above.

Conclusion

Although this chapter dealt primarily with the evolution of politics and economics, it was illustrated that some of the most important scientific findings in biology continue to have resonance in the social sciences. It underscores the intricate tie between science and human nature; that just about everything we do is somehow connected, and that the most robust scientific discoveries are those that are generalisable across many fields of study in describing our place in the universe. De Waal concludes his book by suggesting that Machiavelli reported many of the same observations in his assessment of politics in 16th-century Italy. While there has been a propensity to cover up the unsavoury aspects of politics because they are too insulting to humanity, Machiavelli struck to the core of manipulation and influence.

Tourism, as another form of trade, is a commodity that forces us into situations where reciprocal altruism can be practised. Just as the regular trade practices of aboriginal groups cemented long-term relationships, I see no better vehicle to continue to build cooperation and trust between people than the tourism trade. But this means that, just as it does in bats and chimps, exchanges need to be fair and continual. If not, the system breaks down. So perhaps it is not the servility and menial tasks that separate us into the haves and have-nots, and the various ugly consequences of the division of humanity along such lines, but rather the simple justice that goes along with fair exchanges. Drinks can be served and money exchanged, but there needs to be an element of fairness and respect in the exchange. This means that it might not be so much the type of job that one does that matters, but rather the rewards for doing that job in line with the expectations borne with it.

6 The Business Side of Ethics

The things which ... are esteemed as the greatest good of all ... can be reduced to these three headings: to wit, Riches, Fame, and Pleasure. With these three the mind is so engrossed that it cannot scarcely think of any other good
Spinoza, from *Tractatus de Intellectus Emendatione*, 1677: I, 3

Introduction

The previous chapter provided an historical backdrop from which to examine the current state of business and business ethics, which is the focus of the present chapter. Here, the corporation is contrasted with the individual through an analysis of technology and instrumental reason. Responsibility once again surfaces, but through the vast amount of literature that has emerged on corporate responsibility. In addition, corporate trust and organisational culture are emphasised as vehicles by which to more fully examine the concept of ethics in organisations; while business ethics is discussed later through the Canadian context. The chapter concludes with an emphasis on some ethical responses that have taken place in tourism over the past few years, including fair trade and pro-poor tourism, all of which have linkages to the organisation and to business in general. These provide the opportunity to further examine tourism and development issues introduced in the previous chapter and the prevailing ethical challenges that persist.

Corporatism vs. Individualism

During the late 19th century, private property played a significant role in the development of ideas that led to the emergence of the corporation. Private property served the purpose of tiering society: those who had it were significantly advantaged over those who did not, and this served a dual purpose functionally and morally. Williamson (1989: 26) writes that the onset of liberalism, 'the granting of political and economic equality to individuals who in corporatist minds were manifestly unequal', took away from the natural ordering of society where people were placed into categories depending on their designated status. Private property could thus no longer be used as a vehicle to determine sociomoral hierarchy within society. As a tangible outcome of liberalism, and if properly constructed, the corporation could allow modern man to overcome his moral and spiritual malaise, offering an antidote to the 'spirit of class' through vertical pillaring, internal hierarchies of order and functionally interdependent organisation (Schmitter, 1974).

As the corporate mentality has grown, so too has the expectation that what the corporation happens to be selling is what is most important to us all. This is the case not only with fashion, but also travel. The corporations have us just where they want us:

> Fashion is merely the lowest form of ideology. To wear or not to wear blue jeans, to holiday or not to holiday in a particular place can contribute to social acceptance or bring upon us the full opprobrium of the group. Then, a few months or years later, we look back and our obsession, our fears of ridicule, seem a bit silly. By then, we are undoubtedly caught up in new fashions. (Saul, 1995: 20)

But the corporate mentality does not stop at the private sector. Our public institutions too are run like companies in the name of efficiency. The state, according to Saul (2001), emphasises this daily in its move away from content and consideration towards structure and form. Answers to questions are based on managerial and structural equations rather than humanistic ones, just like corporations. And along with this is the expectation that we act like shareholders, instead of citizens. We need to be clear on this. We are not making purchases as citizens of a country, nor are we owners or shareholders, because we do not own shares and therefore we cannot buy or sell them. It is responsibility that we want, indeed that we need. Saul cites the World Trade Organization's 1997 ruling that Europeans were wrong to ban US beef from cattle that were raised with hormones, as an example of this. The ruling was based on the 'fact' that there was no scientific evidence that the beef presented a health risk. His point is that policy, unfortunately, must be based on scientific fact rather than the will of the citizenry. But what is most troubling about this science is that it is not really science at all because it is initiated under the pretext of commercial interest. So, as Saul notes, food is first a commercial object, second a scientific object and third a matter of social well-being.

Both Fennell (2013c) and Munar (2016) bring this strand of thinking into the tourism studies realm. Fennell (2013c), following Saul (2001) above, argues that universities are now focused more on structure and form over content, such that the neoliberal audit culture of the system promotes quantity of research publications over quality. Munar (2016) contends that the tourism academy ('house' in her words) is so thoroughly enmeshed in a state of bureaucratisation and commercialisation that there is the fear that autonomy and subjectivity are being lost, along with the ability to accelerate knowledge through alternative strategies.

Decision-making in such cases has turned into a form of benevolence that follows its own instrumental pathway. But bureaucrats are often far from benevolent and fair-minded in their public duties. They employ an outward sense of cooperation in principle, but continue to practise a level of self-interest that is designed to benefit themselves and their colleagues, over the needs of the electorate (Buchanan & Tullock, 1982). This has led to a corporatist structure that continually grows in magnitude and self-indulgence, while the citizenry it is supposed to serve grows only more disenfranchised. All under the guise of responsible government.

Given the foregoing, it is no wonder that the collapse of the community as a cohesive unit in the last few decades is symptomatic of a state that has made no bargain with the citizenry to take responsibility for civic order, as suggested by Ridley (1998). With little civic pride, only obedience, it is not difficult to see that, if government treats the electorate as a naughty child, they quite naturally will behave like one. From this, authority has replaced reciprocity and trust, which in turn has given way to top-heavy institutions that are more concerned with using government as a tool, as opposed to having government serve the many as a facilitator and enabler. Heavy government, he suggests, makes us more selfish, not more virtuous.

This leads to the oft-recognised contention that people are ethical, not corporations and governments. And while we can gather up the gall to suggest that corporations are moral, this sentiment can only be tempered with the ultimate modus operandi of the firm: profit. Competition and self-interest rule the day and they make no room on the bus for ethics. Although ludicrous to even suggest, how many corporations have a professional philosopher on staff (acknowledging that many firms have an ethical officer these days)? This is just not rational. Employees are supposed to stimulate productivity, not diminish it, which the philosopher would surely do in opposition to any tightly written job description.

In this we have identified two important but different concepts: self-interest and individualism. The former is driven by our corporatist mentality, the latter by a democratic society that places value in the citizenry as a source of knowledge and diversity of thought. However, as illustrated by Saul (1995), and noted above, our faith in corporatism has denied the legitimacy of the individual. His main argument is that the reliance on the corporatist ideal has forced us to adore self-interest, through passivity and conformity, and the denial of the public good. Economics, Saul (1995) adds, is held accountable for our current state. After 40 years, the whole economics field – its models and theories – has been unsuccessful at making civilisation better. It is not that the advice of economists has not been taken, he says, it is just that it has not proven to be much good. While the corporatist movement emerged in the 1800s as a replacement to democracy, in doing so it inserted the group over the individual as the primary unit of value. The 'me' culture is replacing the culture of civic mindedness. Consumerism is fuelling this change.

In the context of the foregoing discussion, and in our attempts to move away from the corporatist mentality, what news do we have to report for the tourism industry, which is just as corporatist in its thinking, generally, as any other sector? The globalisation of tourism has made the industry 'increasingly competitive and based around a distribution system dominated by multinational companies, and the potential for continued inequalities also is increased' (Tapper, 2001: 353), as noted in the previous chapter. We can suggest the implementation of regional mechanisms to allow the smaller tourism service providers to be advantaged. However, I know this point can easily be refuted. After all, we are dealing with an industry that is fully cognisant of economies of scale – the bigger, the cheaper, and the more profitable the end. But in line with the foregoing discussion, individualism needs to be reflected in our planning efforts, in which quality of life becomes a central issue in allowing people to live their dreams in a profitable manner. It is the softer and more humane side of tourism, what we have termed alternative tourism, but it needs to be made more of a reality than a philosophy. In doing so we can allow tourism to be a positive agent of change, where community has the power to control the type and pace of change. This demands a focus on well-being rather than strictly profit; better accountability of policy makers; better linkages between stakeholders in a transparent forum; a better understanding of systems and networks; and newer values and philosophies that are shared.

Technology and instrumental reason

Jacques Ellul's (1965) *The Technological Society* discusses the erosion of moral values brought on through the development of an increasingly technological civilisation, which seeks to improve the means without careful consideration of the ends. In 1965, when this was published, Rachel Carson had already identified this very point in her treatise on the toxins that were carelessly delivered into the environment to achieve instrumental ends. Technology, it seems, had advanced further than our ability to understand the consequences of its use.

When we place technology ahead of all else it has the tendency to lead to an obsession with planning and procedure and soon after to specialised vocabularies and knowledge that are more concerned with efficiency. New technologies gain the interest of politicians and financiers, as Carson noted, and soon after are leveraged into an instrument of power. Although the initial intention of adopting the technology may have been advancement in one area or another, more typically it ends up limiting the choices of the society in which the technology exists (Ellul, 1965; Saul, 2001). Microsoft is an excellent example of this limited options standpoint, where we are compelled to stagnate and to lose our capacity to think and act as individuals, and where we have few other options to turn to. Bigger software requires bigger and faster machines. If this is not dependency, I do not know what is, and it lends support to the credo: 'You have to serve somebody', and we do serve. The corollary of this is that in all that we do, whether it is business or recreation, we become reduced to a mass conglomeration of reactive machines with diminished levels of freedom.

Taylor (1991) addresses these issues in his examination of three malaises that have eroded societal authenticity. The first of these, individualism, has been important as one of the finest achievements of civilisation *as a right*. The loss of purpose, i.e. no inspiration, or no reason to act strongly for a cause, is linked to a narrowing of vision for society as a whole in the face of individuals who can only focus on themselves. This is the dark side of individualism, Taylor says, which makes us less concerned about others and flattens and narrows our lives. Added to this first 'worry' is the aspect of instrumental reason, which is rationality that is controlled by maximum efficiency and the cold-hearted nature of cost-benefit calculations. This mentality has led to the treatment of just about everyone and everything as nothing more than raw materials for our own instrumental purposes. His third malaise is the political and economic institutions and structures, which work, either consciously or unconsciously, to reduce and restrict our choices through forms of instrumental reason (e.g. technology). This has led to an erosion of political control over our destiny, which, as noted by Taylor, we could quite easily exercise as members of the citizenry.

Perhaps the most damning evidence against technology is in how Saul (2001) characterises it. He feels that, because it has no will, no memory, no purpose and no direction, it constructs utopian visions that take us down dead-end streets. In recognising the pros and cons of technology, researchers have found that great companies are built not on technology first, but rather discipline, knowing strengths, understanding economy and upholding core values (Collins & Porras, 1996). It is, therefore, the people-centred aspects of who we are and what we represent that is perhaps most meaningful, not only to our customers, but to each other as well. Using our example of pesticides as a form of technological progress, we often confuse (or prefer) the value of a green lawn, which is utopian, with the health and well-being of our families and communities.

A contrasting view of this 'damning' perspective on technology is offered by Florman (1996). Florman contends that it is not technology that is the problem, but rather people's improper understanding and use of it. Engineering, and technology, are not cold, impassionless human endeavours that have negatively transformed the world, but rather exist at the confluence of art and science. Technology has the ability to stir our deepest impulses in allowing us to live rich, creative and sensual lives.

Corporate Responsibility

One of the earliest references to corporate responsibility can found in the art of the Dutchman Frans Hals, during the 1600s (Clark, 1969). In Hals' paintings we see individuals, profitable and upstanding members of society, who are prepared to join together in an effort to improve the public realm. They can do this because they have leisure, and they have leisure because they have money. This aspect of community service manifests itself from the belief that upper society and community must forge on together. But as Holland emerged to become the banking centre of Europe at the time, bourgeois capitalism overwhelmed civic responsibility at the hands of self-indulgence; and self-indulgence, as we know, is an agent of self-interest.

This example serves to illustrate the power of capitalism and the notion that the very nature of the market – Holland became the banking centre that it did on the back of the tulip trade – is to operate on the basis of self-interest, and not in regard for responsibility to social, or indeed ecological, externalities (Tisdell, 1989). The Nobel Prize-winning economist Milton Friedman (1970) argued that because the corporation is an artificial being it does not have any responsibilities of a social nature, beyond acting without deception or fraud. In this classic viewpoint, to say that a corporate executive has to act with social responsibility in his capacity as a business person, is asking him or her to act against an employer. Friedman uses the example of the employee who is motivated to purchase pollution reduction equipment that goes beyond the finances of the firm because of his own personal social objective of improving the environment. In this case, the individual is acting not as an agent of the company, but rather from his own social responsibility, and social mechanisms are incongruent with market mechanisms. Mayr (1988) notes this as well in suggesting that corporations have difficulty working under an environmental umbrella because they are unable to keep pace with the competition, which may choose to ignore environmental and social responsibility. Likewise, management is often located considerable distances away from the core of production, so they avoid having to suffer the social and ecological pressures that result from their decisions. Constraints on their behaviour are thus reduced. As such, because commerce has no unique attachment to any particular society, it does not have to be accountable. No accountability comes as a result of no responsibility. The difficulty in accepting corporate social responsibility by the investment crowd has been summarised according to the following five barriers: 'The difficulty in defining exactly what corporate responsibility means, lack of a clearly measurable business case for it, inadequate information from companies, shortage of staff with both financial and corporate responsibility experience, and the short-term focus of financial markets' (Maitland, 2004: 5).

A well-cited example of the tendency for corporations to avoid responsibility is the Birmingham, Alabama, bombing of a church in 1963, resulting in the deaths of four black children. One of the companies at the centre of the race relations was US Steel, whose chairman said:

> When we as individuals are citizens in a community, we can exercise what small influence we may have as citizens, but for a corporation to attempt to exert any kind of economic compulsion to achieve a particular end in the racial area seems to me quite beyond what a corporation should do, and I will say also, quite beyond what a corporation can do. (Beauchamp & Bowie, 1979: 125)

Levitt (1979) too has suggested that the sole purpose of business is to sustain high-level profits. It can only do this, he contends, by doing what it does best (making profits) and by existing in a pluralised society, where there is a division of responsibilities and power between government (which must involve itself in welfare and society) and business. The two cannot share these responsibilities because the end will be a coalescence into a single power that is unopposed and unopposable. In this Levitt believes that business must have narrow ambitions. And if there is anything wrong with the approach of many businesses it is that these are not narrowly profit-oriented enough. His wish is to see business left to its own to embroil itself in conflict, which is, and will continue to be, a central part of human society in giving élan to life. The Hobbesian force of his discussion is reflected in the following:

> If the all-out competitive prescription sounds austere or harsh, that is only because we persist in judging things in terms of utopian standards. Altruism, self-denial, charity, and similar values are vital in certain walks of our life – areas which, because of that fact, are more important to the long-run future than business. But for the most part those virtues are alien to competitive economics. (Levitt, 1979: 140–141)

It is human to seek meaning and purpose in our lives, as Levitt notes above, but the hope for this tends to distort all that we do. Companies understand the human need for meaning by establishing mission statements, which underscore these important values and virtues. They include aspects such as employee well-being, customer satisfaction and community involvement. The bottom line is that there is little empirical data to support the belief on the part of many that corporations actually take these measures seriously when they are matched against the prime goal of profit maximisation (Kay, 2004a).

In the end, Levitt argues that business has but two responsibilities. The first is to obey elementary rules of everyday civility, such as honesty and good faith, and the other is to seek material gain. In accomplishing this, business must fight as if it were at war, 'And, like a good war, it should be fought gallantly, daringly, and, above all, *not* morally' (Levitt, 1979: 141). This underscores, clearly, the self-interested and competitive side of human nature. Is business a surrogate for war; a game that needs to be played out in an arena in full battle-dress? We see it on television these days, where car companies are advertising that by buying a car 'the customer walks out the winner'. Even the law is on the side of profit, by requiring corporations to manage their companies according to the best interests of the shareholders. Those employees who use the corporation's funds to 'indulge their own ethical agendas . . . [violate] their fiduciary duty, a most unethical outcome' (Kay, 2004a: 85).

By the late 1970s and early 1980s, the underlying sole motive of commerce was starting to change as theorists argued that corporations can and should have a conscience and therefore be as morally responsible as an individual (Goodpaster & Matthews, 1982). This meant the insistence that business develop a social consciousness in acting as part of the solution rather than as part of the problem, and further that businesses should lobby government for good environmental legislation, which is the case with some of the responsible tourism operators who have taken it upon themselves to better regulate the industry through responsible practices (Hoffman, 1991). At the heart of the debate on the role of the corporation is its status as a chartered entity – a type of contract in which society permits the company to exist – where the practices of the business will act to do good within society (Beauchamp & Bowie, 1979). It is therefore a societal contract that is developed with a strong link to the broader culture within society. At least, to some, it ought to be this way. Followers of the Friedman and Levitt schools continue to argue that because a corporation is an artificial construction, it cannot be held morally accountable.

This element of social responsibility in tourism has been addressed by Walle (1995), who points to a dichotomy that exists in business ethics theory as established by the work of Friedman and Davis. In the Friedman school, the modus operandi of business is to generate profits, as noted above, and as such business leaders have no responsibility regarding social policies and strategies (within the law). By contrast, the Davis school advocates more of a socially responsible behaviour for business, which will, in turn, lead to greater profits and prevent government intervention through the generation of positive publicity (Table 6.1). Through these extremes, Walle argues that tourism, because of its uniqueness, cannot follow the universal or generic strategies of mainstream business, which focus on the organisation and its customers (e.g. manufacturing). Instead, Walle argues that:

> Tourism is not a generic industry since it uniquely impacts on the environment, society and cultural systems in ways which require a holistic

Table 6.1 Ethical orientations: A comparison

	Social obligation no. 1 (Friedman)	*Social responsibility no. 2 (Davis)*	*Social responsiveness (extension of no. 2)*
General overview	Legal and profitable.	Current social problems responded to.	Future social and/or environmental problems are anticipated/ addressed.
Choosing options	The sole consideration aside from profit is legality.	Decisions respond to social issues that overtly need to be addressed.	Decisions based on anticipation of future needs and/or social problems even if they do not impact or are caused by firm.
Strategies evaluated with reference to	Is the strategy legal? Is the strategy profitable enough?	Has the organisation responded to problems and issues that have emerged as significant?	Future problems are addressed even if the organisation is not directly causing them.

Source: Adapted from Walle, 1995.

orientation within a broad and multidimensional context. Contemporary business ethics, however, has been slow to embrace such a holistic perspective. Historically, the focus has been on the organization and its customers. Impacts on third parties (externality issues) have often been ignored. (Walle, 1995: 226)

His initial conceptualisation on social obligation, responsibility and responsiveness is reworked to illustrate how such orientations might be applied to tourism's uniqueness (Table 6.2).

Other researchers have criticised the Friedman and Levitt schools for their lack of inclusiveness in regards to stakeholders in business (Robson & Robson, 1996). Stakeholder theory has been used as a normative planning tool in tourism by researchers who observe that all stakeholders have intrinsic value and therefore have the right to be treated as an end in and of themselves (Sautter & Leisen, 1999). Stakeholders' rights must not supersede the rights of others, and it is the task of management to select activities that provide optimal benefits for all involved. In following this approach, tourism organisations can be both smarter and more sensitive to the needs of others.

Still, others point out that stakeholder theory fails because of the sheer number of stakeholders who might wish or need to be part of the planning process, and who will also have vastly different objectives and values from

Table 6.2 Special ethical considerations for tourism

Tourism's perspective	*Social obligation no. 1 (Friedman)*	*Social responsibility no. 2 (Davis)*	*Social responsiveness (extension of no. 2)*
Progress is not inevitable or inherently beneficial	Since the concept of progress is not universal or inevitable, we should not place an over-reliance upon it in our strategies and tactics.	Tourism has a responsibility to encourage development that meshes with the local environment and culture, not in accordance with a universal concept of 'progress'.	Since 'progress' leads to concomitant changes in culture and the environment, tourism strategy should be appropriate and mitigate its impact.
Tourism can be undermined by pressures of the industry	Changes wrought by tourism might undercut the industry; such potential should be prevented and mitigated when doing so is a good tactic.	Tourism causes negative impacts and pressures on people and the environment that should be mitigated.	The industry has both practical and ethical reasons for responding to impacts on the environment and local people.
All relevant stakeholders need to be considered when strategies are forged	Government regulation and loan conditions might demand a response to the needs of all relevant stakeholders.	Tourism should respond to the needs of various stakeholders that are impacted on by the industry.	The industry should anticipate future impacts from various sources and respond in proactive ways.

Source: Adapted from Walle, 1995.

their counterparts (Robson & Robson, 1996). What constitutes success, therefore, is open to debate because of an extensive communication network that is unmanageable and seen to be more of a constraint than a catalyst. The feminist perspective is a case in point, where women do not have as much of a stake in the new economy (e.g. due to low pay and part-time work), because of the inequalities in power and knowledge between groups. Women are unable to have a full stake owing to the constraints that exist in the social world and that prevent their moral consensus. In tourism, and other sectors, this disparity prevents us from initialising changes in the ethical performance of businesses.

Responsibility in tourism has been fleeting for other reasons. In a study seeking to see what factors were important to tour operators in adopting a more responsible stance now and in the future, it was concluded that control, finances and scale were oft-cited reasons (Miller, 2001). Tour operators felt that because of their positioning in the middle of the tourism industry, they had little control over what was happening regarding supply and demand. However, operators noted that although they knew what to do to act responsibly, they did not have the resources to put these actions into place. Small service providers felt that it was the bigger tour companies that had the most power and the most advantage over instituting change in the industry. Furthermore, service providers felt that consumers held potential in instituting more responsible behaviours in tourism. None of the organisations interviewed mentioned that they could be more responsible out of altruism alone. These findings resonate well with Holden's (2009) conclusion that environmental policy has had little effect on the workings of the tourism market. Interestingly, in a study on the perceptions of tourists, entrepreneurs and residents in relation to the responsibility for environmental impacts of tourism in Greece, tourists felt that entrepreneurs and locals were more responsible for impacts, while local people considered themselves to be more responsible for ecological impacts, relative to the other two groups (Kavallinis & Pizam, 1994). Responsibility is therefore subject to justification and blame.

Studies on corporate social responsibility have expanded into specific sectors of the tourism industry. This reductionist approach is thought to be essential in pin-pointing the causes and concerns of these specific industries in tourism. Examples include Ravinder (2007) on the aviation industry; and Holcomb *et al.* (2007) and De Grosbois and Fennell (2011) who investigated Corporate Social Responsibility (CSR) principles and practices of the hotel industry. In this latter study, De Grosbois and Fennell looked specifically at the carbon footprint of the 150 of the largest hotel groups in the world and found that carbon reporting among the sample was scarce, with poor calculations and little consistent means for comparison of carbon footprint indicators. Yet, although expansion in research on CSR is evident, Coles *et al.* (2013) argue that much more critical engagement with the concept is needed. Work has emerged on implementation, the economic rationale for being more responsible, and on the social implications of CSR. Needed, the authors observe, is greater conceptual and methodological sophistication in research to meet the needs of management and governance in the tourism industry.

A potentially new and fruitful avenue of research and practice for tourism scholars in the area of CSR is the benefit corporation (BC). These for-profit corporations have as their mandate promotion of the public good. Cummings (2012) writes that new legislation in several US states has allowed these corporations, which are identical in almost every way to conventional for-profit organisations, to pursue their social enterprise agenda and be protected by law. This means that the corporation is free to sacrifice some of its profit to achieve any type of social good it deems important. For example, the directors of a benefit corporation may wish to power their offices using alternative energy sources, and if the cost of these alternative sources exceeds the cost of conventional fossil fuel sources, they would still be able to use the newer technologies (Loewwnstein, 2013). Loewwnstein argues that this places the directors of BCs in an envious position: they are able to spend other people's money to further their social and environmental priorities.

Trust and Culture in Organisations

Like communities and other social groups, corporations are dependent on trust for ethical guidance (Fukuyama, 1995). The creation of hierarchies within the organisation – again, not unlike society in general (as observed earlier in this chapter) – are seen to be essential, because they allow individuals to develop and uphold the trust and values of the unit, acknowledging that all people cannot be trusted at all times to live up to the ethical standards of the unit. Free-riders, as we saw in the theory of reciprocal altruism, will often tax the system without giving anything back to the common cause.

As a culturally determined virtue, trust is based on deep ontological beliefs, such as the nature of God, or the nature of human–environment relationships, in addition to secular norms, such as codes of ethics and professional standards (Fukuyama, 1995; Malloy & Fennell, 1998a). These beliefs are consensual and act to bind the group by increasing the sense of trust and reducing the reliance on other more formal laws and regulations. Social capital (people forming the glue that binds the unit together) required to build such a moral unit cannot be purchased or invested in through conventional means (like the market or an education), but rather must be assembled through 'habituation to the moral norms of a community and, in its context, the acquisition of virtues like loyalty, honesty, and dependability' (Fukuyama, 1995: 26–27). Trust may be envisioned as a form of tax. Those who have it do not need to pay it, while those units that have widespread distrust must be prepared to pay heavily.

The business literature supports the notion that different cultures use trust differently in how they conduct themselves and their commerce. North America and Western Europe are two regions referred to as individualistic cultures, defined as those societies where people are concerned more with themselves and their immediate family than with others. By contrast, collectivist societies are those in which there is a deliberate move towards association in groups that look after the individual in exchange for his or her loyalty (Robertson & Fadil, 1999). Individualistic cultures are marked by autonomy, competitiveness, self-sufficiency and achievement, and the desire for wealth, which is positively related to unethical actions (see Hegarty & Sims, 1979). Citing the work of Frank (1988) on altruism and trust, Bowie (2001) observes that the altruistic person (one who will not behave opportunistically even if he or she can get away with it) is the best person to make a deal with. Altruism is adopted by these individuals not because it pays but because they are committed to it. In reference to his commitment model, as noted in Chapter 2, Frank writes that:

> For the model to work, satisfaction from doing the right thing must not be premised on the fact that material gains may later follow; rather it must be intrinsic to the act itself. Otherwise a person will lack the necessary motivation to make self-sacrificing choices, and once others sense that, material gains will not, in fact, follow. Under the commitment model, moral sentiments do not lead to material advantage unless they are heartfelt. (Frank, 1988, cited in Bowie, 2001)

Bowie hypothesises that labour–management relations in the US are opportunistically oriented, such that any reforms initiated are not based on employee concern but rather on profit. What is missing is the committed altruism that is a foundation for trust and that has proven to be an effective management aid in collectivist societies.

Organisational culture (OC) is defined as what is and what is not appropriate behaviour within the organisation (Trevino, 1986). 'Appropriate' can apply both to behaviour that is grounded in tangible aspects, such as codes of ethics, as well as the intangible aspects, including leadership and trust. As regards the former, a code of ethics within an organisation can tell us much about the organisational culture of that firm, in that it acts as a barometer regarding ethical conduct (Singhapakdi & Vitell, 1991). However, evidence suggests that it is the intangibles that more readily define organisational culture and that may make the most important contribution to appropriate behaviour in the organisation (Peters & Waterman, 1982). This is further supported by Guy (1990), who notes that organisation is not created as much by enforcement (e.g. codes of ethics as the letter of the law) as by peer pressure. In judging how ethical a company may be, we might look at how it treats its people from day to day, and person to person (Enderle, 1987; Freeman, 1990), acknowledging that rules established by the market, organisational policies and law are not always ethical. We may instead be guided by sound leadership and trust in our efforts to alleviate work stress and minimise employee turnover (Jose & Thibodeaux, 1999; Ross, 2003). And in regard to leadership, although there may be some initial complexities in operationalising it, in the end it affords us an axis from which to streamline our efforts in being moral.

In relying on these intangible aspects in our organisation, we have opened ourselves up to an appreciation of what values can mean to individuals and the firm. So ethics and values are not only personal issues but also organisational ones (Driscoll & Hoffman, 1999). Just as we can be affected by the actions of an unethical person, so too can workers be affected by the unethical actions of the organisation. Correspondingly, values help define the core of people, including what they love or dislike, and the various sacrifices they make in attaining their goals (Hunt *et al.*, 1989). Within an organisation values are important in conveying a sense of identity to its members and to the external activities of the firm. The work of Schein (1985) is salient in this regard; he noted that corporate values, which underlie corporate culture, define the standards that define the internal and external nature of organisations. Underlying values are corporate ethical values, which are essential in establishing and maintaining standards and the right course of action. When the ethical standards of an agency are upheld, organisational success can be enhanced (Hunt *et al.*, 1989). This has been corroborated by Peters and Waterman (1982) in their work on excellent companies; they found that all had a commitment to well-defined sets of values that were shared throughout the organisation.

Although corporations are usually bent on structuring values that maximise shareholder profits, the most successful companies are those that do not set out with profit maximisation in mind. In referencing the theory of obliquity, or the idea that goals are often best achieved when pursued indirectly (characterised by systems that are complex and not well understood), it has been said that we do not maximise things such as happiness, shareholder value, the gross national product and so on, simply because we do not know how and never will (Kay, 2004a). Those entities that promise this often fail to succeed because, although they are the most profit-oriented, they are often not the most profitable. Profit, happiness and success in general cannot be pursued, because of our ignorance of the complexity of systems, but rather must ensue through other means (see Frankl in Chapter 3). This takes place through a focus on values other than strictly profit, and may include such things as ensuring that staff are happy, pouring money and resources into the community or nature and so on. Unhappy businesses resemble one another because of the focus on profit. It is the happy businesses that do not resemble each other because their success is generated in different ways and is dependent on the values that each put forward in place of profit (Kay, 2004a).

It is interesting to note, especially in the context of the foregoing discussion, that success is also a product of natural selection in the face of complexity, as observed by Kay (2004b). That is, those companies most able to adapt to their environment and changing conditions are the ones that prevail. Kay uses the oblique relationship between intention and outcome in reference to Dawkins' metaphor of the selfish gene. As suggested earlier, it is not the gene itself that is selfish or contains motives, but rather the ability to withstand various changes and pressures within the environment – adaptation – that enables the gene to survive and reproduce itself. In reference to altruistic themes brought forward in the same evolutionary fashion, Kay notes that we have evolved to detect the instrumental behaviour of others. He uses the example of the person who buys others a drink in the hope that they will buy his mutual funds, versus the opposite situation of one who buys us a drink because he or she is a friend. The outcome is the same – a drink – but the intentions are very different. The point of this is that honesty and purity in human relations, in things like business, are there as part of the human condition, but may be masked through other motives and desires. It is when we value profit to such

Table 6.3 Theoretical ethical climate types

	Locus of Analysis		
	Individual	*Local*	*Cosmopolitan*
Egoism	Self-interest	Company profit	Efficiency
Benevolence	Friendship	Team interest	Social responsibility
Principle	Personal morality	Company rules/procedures	Laws/professional codes

Source: Victor and Cullen, 1988.

an extent that it obliterates our other qualities that we often find that the reverse of what we wanted in the first place has transpired.

In tourism research, organisational culture is a topic that has received sparse attention, given the interest scholars have shown in other fields, such as business. In perhaps one of the earliest examples of organisational culture and tourism, Upchurch and Ruhland (1995) focused on advancing the hospitality industry's understanding of ethical work-climate types through an analysis of the normative ethical theories of egoism ('an individual should follow the greatest good for oneself'), benevolence ('actions that are delivered on a fair and impartial basis, are based on maximizing the good, and follow impartial distribution rules'), and principle (which 'is a theory suggesting that decisions are based on rules. Outcomes or actions should be based on the merit of the rule') (Upchurch & Ruhland, 1995: 37). These were typologised through Victor and Cullen's (1988) measure of ethical climate within organisations, which examines three levels of analysis (Table 6.3).

Respondents in this study were Missouri lodging operators who filled out the Likert-style ethical climate questionnaire, designed to measure the three ethical theories. The authors found that benevolence was the most frequently perceived ethical climate type present in the lodging respondents (which the authors say is consistent with the literature). Benevolence, therefore, indicated a certain responsiveness to the clientele in a socially oriented manner. The local level of analysis was found to be the primary determinant of ethical decision-making in the organisation.

In other tourism research on organisational culture, Malloy and Fennell differentiated between ethical and non-ethical work environments in the ecotourism industry, based primarily on the work of Schein (1985). Schein has illustrated that culture sets certain standards of acceptable behaviour within organisations. Individuals are thus subjected to a process of socialisation based on normative and value-based behaviours of the unit. In juxtaposing the work of Schein with Kohlberg in a framework of moral development, Malloy and Fennell (1998b) illustrated that, while most tourism enterprises exist at what was referred to as the market level, further moral development might only take place at the hands of an enhanced organisational environment that sets standards based on social and ecological criteria, and beyond the profit-driven market level. Rules for profit maximisation are a means to escape punitive action, while the new standards are the basis for a climate of benevolence (more on this work in Chapters 10 and 11).

Business Ethics

In Chapter 5, it was suggested that the propensity to involve ourselves in commerce has been a strong and lasting element of human nature. This is demonstrated by an extensive history of questions that have been directed towards the ethical practices of business for millennia. For example, the Code of Hammurabi – created 4000 years ago in Mesopotamia – emphasised basic principles around honest pricing; Aristotle gave lectures on the vices and virtues of merchants and tradesmen; and the New and Old Testaments referenced proper ways to conduct business (Hoffman *et al.*, 2001). But the need to question business practice at all suggests that there is not universal acceptance of its rightfulness as part of our character. Adam Smith (1759/1966), for example, held a very low opinion of businessmen and those in lofty positions of power who were foolishly worshipped by the masses, which he noted often translated into imitation. Some moral thinkers of Greek and Roman antiquity were hostile to business because of the belief that commerce was tied to fraud and avarice. Those who practised it were seen to be corruptible and barbarian, and to hold manners that were well below the standards of what was required of an ethical civilisation (McHugh, 1988).

If ethics is definable as what is good or right for people (and the natural world, as we shall see in the next chapter), business ethics can be seen as what is good or right in business. The development of the field of business ethics began in the 1970s in a more formalised fashion, and later became institutionalised in the mid-1980s as an interdisciplinary field (De George, 1987). It took its lead from medicine, which had been immersed in a number of timely social issues throughout the early and mid-20th century. The emerging ethics of medicine provided the opportunity to move the debate on ethics beyond a standoff between absolutists (i.e. those who viewed ethics from the perspective of unquestioned principles and authoritative commands), and relativists (i.e. those who held the view that ethics was subjective and based on attitudes and feelings, brought forward by the social sciences). The debate took shape (see also Chapter 3) in the following ways:

> (1) In place of the earlier concern with attitudes, feelings, and wishes, it substituted a new preoccupation with situations, needs, and interests; (2) it required writers on applied ethics to go beyond the discussion of general principles and rules to a more scrupulous analysis of the

> particular kinds of 'cases' in which they find their application; (3) it redirected analysis to the professional enterprises within which so many human tasks and duties typically arise; and (4) it pointed philosophers back to the ideas of 'equity', 'reasonableness', and 'human relationships', which played central roles in the *Ethics* of Aristotle but subsequently dropped out of sight. (Toulmin, 1986: 266–267)

Applied philosophy (that which is 'concerned with clarifying moral issues and determining how general principles can be applied to concrete cases' (Fox & DeMarco, 1986: 11)) found a role to play in the day-to-day issues in medicine and other fields, thereby coming to the rescue of theoretical ethics, which had little role to play outside the lecture theatre.

The application to business has led to a great deal of interest in both the applied and theoretical contexts in North America (Dunfee & Werhane, 1997). The need for such an approach arrives on the back of scandals in the latter part of the 20th century and beginning of the 21st involving Enron and WorldCom Inc., sounding the alarm that more needs to be done to brace against corporate behaviour that has led to hundreds or thousands of investors losing billions of dollars. In a US study, *2002 Report to the Nation: Occupational Fraud and Abuse*, it was found that: (1) fraud costs about 6% of revenue, or $4500 per employee; (2) over 80% of fraud involves asset misappropriation; (3) fraud was most commonly detected through a tip from an employee; (4) organisations with hotlines were able to cut fraud losses by 50%; and (5) smaller businesses are most vulnerable to fraud (Verschoor, 2003). In response, many prominent organisations have started to distribute business ethics awards, sanction ethical mutual funds, establish tax breaks for acting responsibly, run conferences on corporate ethics, and hire and train ethical officers and ombudspersons. Most recently, i.e. after the scandals noted above, there has been an even stronger push for corporate social responsibility through a number of new federal laws in the US (see the section on corporate social responsibility above on benefit corporations).

The pressure to find common moral ground in business has also been an issue in the business schools. For example, in a recent study of MBA students in the UK, USA and Canada by the non-profit Aspen Institute, on issues related to social responsibility and corporate mismanagement, results indicate that only 22% of respondents (about 1700 students) felt their business school was doing enough to prepare the next generation of business graduates to manage value conflicts. Furthermore, 19% of the sample said that students were not being prepared at all. Almost half of the respondents, in perhaps a more damning indictment, said that the priorities communicated in their MBA programmes may in fact contribute to corporate misconduct (Harding, 2003). This was attributed to the feeling among some academics that the teaching of ethics and corporate responsibility is difficult in business schools because the focus on telling students only what is right or wrong is felt to be patronising and irritating. Instead, a model that focuses on value judgements and decision-making is felt to better prepare students to make good decisions based on critical thinking. The criticism of the focus on one type of knowledge in business schools has prompted some theorists to suggest that business is better suited to trade schools and community colleges than the universities, where, in the case of the latter, it has steadily eroded the liberal arts emphasis in favour of failed models and theories that prevent us from thinking. The type of organisation that has gone into business faculties is behind the Nobel Prize for economics, which Saul says emerged only because of financial backing from a bank (Saul, 1995).

The problem with business school programming has been tackled by Belhassen and Caton (2011) in the context of tourism. These authors contend that it is only, or perhaps primarily, through the embrace of critical pedagogy in tourism education that forward progress will be made in the spheres of individual freedom, social justice, and business productivity. Tourism education success should be measured not only on the basis of technical skills passed on to graduates (instrumentalist terms), but also in terms of how they might be 'moral architects in their occupational domain' (Belhassen & Caton, 2011: 1394) in underscoring the value of education from an intrinsic perspective.

In addition, despite the energy placed into business ethics over the last few decades, it is by no means an essential ingredient in guaranteeing success. This is emphasised by Martin (1998), who illustrates that most of the companies identified in Peters and Waterman's classic *In Search of Excellence* (1982) had gone out of business within 10 years. And, while there are countless examples of companies that have done well by acting ethically, the prevailing norm seems to be that ethics is more often a barrier to realising business goals than a catalyst. So, in order to stay afloat, Martin reports, it is essential to cut corners ethically, especially because honesty does not pay (Bhide & Stevenson, 1990). Those in business who keep their word do so because they want to, not because they have to or need to. The ethics-as-constraint world view is described in the following mind experiment regarding a sceptic (Beversluis, 1987: 83):

(1) We all have a right to economic survival, i.e. a right to survive in business.
(2) This right implies that we have no obligation to do anything that is incompatible with surviving in business.
(3) One cannot survive in business if one is ethical, i.e. if one takes everyday senses of honesty, law-abidingness, fair-play, etc. into one's own dealings; one has to 'play the game'.
(4) Therefore if by 'business ethics' we mean taking everyday senses of honesty, law-abidingness, fair-play, etc., into our business dealings, we

have no obligation to do so, and hence there is 'no such thing as business ethics'.

Even when there are solid grounds for taking the moral turn in business rather than sticking by strategic or instrumental reasoning (refer back to Habermas' discourse ethics in Chapter 3), it is still acceptable to remove morality from the equation. Theorists, including Freeman (1999), refer to this phenomenon as the 'separation thesis'. This view postulates that because it is the main purpose of businesses to make a profit, all other goals should be subordinate. As such, goals of conservation or community benefit and development are recessive to the more dominant need to satisfy shareholder interests. Morality is thus a constraint on the proper functioning of the firm.

The principal response by the business community to avoid the type of thinking that so overwhelmingly prevails has been the development of centres or institutes of business ethics. One such centre, The Canadian Centre for Ethics and Business Policy (1999, as cited in Fennell, 2000b), is volunteer-driven and comprised of organisations and individuals dedicated to developing and maintaining an ethical corporate culture. This independent centre works with its own contributors and other business-oriented groups involved in ethics, for the purpose of developing programmes and addressing current business ethics issues, with the following broad goals:

(1) To encourage organisations to take into account the ethical dimension in making their business decisions and developing their policies and practices. To explore and promote the role of ethics in the conduct of organisations and of individuals who work in and with them.
(2) To promote ethical values by encouraging and contributing to public, professional and organisational awareness and knowledge.
(3) To serve as an information resource centre, catalyst and reference point for Canadians.
(4) To encourage constructive discussion and debate by forms of communications as appropriate.
(5) To network with other Canadian (and North American/International) organisations with similar objectives and goals.
(6) To encourage and support research into issues of corporate ethics, practical tools and techniques.
(7) To assist organisations in their development of a culture in which ethical issues are incorporated into their decision-making process.
(8) To support business goals and promote the view that ethical business practices are consistent with such goals.
(9) To identify issues that merit consideration for their ethical dimension.

The prevalence of these units in business led the author to conclude that it is not so much a question of *whether* centres of applied ethics will be incorporated into the mainstream of tourism research and education (acknowledging the centres of hospitality ethics as noted in Chapter 1), but rather *when*. The 'when' may not occur as quickly as we wish if we are to emphasise the separation of ethics and business as highlighted by many tourism pundits in the 1990s, even in full view of the sustainable development paradigm. Tourism business is seen as placing ethics on the bottom of an extensive list of priorities. In making this point, Haywood has suggested that: 'Business and society are still seen as separate from each other, and the language of rights and responsibilities, which attempts to link the two, remains irrelevant to the world of practicing managers' (1993: 235). But although there is nothing wrong with self-interest from the economic side of things, we must be clear that it cannot take the lead in a decent society. It must be led by society (Saul, 2001).

Social Status, Consumption and Marketing

Citing John D. Rockefeller, who, in 1905, stated 'I believe the power to make money is a gift from God. ... It is my duty to make money', Kagan (1998: 154) suggests that, although unbridled capitalism has lost some of its virtue through recognition of misery in urban peasants, wealth has become a key symbol of virtue in contemporary America, where one feels obligated to display it at every turn. Its importance is reflected in the fact that we naturally seem to gravitate to people who have more of it.

Veblen (1899/1953) was the first to comprehensively describe social status in the context of leisure in his book *Theory of the Leisure Class*. He differentiated between conspicuous consumption, conspicuous leisure and conspicuous waste in suggesting that those who are financially well-off have the ability to use their wealth as status. This could take many forms, including the purchase of cars (big impressive ones, or smaller compact ones that are less environmentally harmful), houses, art, travel, wine, opera tickets or books – all of which may be used to demonstrate one's financial loftiness. In the case of the latter, a carefully presented set of books in one's house is a beacon towards illustrating the well-read and articulate nature of the inhabitant. The same applies to tourism, where regular and extended trips abroad demonstrate affluence, providing membership to the upper end of the social continuum. The more of these we have, the better chance we have of reaching social exclusiveness.

However, it is more than just the volume of travel and books that admit one to the club it is also the quality of these possessions. So, travel to the mass tourism destinations (where the price is lower and the product is more homogenised) is not nearly so impressive as an extended trip to Provence to take in the culture and environment. Homogenisation, therefore, is an enemy to those who seek their identity through high culture. New and different is appealing, and therefore more expensive, which motivates the producer or developer to continually brandish new products in time and space. The problem for the developer, as alluded to above, is the competition that makes the unique commonplace – a reason why tourism service providers cannot sit idle while others continue to perfect their programmes through effective planning techniques (Fennell, 2002).

Tourism is thus a positional good (a good's value is dependent upon its desirability ranking) in the same way as clothes, houses or cars. However, the consumption of tourism presents a problem in that those who the tourist most frequently interacts with (at home or at work) are unable to see the individual consume the good (Schor, 1998) *in situ*, or in the field. This makes the purchase of shirts, hats or other souvenirs so important in informing people on the home front. And, just like the choice of clothes or a car or house defines the person (e.g. 'the clothes make the man'), 'we are *where* we travel', or, more philosophically: 'I travel, therefore I am.' Consequently, the choice of what to buy or where to travel helps to establish group identity (like the choice to purchase a Harley-Davidson motorcycle).

Why does the tourism industry continue to develop new forms of tourism? At one count in the early 1990s, there were 90 different types of tourism in existence (Boyd, 1991). This number must surely be 110 by now. Any one for 120? The point is that tourism is no different from society as a whole, which is bent on the development and perfection of goods and services for sale. As noted above, if we cannot improve upon a destination or if competition is too severe in the area, we have to move on and develop new attractions elsewhere (Butler, 1980). The same is true with board games, cars, movies and anything else we develop for sale. The basis of this constant improvement is profit and therefore self-interest. We do not make things for the intrinsic value they bring us, but rather for the extrinsic reasons: profit and status.

But two things emerge from this. The first is that, although wealth brings the demand for more, it also furnishes the inability to enjoy what we have (Blackburn, 2001). Second, and just as importantly, there is no good news for those who feel that by occupying a higher social bracket they are more ethically advanced. As Pinker (2002: 415) has observed, 'the circuitry for morality is cross-wired with our circuitry for status'. And, although the super-consumption characteristic that plagues us today continues to prevail, it does not appear to have a biological explanation, as yet (Ehrlich, 2000), although it links quite nicely to self-interest as a foundational characteristic. Even so, science tells us that having more possessions does not make us happier. A study conducted in the mid-1990s found that, compared to people in 1957, even though Americans in 1993 have many more possessions than their counterparts in 1957 the number of people reporting that they were 'very happy' stayed the same (Myers & Diener, 1995). That is, beyond a threshold point of personal possessions related to physiological and psychological needs, more does not make us happier (see Kasser, 2003 for a synthesis of dozens of studies on the negative correlation between consumerism and subjective well-being).

The culture of consumption in Western society takes a different form in the work of McKibben (1999), who recounts a story of how blackfly season in the Adirondacks of the USA has had an adverse effect on the tourism industry in his backcountry town. In order to rid the region of blackflies, the community had thought of using the bacterium *Bacillus thuringinsis* in the streams where blackflies breed, thus reducing their population and increasing tourism. While the cost per individual to control the blackfly population would amount only to $56, some felt that the disruption of streams would have an adverse effect on trout and bird populations, both of which feed on the flies. Others commented that blackflies were simply part of the fabric of the community, that made it, not unique, but a challenge to live in such areas, which was nevertheless something that needed to be preserved.

McKibben writes that the challenge is premised on one main force: that we are not the most important thing on earth. But we are meant to be, at least through the campaigns by major corporations that would have us think that 'This Bud's for you', or that I eat at 'my McDonald's'.

All things, in his words, orbit our desires because we are the heaviest object in the universe. We are thus defined by our patterns of consumption, with the inability to pull ourselves away from our desires. We cannot see ourselves behave as anything but consumers in our haste to be as comfortable as possible. And so it follows that some of the residents of McKibben's community want to be able to consume bite-free air without annoyance. But McKibben's philosophy is one that hinges on a shift to voluntary simplicity, where we can accept winter's challenges, blackflies, bad roads, power outages and so on, as pleasurable commodities that enable us to develop a different self-image, with the end result of happiness. In a world where consumption threatens the very basis on which we depend, the acceptance of the world on its own terms is itself virtuous. So blackflies, left alone to do what blackflies do best, are a simple (and itchy) reminder that we are not the centre of the world.

Unhappily, we have been taught that what McKibben proposes is irrational because it does not maximise our gains and minimise our losses in our efforts to seek pleasure. But in the words of Orr (1999), we freely defend our rights as consumers and at the same time deny our rights as citizens, with no internal drive to force political institutions to harmonise with the cycles and systems of the earth. How can we, he observes, if we have been manipulated by a system that is geared towards profit and pleasure. The system has developed according to a number of distinct stages:

> The first step involved bamboozling people into believing that who they were and what they owned were one and the same. The second step was

> to deprive people of alternative and often cooperative means by which they might fulfill basic needs and obtain basic services. The destruction of light-rail systems throughout the United States by General Motors Corporation and its co-conspirators, for example, had nothing to do with markets or public choices and everything to do with backroom deals designed to destroy competition for the automobile. The third step was to make as many people as possible compulsive and impulsive consumers, which is to say addicts, by means of daily bombardment with advertising. The fourth step required that the whole system be given legal standing through the purchase of several generations of politicians and lawyers. The final step was to get economists to give the benediction by announcing that greed and the pursuit of self-interest were, in fact, rational. (Orr, 1999: 142)

In tourism, there seems to be the willingness to allow our less fortunate counterparts in LDCs to 'develop' by selling off their dignity and the integrity of the natural world for the benefit of those who lack for nothing. As noted above, our social existence is marked by huge variations in status, power and wealth. While some people have achieved power, others are powerless to control their fates. Furthermore, while some have wealth and have been able to maintain this wealth within the family through successive generations, others are chronically poor. The poorest 20% in the most developed countries have only 5% of the total income, whereas the richest 5% have 25% of the income; while just 2% of those in the USA control upwards of 75% of the corporate stock (Lewontin, 1982). In 2017, Oxfam reported that eight men had fortunes that equalled the wealth of the bottom half of humanity, and that the gap between rich and poor was widening (Oxfam International, 2017). This is compounded by globalisation because of the relative magnitude of the comparisons we make between the developed and the less developed. We are now measuring ourselves at a different scale, as opposed to our community or nation. This has been referred to as competitive acquisition and it is intensified through the media and by the advertising industry (Schor, 1998) as suggested above.

In a previous chapter it was noted that, while many enjoy the benefits of travel, more still can only fantasise about it. According to Townsend (1987), this is one of many forms of deprivation, which he defines as 'a state of observable and demonstrable disadvantage relative to the local community or the wider society or nation to which an individual, family or group belongs' (Townsend, 1987: 125). He stipulates that deprivation has more to do with conditions (e.g. physical and environmental circumstances) than resources, the latter of which may be more attributable to poverty. But while the latter garners a wider audience and sympathy among policy makers, it may be no more important as a basis by which to gauge social conditions than deprivation. Townsend operationalises deprivation according to the following 13 types, organised according to material and social deprivation. For the purposes of brevity, only certain examples will be included for these items:

Material:

Dietary. No fresh fruit most days.
Clothing. Inadequate footwear for all weathers.
Housing. Housing not free of infestation.
Home facilities. No television; no radio; no central heating.
Environment. Nowhere for children to play safely; industrial air pollution.
Location. No recreational facilities for young people or older adults nearby.
Work. Poor working environment (polluted air, dust, noise, vibration and so on).

Social:

Lack of rights in employment. No paid holiday.
Family activity. No days staying with family or friends in previous 12 months.
Lack of integration in community. Being alone and isolated from people.
Lack of formal participation in social institutions. Did not vote at last election.
Recreational. No holiday away from home in last 12 months.
Educational. Fewer than 10 years' education.

Although travel may not be a requirement for survival, it appears as though it is one of many essential aspects of life that may help us avoid being or feeling deprived. As of 1997, over 800 million people lived in what is referred to as absolute poverty, as noted by Singer (1997). In recognising the massive inequalities within the global village, he feels that the only way to alleviate the poverty problem is to insist that those in the affluent nations be morally obligated to assist the starving, by transferring wealth and food to the LDCs. By giving 10% of per capita income to the impoverished we can improve the standard of living, leading to reform in education and economic security. Singer's most forceful point is that by not giving in this way we are permitting the disadvantaged to suffer malnutrition, ill health and death. And allowing someone to die is not that different from killing them.

Tourism represents an opportunity to examine the intersection of the primitive with the modern. This has been documented by MacCannell (1992), who illustrates how the gap between the past and present can be bridged by tourism, which places modern people in face-to-face encounters with the 'primitive'. We know that these individuals are not savages, but what is savage to some is the staged authenticity that goes along with allowing tourists to view sacred shrines and ritual performances at regular times during the week. The commercialisation of these performances, according to MacCannell, has the potential to have long-term economic impacts for such communities. But the fallacy of this interaction is the utopian belief that profit takes place without exploitation – pretending we can get something for nothing:

> It is his coming into contact with and experience of the ultra-primitive which gives him his status. But this has not cost the primitives anything. Indeed, they too, may have gained from it. Taking someone's picture doesn't cost them anything, not in any Western commercial sense, yet the picture has value. The picture has no value for the primitive, yet the tourist pays for the right to take pictures. The 'primitive' receives something for nothing, and benefits beyond this. Doesn't the fame of certain primitives, and even respect for them, actually increase when the tourist carries their pictures back to the West? It seems to be the most perfect realization so far of the capitalist economists' dream of *everyone* getting richer together. (MacCannell, 1992: 29)

MacCannell responds by saying that this is impossible. This also holds true in the West, where everyone does not get richer together – merely a fortunate few who seem to have preferred positions within the economy. However, it is not just the marginalised populations of the world who are under the spell of tourism utopias. The tourism industry attempts to construct the value of intimacy in its attractions (e.g. Disney) in a business that, at its root, is perhaps the opposite. They entertain us with what they have created, but what they have created is precisely what we have lived. They have taken from society, repackaged it and thrown it back at us with a cost. It only works because it is precisely what the mass tourist wants (see also Cohen, 1972; Urry, 1990).

But is what the mass tourist wants remotely the same as what the alternative tourist wants? Some might say no, because the latter chooses to define themselves as the antithesis of the mass tourist. Instead of learning about destinations in magazines and brochures, the alternative tourist chooses to inform themselves through other methods, such as travel guides (Mowforth & Munt, 1998). This is addressed by Silver (1993), who says there is really no way to advertise for the alternative traveller because of the dislike for the commodification of people and places that typically takes place through conventional means of advertisement. To confuse things further, a third market has emerged – the chic traveller – as noted by Silver, who positions themselves somewhere between the mass tourist and alternative tourist. Advertisers have learned that there are places that cannot entirely fulfil the wishes of alternative or mass groups, either because more tourists are wanted in places like Hawaii, where there exists a mature industry, or in places like Papua New Guinea, which wants more tourists but does not wish to develop a mass tourism industry. In the case of the latter, affluent travellers can enjoy primitive conditions but in more luxurious surroundings that are more comfortable and safe. Here the experience is advertised as being more authentic as well as more ethically responsible to a sophisticated, wealthy and more highly educated clientele. For the tourist this means that they are not a mass tourist, because this would be unacceptable, but rather somebody who can feel good about his or her vacation because it supposedly has less impact on the culture and natural resources of the destination.

But are the changing lifestyles of the tourist population, as noted by Krippendorf (1987) and others, a reality or a ploy conjured up by the marketers to make us feel different or important, and another way to make more money? For tourists, ethical holidays have been purchased for reasons other than altruism. The decision to buy alternative or eco is based on getting a better product through smaller group sizes and the opportunity to access unique areas. As for service providers, just as many operators said they would be advantaged as those who were undecided about the value of ethics in their product as an advantage over their competitors (Weeden, 2001).

Indeed, Wheeler (1995) addresses the paradoxes in marketing and contends that it is the area in business most often cited with ethical abuse, especially as marketers are the first to exploit fads and fashions. In tourism there is the tendency to induce the host community to use its natural assets (beaches, forests and streams) in the development of the industry. In the transformation of the region into a tourist destination, the original resources become less of an asset, and less valued, in the acceptance of economic pressures for large-scale development. The role that marketing plays in this process is to attract greater numbers of tourists in line with supply. If we were truly concerned about the impacts of the industry, and how these are marketed, Wheeler suggests, our developments would be small-scale, locally controlled, less seasonal, slower-paced, and socially and environmentally benign. We might also expect that the marketers, who may market products they never themselves see or experience, be witness to the impacts they help create.

Marketing studies of the kind noted above are important in gaining a sense of the advantages of ethics in tourism. But the very nature of the fact that we use ethics in tourism as a competitive advantage is problematic. In doing so we lose out on the intrinsic basis of ethics to be good for its own sake. As such, ethics then becomes an afterthought or secondary to the real job of selling for profit. We could market ethically not for the purpose of making more money, but because we think it is the right thing to do regardless as to whether it makes us more money or not. As a business you are fooling yourself as well as your clients.

During the late 1980s and early 1990s, the green marketing movement escalated in tourism as part of the new sustainable development platform. This was not isolated to the tourism industry only, but also many other sectors. Termed the 'enviropreneurial' approach by Peattie (1999), marketing is blended with environmental concerns in the search of innovations and opportunities. This prompted theorists to suggest that corporate environmentalism is an imperative for all businesses, with the expectation that those who move first are those who should be advantaged over those who do not (Polonsky & Rosenberger, 2001). Tourism Concern (1992) is one group that has developed a list of different stipulations in its guide to appropriate marketing. These include the employment of guides who portray societies honestly and dispel stereotypes; the promotion of tourism appropriate to the capacities of the destination; the avoidance of the imposition of Western mores on regions with different values; the attempt to dismantle racial, sexual, cultural or religious stereotyping within the industry; and the marketing of holidays that correspond to the tourist product and experience offered.

Just how effective is the push behind green marketing? Knowing what is good for the natural world has always been a difficult task (Miller & Szekely, 1995). This makes linking such knowledge to business practice all the more challenging. This is in part because the hue or 'greenness' of a business is a function of the degree to which these businesses incorporate the principles of sustainability or other ecological perspectives (Johnson, 1998). Those who wish to be green on the basis of demand, Johnson says, and without the proper philosophical underpinnings, have short-lived success in this regard:

> There frankly is little support for the notion that green . . . marketing translates into business organizations and practices that yield significantly improved, sustained environmental performance. More often than not, 'green' consumer preferences are insufficiently focused, or focusable, in terms of being able to induce fundamental changes in practices and stakeholder relations that yield indefinite commitments to enhanced environmental performance. Typically, green marketing aimed at capitalizing on green consumer sentiments is fleeting, leading to tendencies toward the eventual subversion or abandonment of green production practices. (Johnson, 1998: 264)

This itself is a form of instrumental reason that is led by a system that is largely uncertain of its inputs and outputs along a supply chain that crosses international borders, rendering detection of eventual effects extremely difficult (see Keating, 2009 for a discussion of ethics and the supply chain in reference to Chinese travel to Australia).

Empirical work on ethics and marketing is considered essential in pushing forward a more responsible tourism industry, but few studies have been undertaken. Yaman and Gurel (2006) investigated Australian and Turkish marketing executives and found that there were key differences. Turkish marketers were more deontological in their reasoning as well as more relativistic, i.e. more tolerant towards the ethical judgements of people from other cultures, than their Australian counterparts. Hudson and Miller (2005) developed a model for environmentally responsible marketing and applied it to the business practices of the firm Canadian Mountain Holidays. The authors report challenges on the balance between environmental action and the communication of activities.

Scholars have also applied postcolonial theory in efforts to more firmly emphasise the tourism representation discourse. Echtner and Prasad (2003) identified three prominent myths in Third World marketing: the myth of the unchanged, the myth of the unrestrained, and the myth of the uncivilised. This type of marketing continues the long and sustained trend of attitudes, images and stereotypes that reflect the colonial legacy. This representation of cultural Others is so firmly entrenched that even non-profit brokers with humanitarian missions cannot break free. Caton and Santos (2009) investigated the company Semester at Sea, and found that their promotional tactics reinforce colonialist stereotypes of non-Western people. The company's brochures and website images 'tended to essentialize and exoticize hosts by depicting their cultures as primitive or backward, stagnant, and dependent on the West for advancement' (2009: 201). As observed by Echtner and Prasad (2003), attention needs to be directed at counter-colonial discourses for the purpose of resisting these types of representations.

Ethical Responses in Tourism

In the previous chapter the example of multinational domination of the tourism sector was used to demonstrate the magnitude of structural inequalities within the industry. In this chapter it is the purpose to examine some ethical responses that have emerged to generate more equality in the context of the prevailing business-centred model. These include fair trade in tourism and pro-poor tourism.

Fair Trade

An area of tourism that has recently emerged with an ethical imperative is the fair trade movement. It developed with the intent of redressing 'unequal trading by promoting fair trade in commodities with small producers in the South, enabling them to take control over production and marketing and challenging the restrictive power of transnational corporations' (Cleverdon & Kalisch, 2000: 171). This is reflected in the following definition by the International Federation for Alternative Trade, who declare that Fair Trade is designed to:

(1) Support efforts of partners in the South who by means of co-operation, production and trade strive for a better standard of living and fairness in the distribution of income and influence;
(2) Take initiatives and participate in activities aimed at establishing fair production and trade structures in the South and on the global market. (Cleverdon & Kalisch, 2000: 173)

In relation to tourism, the authors suggest that a focus needs to centre on who benefits, by how much and in what ways. Central to the concept in tourism are issues such as the distribution of benefits, access to capital, ownership of land and resources, stakeholders, transparency of trading operations, respect for human rights, ecological sustainability, the development and respect for laws, and the raising of prices in evening out benefits. As regards the latter, whenever prices are raised, given the alternative of other options that deliver similar products, there is the inevitability of a niche market (e.g. oranges are of the organic and non-organic variety, with a niche market developing that favours the former). There are questions, therefore, that need to be raised about the ethics that apply to a narrow niche only – perhaps 2 to 3% of the overall market. The tourism watchdog, Tourism Concern, would like to see fair trade emerge both as a niche market and for the industry in general (Boyd, 1999). Tourism Concern also feels that corporate social responsibility is a necessary precondition for realising fair trade in tourism. This is one of five targeted areas set out by this organisation in their efforts to generate this – the other areas include local community, consumers, national government policy and planning, and international agreements (Kalisch, 2000).

One type of tourism that is representative of the breakdown in fair trade is conservation-based tourism in the LDCs. Researchers suggest that fair trade principles could be used in and adjacent to protected areas in these countries – resulting in diversification and the overall improvement of people's lives and resources (Goodwin & Roe, 2001). At present, most tourism fails these groups because of the inability to access the market, lack of start-up capital and lack of knowledge, among other factors. This has compelled Goodwin to observe that fair trade, far from being altruistic, appears to be motivated more by the desire of consumers to feel good about themselves. And this (feeling good), Goodwin notes, is 'one of the main drivers of responsible tourism' (Goodwin, 2003: 273).

Pro-poor tourism

Tied to the concept of fairness and justice in the tourism trade is the emergence of pro-poor tourism, a movement that is supported by PPT (Pro-poor Tourism) in the UK. They define pro-poor tourism as tourism that generates net benefits for the poor through strategies that aim to unlock opportunities, rather than to expand the overall sector (PPT, 2002). PPT pundits submit that it differs from other forms of sustainable tourism and alternative tourism because of its focus on: (1) tourism in the South; and (2) poverty as a social problem, rather than on concerns that are mostly ecological (as in ecotourism). The group establishes that PPT requires good planning and committed governments in order to institute change. This goes along with the stimulation of economic linkages through the cooperation of many interested stakeholders who hold the benefits of the community as most important. These are articulated in the following strategies for pro-poor tourism (Roe & Urquhart, 2002: 5–6):

(1) Strategies focused on economic benefits
- *Expanding business opportunities for the poor.* Small enterprises, particularly in the informal sector, often provide the greatest opportunities for the poor.
- *Expanding employment opportunities for the poor.* Unskilled jobs may be limited and low-paid by international standards, but are much sought after by the poor.
- *Enhancing collective benefits.* Collective community income from tourism can be a new source of income, and can spread benefits well beyond the direct earners.

(2) Strategies focused on non-economic impacts
- *Capacity building, training and empowerment.* The poor often lack the skills and knowledge to take advantage of opportunities in tourism.
- *Mitigating the environmental impacts of tourism on the poor.* Tourism can lead to displacement of the poor from their land and/or degradation of the natural resources on which the poor depend.
- *Addressing social and cultural impacts of tourism.* Tourists' behaviour, such as photography and Western habits, is often regarded as cultural intrusion. Sex tourism exploits women. Tourism can affect many other social issues, such as health care.

(3) Strategies focused on policy/process reform
- *Building a more supportive policy and planning framework.* Many governments see tourism as a means to generate foreign exchange rather than to address poverty. The policy framework can inhibit progress in PPT; reform is often needed.

- *Promoting participation.* The poor are often excluded from decision-making processes and institutions, making it very unlikely that their priorities will be reflected in decisions.

Box 6.1 Tropic Ecological Adventures, Ecuador

Tropic Ecological Adventures is a small for-profit company that was established with the specific objective of demonstrating the 'viability of environmentally, socially and culturally responsible tourism' as an alternative to oil extraction in the Ecuadorian Amazon. It operates tours to natural areas in Ecuador, including the Amazon, usually for small high-paying groups. It has links with several communities, of which two are the focus of the case study: Tropic has worked with the Huaorani people to develop a joint initiative, bringing tourists into the community for overnight stays and to experience the Huaorani culture and lifestyle. It was marketing the long-established Cofan initiative at Zabalo, though it has recently been forced to suspend these operations due to security issues in this area near the Colombian border.

Although Tropic found that its community-based programmes were less profitable and less marketable than some of its other activities, it has managed to successfully address this problem by coupling them with more mainstream packages such as visits to the Galapagos Islands. Unfortunately, however, a decline in tourism in the Ecuadorian Amazon in 1999 and 2000, following kidnappings and political upheaval, has heightened competition among tour operators and driven down prices, which has undermined Tropic's impact-minimising approach of bringing in small groups of high-paying tourists. A further setback arose from the Civil Aviation Authority's decision to close down the airstrip at the Huaorani site (due to poor maintenance – a community responsibility). However, a flight service has been re-instituted, and ecotourists now take a 45-minute flight from the small town of Shell to the grass airstrip at the Huaorani village of Quehueri'ono, after which guests board a dugout canoe for the final trip to the lodge. A major aspect of the ecotour is the damage that oil pipelines have done to the local environment and hunting grounds (Vegan Magazine, 2013).

Source: PPT (2002).

- *Bringing the private sector into pro-poor relationships.* Locally driven tourism enterprises may require input to develop skills, marketing links and commercial expertise.

Box 6.2 Pro-poor tourism around South Africa's Addo Elephant National Park

A recent study centred on a participatory analysis of the tourism trading system in the area around the Addo Elephant National Park (AENP) in South Africa's Eastern Cape province. A key aim was to promote the integration of emerging community tourism initiatives with 'mainstream' tourism, specifically as an anti-poverty strategy. This required exploring how to maximise the linkages between the different components in the tourism system, which included government service providers, a range of existing tourism businesses, the South African National Parks Board (SANP), tourism marketing organisations and poor communities living around the borders of the park. Through a multi-stakeholder dialogue process, it became clear that community tourism projects, such as drama groups, choral groups, and arts and crafts groups in the Addo area, have the potential to add value to the tourism system through diversification of the mainly wildlife-related tourism product. And, on the other hand, the role of the private sector in tourism partnerships is key for effective marketing and business skills development. The process of dialogue may ultimately lead to local standard-setting and a *locally developed sustainable tourism brand*, with a strong anti-poverty focus.

Source: Roe and Urquhart (2002).

The authors suggest that private companies, governments, civil society and donors can play a strong role in PPT through the provision of technical advice to the poor; the revision of regulations that impede the poor from developing their enterprises; support for campaigns that enhance pro-poor objectives; and the promotion of pro-poor tourism within the international agenda (see also Scheyvens, 2007). The preceding benefits, impacts and strategies for reform highlight the many inequalities that exist between parties of the North and South. But these inequalities have remained steadfast for years, and are symptomatic of the politico-economic structure of the industry, as identified in Chapter 5. Although it is difficult to see how this pattern will ever change, a further examination of this issue will take place using normative ethics, in Chapter 11. Boxes 6.1 and 6.2 illustrate how pro-poor strategies have helped in two different world regions. In an attempt to provide a balanced appraisal of the significance of the pro-poor tourism platform, there is perhaps an element of paternalism on the part of industry and other stakeholder groups that may be seen as distasteful. Although the basic tenets of the pro-poor tourism platform are sound in principle, knowing what is good or right for tourism in the lesser developed countries, and instituting measures from the most developed countries, may seem to some to be patronising. Harrison (2008) shares some of these concerns in suggesting that the pro-poor tourism platform has become too heavily weighted to the community-based tourism realm. This has emerged because of an absence of strong theoretical and methodological strategies. Harrison argues that it is now easier to discuss what PPT is not than what it is.

Conclusion

The business literature is perhaps the most fertile ground for knowledge on the application of ethics in social science. This chapter sought to highlight some of this information through work on corporatism vs. individualism, corporate responsibility, organisational trust and culture, and business ethics. It was argued that the organisation has replaced the individual as the moral touchstone in society. In this, it becomes a much easier task to divest the corporation of any wrongdoing than any specific individual or small group. This has led to the denial of the legitimacy of the individual, in favour of the group as the primary unit of value. Technology has served to reinforce this change. Although corporate responsibility was discussed earlier in the book, it was given a higher priority here. This included two polarised arguments on the role of profit and ethics in the climate of the firm. Business ethics was also a topic of discussion, with the suggestion that tourism should venture down the same path (i.e. the development of centres of tourism ethics). Finally, ethical responses in tourism were illustrated, including marketing, fair trade in tourism and pro-poor tourism, along with many of the advantages and disadvantages of these perspectives.

7 Ethics and the Natural World

In the world of values, nature in itself is neutral. . . . It is we who create value and our desires which confer value. . . . It is not for us to determine the good life, not for nature – not even for nature personified as God
Russell, 1957: 55–56

Introduction

Holden (2003) has observed that the writing of a paper on tourism and environmental ethics in the past was highly unlikely. The fact that they are being written today represents a significant change in our thinking on tourism–environment interactions. But as the previous discussion served to illustrate, if we do consider ethics at all in tourism, it is usually in the context of the environment as *la seule morale*.

The focus of this chapter is on a range of ethical issues that have been identified in the environmental literature. Keeping with the approach used in the book, many issues will be addressed that fall outside the tourism literature. After a brief overview of ecosystems and ecosystem services, the chapter examines the myth of stewardship, environmental values, virtues and costs, and environmental justice and rights; it then concludes with an example of a centre of environmental ethics.

The Myth of Stewardship

The discussion in Chapter 2 on the noble savage pointed to the belief that traditional societies lived peaceably and in harmony with the natural world. This is especially true of societies that were more sedentary and reliant on resources like fishing and horticulture (Gadgil *et al.*, 1993). For example, Indonesia, India and China, where the development of integrated systems helped in the production and maintenance of resources over the long term. Once empowered and assured of a stake in these enterprises, local people were more responsible in their treatment of such resources. But these practices are seen to be more the exception than the rule. Instead, research points to the fact that indigenous people were far from ecologically conscientious in their actions, knocking over a house of cards that was built on the belief that many of our ecological problems were derived solely from the Industrial Revolution (Attwell & Cotterill, 2000).

In his book on dismantling the fantasies of environmental thinking, Kauffman (1995) identifies a number of cases in which traditional societies held less than romantic views of the natural world. In many of these cases, archaeologists have found evidence of massive overkills of many of the large species including bison, elk and wolves, near the sites of ancient tribes. This is corroborated by other researchers, who note that the largest species and the biggest members of the species were often taken as game, despite the fact that these species were the ones that were least able to respond from over-harvesting (Winterhalder & Lu, 1997).

It also appears that this overuse of resources was not isolated to any one region. The human prehistoric communities of the Pacific islands decimated species of land birds on islands like Hawaii, New Zealand and Easter Island. The loss of avifauna biodiversity runs to a magnitude of 2000 birds, representing some 20% of today's diversity of bird species (Steadman, 1995). For example, human arrival in the Hawaiian Islands approximately 2000 years ago led to the extinction of 60 species. Once people occupy an island, predation, habitat loss and the introduction of other predators and pathogens become responsible for extinctions in somewhere between 100 and 1000 years in these insular regions. This suggests that humans have always found it difficult to manage stocks in a sustainable way – a chronic problem that traces its way to present times. This has been verified by Diamond (1993), who illustrates the following by way of observation of the mannerisms of New Guinean villagers:

> Like other humans throughout the world, New Guineans kill those animals that their technology permits them to kill. The more susceptible species become depleted or exterminated, leaving less susceptible species which people continue to hunt without being able to exterminate them. When technology improves, as it did in New Guinea with the arrival of the bow and arrow, or with the arrival of dogs a few thousand years ago, or with the arrival of shotguns within the present generation, some species that have been able to survive previous hunting technology become susceptible and disappear. (Diamond, 1993: 268)

The same pattern of overuse took place in the Amazon (Hames, 1990). Findings indicate that none of the villagers maintained a conservation ethic, as individuals were found to take any type of game in any region, despite the fact that such regions were found to have diminished stocks. The high regard for protein in the diet of Amerindian groups is sufficiently valuable to intensify their efforts to capture a species of game even in the face of its depletion. In other studies, researchers found that the Yuqui of Bolivia frequently kill pregnant monkeys to eat the foetus, considered to be a delicacy (Stearman, 1994). The Piro of Peru demonstrated no restraint in killing large game adjacent to their reserve, despite the fact that these species were severely depleted (Alvard, 1994). In India, resource exhaustion has taken place due to the heterogeneity of cultures, pressures from large human and livestock populations, and the resource demands from industry, leading researchers to conclude that no segment of society is motivated to use the resources of the country in a sustainable fashion (Gadgil, 1991). More recently, researchers have discovered that the practices of prehistoric Inuit hunters in northern Canada transformed pristine Arctic lakes into moss-choked waste disposal sites through massive whale slaughters over a span of four centuries (from 1200 to 1600). The lakes have yet to return to their original condition (Calamai, n.d.).

In perhaps one of the most comprehensive studies of conservation and traditional societies using the 186-society Standard Cross-cultural Sample, Low (1996) discovered that: (1) people in traditional societies did not express a conservation ethic; (2) there was little pattern to the expression of a conservation ethic, even when different environments were taken into consideration; (3) sacred prohibitions did not prompt conservation, but rather accelerated use of resources; and (4) traditional societies frequently caused environmental degradation. Perhaps most indicative of the behaviour of traditional societies from these findings is that conservation was only favoured when there was a quick and clearly detected feedback that had a positive impact on the familial unit or on the individual (refer back to the theories of inclusive fitness and reciprocal altruism).

The overuse of resources, however, is not exclusive to humans. In the animal world the survival of one population often takes place at great cost to others. For example, foxes have been found to kill many more gulls than they need (Kruuk, 1976), and the same can be said of billfish and certain fish schools (Bigelow & Schroeder, 1953). Compounding this has been the practice of interpreting the actions of animals from a romantic standpoint. While courtship and grooming among animals could be viewed as positive human virtues, other acts deemed destructive or disgusting, by human standards, could be justified only through name-changing, despite the fact that these actions (e.g. rape in turtles (Berry & Shine, 1980); masturbation in mammals (Beach, 1964); and murder and cannibalism in all animals except the vegetarians (Polis, 1981)) were essential in the maintenance of the population (Williams, 1988). Until the theories of inclusive fitness and reciprocal altruism came about, there was no consistent theoretical way to describe the behaviours of animals in social situations (Williams, 1988). This led Williams to argue that we must rebel against the selfish genes, and nature selection in general, for the betterment of the human condition. It is natural for animals to behave the way they do, but as moral beings we have the ability to transcend our predispositions. This is precisely the perspective adopted by Huxley in his *Evolution of Ethics*, where he argued that:

> If the cosmos is just 'and of our pleasant vices makes instruments to scourge us,' it would seem that the only way to escape from our heritage of evil is to destroy that fountain of desire whence our vices flow; to refuse any longer to be the instruments of the evolutionary process, and withdraw from the struggle for existence. (Huxley, 1894/1968: 63)

Although we appear to be conditioned to be self-interested, evolution also gave us the capacity to think, reason, learn and communicate, which has been essential in our abilities to cooperate for our long-term viability. The development of ethics has thus furnished us, in Huxley's words, 'with a reasoned rule of life' (Huxley, 1894/1968: 52). This means that methods for the struggle for existence of the tiger are incongruent with what it means to be an ethical human.

The fact that native communities are not the conservationists we thought they were shows a consistency in the human condition towards the overuse of natural resources for our own self-interested ends. We may have inherited a propensity to avoid restraint from our ancestors, but worlds away in the technological society of the 21st century, the game and the stakes are a great deal more complex. The work of William Rees (1992) on the ecological footprint has captured the essence of this argument in contemporary times. (On balance, see Higgins-Desbiolles' (2009) case study on Camp Coorong, Australia, where an indigenous community is teaching traditional ecological values for the purpose of transforming consciousness among ecotourists).

Ecosystems and Ecosystem Services

As biology became more systematic and reductionistic during the late 1800s, thanks to the work of Darwin and others, scholars began to look away from the predominant field of natural history, which was felt to be less legitimate than the newer sciences. Natural history was resurrected during the early 1900s through the recognition that biology had failed to comprehensively examine the science of relationships, essential in the development of other aspects of biology. Ecology (Ernst Haeckel coined the word ecology, meaning *oikos* or home) afforded this in laying the groundwork for the study of the interactions between organisms and their environments. This takes place on three levels: (1) the ecology of individual organisms, for example, behaviour related to habitat selection; (2) ecology of groups of individuals or populations; and (3) community and ecosystem ecology, which examines the organisation of distinct communities and ecosystems (Brewer, 1979).

Everything we need to sustain life, including the biotic factors (living) and abiotic factors (non-living aspects, such as temperature, moisture and light), exists within a thin layer of air, soil and water, which spans only 15 km. While the systems that developed from the unique relationships that evolved from these elements and agents have remained fairly stable over time – acknowledging that ecological change is a constant – the rate and extent of change that has come about in the last 300 years, and particularly the last 60 years, has been monumental. I would like to use the word 'unprecedented' here, but it would fail to acknowledge catastrophic events such as comets, earthquakes and other natural forces, which we are not exactly clear on. These changes have occurred primarily as a result of population growth, which in many areas is unsustainable even for the immediate future, but also because of technologies and human lifestyles that damage the ecological systems more and more each year.

One of the main dislocating features of this growth and technological advancement is that people are unable to appreciate the importance of the services that the ecosystem provides (Daily, 1997). Children in urban settings think that food is grown at grocery stores and that water simply appears from a tap, without any further understanding of their origin. The parody of this is that the functions and services that the earth's ecosystems provide for humanity have not been taken into account in commercial markets, despite the fact that all of our commerce, indeed our very survival, relies on these services (Costanza *et al.*, 1997). Such services include gas regulation, climate regulation, disturbance regulation, water regulation, water supply, erosion control, soil formation, nutrient cycling, waste treatment, pollination, biological control, refugia (habitat), food production, raw material, genetic resources, recreation and cultural aspects. Unhappily, exact figures on the monetary importance of these services have been left off the table in efforts to devise proactive policies.

Costanza *et al.* (1997) estimate that the annual value of these services is $16–54 trillion, which is about 1.8 times the current global GNP. Given the level of pressure and stress that continues to make these services scarcer, it is estimated that the value will increase to the point where some may hold values to infinity due to their critical state. The study underscores the importance of these services for our overall welfare and the obvious need to maintain the resources of the planet in order that we may continue to survive.

Ecological Values and Justice

While the emergence of humanism during the Age of Enlightenment had many societal benefits, some researchers have identified it as a root cause of today's environmental crisis. Ehrenfeld (1981) makes reference to the 'arrogance of humanism', at the core of which rests the notion that human reason alone can ameliorate the many problems that humanity faces. This supreme faith in the ability to rearrange the affairs of humans and nature in a logical order has progressed without an adequate understanding of the complexities of individuals, society and the natural world.

The arrogance of humanism can even be seen in those who purportedly loved animals. One of the leading ornithologists of the late 1800s and early 1900s was Charles Wilson Peale, who, in his diaries, depicts his interactions with the Carolina parakeet, one of the most beautiful birds of North America. He describes graphically his method of separating the birds from the trees by way of shotgun:

> At each successive discharge, though showers of them fell, yet the affection of the survivors seemed rather to increase; for, after a few circuits around the place, they again alighted near me, looking down on their slaughtered companions with such manifest symptoms of sympathy and concern, as entirely disarmed me. (Bryson, 2003: 474)

Peale was a lover of birds, but at the same time saw nothing wrong with slaughtering them en masse for no other reason than personal interest in making observations about their physical characteristics and behaviour. This prompted Bryson to note that, for the longest time, until rather recently in fact, those most interested in living things were those most likely to kill them off. His point is that, if some higher being was to assemble an organism with the sole purpose of looking after and monitoring life on the planet, it would not choose humans.

This sentiment about nature and the associated humanism can be seen more of late, and even in some of the most prestigious of journals. In his article in *Science*, Dubos writes: the 'richnesses of nature are brought to light only in the regions that have been humanized'. In reference to his home in France, he observes that 'The hills have such low profiles that they would be of little interest without the venerable churches and clusters of houses that crown their summits' (Dubos, 1973: 770).

Perhaps one of the most disturbing findings in the research on people and the environment is the polarisation between environmental values and environmental behaviour. When asked how we feel about the natural world, people typically say they value it and recognise the need for its preservation. But these feelings do not always translate into environmentally responsible behaviour, especially when two or more groups with competing values are involved, and when the livelihood of those in question is at stake. The rainforest, and those who make it their home, is salient in this regard. The World Wildlife Fund (WWF) argues that the rainforest will be gone by 2020 if logging is not kept in check. Seeing no other prospects for jobs, local people are confused as to why international agencies wish to put a stop to the logging. In fact, *The Economist* published the local sentiment of the deputy editor of the Cameroonian magazine, *Africa Express*, who in this case complained that:

> You destroyed your environment and got developed. Now you want us to stop doing the same! What do we get out of it? You have your televisions and your cars but no trees. People want to know what they gain by conserving the forest. (Anon, 1999: 55)

According to the director-general of the WWF, Clause Martin, and described in the same publication, what you do get are impoverished landscapes that are abandoned because of their lack of productivity, thus contributing to more poverty than before the loggers came to town. And often what is not explained is that logging concessions are not handed out to responsible logging companies, but rather to 'associates' of the current governmental regime, where some make very nice gains at the expense of others and the rainforest (recall the theory of inclusive fitness). The primary industry model in lesser developed countries has been explored by Carriere (1991), who notes that environmental destruction is the result of a constellation of political issues and forces that led to the domination of the export industry with very few linkages with the rest of the economy (as noted in Chapter 5). It amounts to short-term profits, continued ecological impoverishment, and unsustainable peasant livelihoods, where populations are increasingly pushed on to more marginal lands and are reliant on wage labour for survival instead of subsistence agriculture. Peasantry is thus forced to over-exploit these lands in order to survive. The natural resource degradation in these regions is not typified by land mismanagement or land abuse, Carriere observes, but rather by the model of accumulation and associations based on power. Sustainability is thus unrealistic because of the entrenched sociopolitical system that works to prevent it. This is why it is suggested that we need responsible governments who are willing to institute formal shifts in political and social policies at the grandest scale. Along with this include citizens who are empowered, institutions that are enlightened (and allowed to be enlightened), environmental reform and the reconstruction of corporate mentalities (Hammond, 1998). As such, the proper value of resources for trade will not be balanced unless governments and other groups are prepared to develop new and innovative ways by which to reward stakeholders. In Mexico, for example, the Cebadillas community will receive $250,000 over 15 years to preserve the habitat of the Western thick-billed parrot. Six environmental organisations have come together to make this initiative successful through a financial programme that was easily developed (Gullison *et al.*, 2000).

There is consensus in the literature supporting the notion that environmental concern can be classified on the basis of three distinct value scales, as outlined by Stern and Dietz (1994). The first, egoistic, presumes that the values that people attach to the environment have a direct effect on themselves personally. It follows that these individuals may be opposed to environmental protection, for example, if the personal costs are perceived to be too high. The second scale, social-altruistic, is based on the work of Schwartz (1970), who illustrated that people have a personal moral obligation to act in situations that may have an adverse effect on others and that they may be able to prevent. Energy conservation is an example of this value orientation that has costs and benefits to other human groups (Black *et al.*, 1985). The third scale of value orientation is the biospheric realm, which refers to a global concern for the costs and benefits of action or inaction to entire ecosystems, or to the planet as a whole. The moral imperative built into this scale is analogous to that which has been discussed at the social-altruistic level. Stern and Dietz hypothesised that people commit to action when proenvironmental personal norms are activated by beliefs that an environmental condition has adverse consequences for self and close kin (in the egoistic value orientation), for other humans (in the social-altruism orientation), or for other species or systems (in the biospheric orientation). They found that, as stated in the literature, environmentalism is tied to certain values, which links environmentalism to the biospheric-altruistic orientation and, inversely, egoistic value orientations. Biospheric altruism is said to result more from selective processes brought on by culture than from natural selection (genes).

In related research, researchers have argued that there are three codes of moral thought that can help to classify the moral intentions we maintain across cultures (Shweder, 1990). The first of these, the ethic of autonomy, makes reference to the self as an overriding preference for moral regulation along the lines of one's choice, autonomy, control and fairness as cardinal virtues. Western secular society typifies this perspective. The second moral domain, ethics of community, examines the individual as part of a broader set of collective relationships. This code requires the agent to take on virtues, such as duty, respect and obedience to authority, consistent with one's age or social status. The final domain, ethics of divinity, has a broad spiritual significance that is linked to purity and sanctity. Behaviours and acts that are not consistent with this spiritual purity are deemed disgusting and are therefore condemned, including pollution and hunting.

The classification of values on the basis of scale, as above, provides a template from which to examine a wide range of different environmental issues. For example, the debate over externalities, such as air, climate, wildlife and genetic diversity, is irrelevant to the calculations of economists because these externalities do not hold value (Suzuki, 1994). Even more broadly, is the fact that an externality can occur when the welfare of one is directly influenced by the actions (e.g. pollution) of other individuals or groups (Romp, 1997). In essence, economy is said to rest outside ecology as a separate entity, when it should be nested within an ecological framework. Our efforts to control pollution are a case in point – industrialised nations of the North continue to generate increasing volumes of hazardous waste that they continue to send to the lesser developed nations (Pellow *et al.*, 2001). The shifting of the burden to these regions has resulted from more stringent environmental regulations that have driven up the costs of waste disposal, as well as more relaxed legal frameworks in these LDCs. While economists describe these transactions as economic efficiency (many in the LDCs liken this to 'toxic colonialism' or 'garbage imperialism'), theorists refer to such forms of resource management as instrumental reason (Millar & Yoon, 2000).

Pellow and his colleagues further suggest that a growing number of scholars have become concerned with the impacts of environmental pollution across class and race. It is not just an issue that affects LDCs, but one that has been documented in the US regarding the upper and lower classes. That is, there is a predominant finding pointing to the fact that environmental racism is a very real problem in the USA. Citizens are not treated equally because locally unwanted land uses (LULUs), such as lead smelting, coke ovens in steel mills, sewage treatment plants and so on, are located in communities of colour, where the inhabitants are made to suffer a disproportionate share of the long-term exposure. This phenomenon takes place because these groups are less powerful than other groups.

It is not a far stretch then to envision the same scenario in a tourism context; something we might call tourism development racism. Here the displacement of local people from their traditional means or practices in support of a tourism industry takes place where the latter group provides little benefit to the former. In adapting Pellow's work on indicators of environmental inequality as racism, it is suggested that the following takes place with regard to tourism development:

(1) Widespread unequal protection and enforcement against tourism development in areas where inhabitants maintain traditional means or practices.
(2) Disproportionate impact of the development on the poor and residents of colour.
(3) The creation of unsafe and segregated housing.
(4) Discriminatory transportation systems (if at all) and zoning laws.
(5) The exclusion of the poor and people of colour from environmental decision-making in government and corporations.
(6) The neglect of human health and social justice issues by the established environmental movement.

These authors are quick to note that environmental inequality is a social process, which affects and involves many different groups across space and time. According to Hampton (1999), environmental equity is often hampered because of a lack of public participation. Through Hampton's work, we get a sense of the countless examples of people who are living at or below the poverty line and who are subjected to larger amounts of environmental pollution because of a lack of participatory methodologies that cater to needs of these groups. This means that it is the advantaged segments of society that often have better access to decision-making processes than their disadvantaged counterparts. Citing the principles of public participation from the International Association for Public Participation, Hampton (1999: 166–169) offers the following principles in generating equity, justice and fairness for all groups, many of which echo the pro-poor tourism platform, as described in Chapter 6:

The development of the participation process. The public should be involved in defining the process of participation and it should be agreed upon by a planning agency and participants. Although this may lengthen the consultation process and increase costs, it is essential for the development of an acceptable methodology.

Participation throughout the decision-making process. It is generally considered that participation should start early in the decision-making process.

Involvement of interested and affected parties. All interested parties should be identified and involved in the participation process, which should actively seek out and facilitate the involvement of those potentially affected by an issue.

Expression of public values. Participatory democracy is more likely to lead to justice than representative democracy, as justice is constructed through social interaction and the compatibility of supporting values. The values of all participants need to be made explicit.

Provision of information. All information should be freely available to the public, particularly when this information is relevant to the public's understanding of a decision.

Credibility of the process of participation. It is important for the participation process to be credible to participants. The process should therefore communicate the interests of participants and they should also know how this will take place.

Extent of influence on decision-making. The extent to which the process of public participation can promote environmental equity is dependent upon the degree to which the preferences influence the final decision.

Blackstone (1985) has suggested that we would have far fewer environmental issues if we could only see it as a human right to have a liveable environment. He views it as unfortunate that individuals see it as their right to have equity, happiness, property and liberty, but not the right to have a healthy environment. Because the rights of humans come before the rights of the environment, it is important to have social institutions and policies in place to make sure that the manipulation of the natural world is in the best interests of the public and not individuals. While Blackstone acknowledges the importance of freedom as part of a civil society, it is also necessary to restrict certain freedoms in line with the rights of humanity in general. For example, the group 'Wise Use', which lobbies for expansion in logging, mining and ranching, may be seen as the other side of the debate on the preservation of our natural heritage. They argue that environmentalists stand in the way of the individual rights and freedoms that are central to the American traditions of manifest destiny and the frontier mentality (Sachs, 1995).

The application of environmental justice has been described by Beim (1998: 160–261, 271–272) through a problem/solution scenario in reference to incinerators and the NIMBY (not in my backyard) phenomenon, as outlined in Box 7.1. Although it is a non-tourism scenario, the application could just as easily be changed to reflect an over-abundance of waste from hotel

Box 7.1 Environmental justice problems and solutions

Problem

The passage of a number of environmental laws in recent times, coupled with educational campaigns on the part of environmental groups and a supportive citizenry, have resulted in a substantial reduction in the quantity of hazardous waste produced. It is not likely, however, that the production of hazardous waste can be completely eliminated. As long as hazardous waste is generated, it will be necessary to dispose of that waste as safely as possible. Incineration is an effective method of hazardous waste disposal and it has become safer than ever with the passage of the US Clean Air Act of 1990. Yet, local communities (often with the help of vocal environmental groups) have so successfully opposed construction of hazardous waste incinerators in their areas that no new facilities have been built in the US in many years.

Solution

On the issue of opposition to the construction of the hazardous waste incinerator, it is not unethical to protect one's self, family, and community from a risk. This is especially true concerning the risk from, for example, a hazardous waste incinerator, over which one has so little control. Even with the best design and pollution controls, no one can absolutely guarantee that the incinerator will have zero emissions. Nor can anyone absolutely guarantee that the low level of emission will not pose a hazard to anyone. Although the risk may be estimated to be as low as one excess cancer or other disease in each one million exposed people, this is no consolation if that one affected party is you or a member of your family.

It would, however, be unethical to avoid your risk by shifting it to someone else. The manufacture and use of products, such as furniture, rugs, clothes, cars, and electronics, have become an integral part of our way of life. Wastes are produced by the manufacture of all these products. As long as the products continue to be used, there is a moral obligation to share the risk that results from the disposal of the waste generated in making these products.

Sometimes, widespread opposition to hazardous waste disposal results in industry finding creative ways to eliminate the production of waste that they cannot dispose of. Those who are successful in not advocating the NIMBY policy can then claim the highest of ethical behavior since all society had benefitted from the improved production processes. All too often, unfortunately, industry simply moves its operations to a state or country where environmental restrictions are less severe. In addition to possibly shifting the risk to someone else, the NIMBY policy also causes the loss of jobs. Since one cannot ascertain whether opposition to the hazardous waste incinerator will lead to the elimination of the production of the waste or the shifting of the risk to someone else, this behavior cannot be classified as ethical.

On the issue of problems raised by such opposition, a possible (but not perfect) solution may be a national hazardous waste incinerator siting program. Each region would be required to accept its share of incinerators. A government agency, or some other neutral body, could evaluate all potential sites in each region based on factors such as availability of facilities, accessibility, number of people affected, etc. A lottery system could be used to make the final choice among the top, roughly equal sites. To mitigate, somewhat, the risk imposed on the selected site, benefits, such as free health care or reduced taxes, could be provided to the people in the selected area. In this way, the risks and alternatives would have to be squarely faced by virtually everyone. It would not be possible to simply evade the problem by shifting it to someone else.

Source: Beim, 1998.

and resort development, or from large-scale hotel development itself, and the complications of this for local people.

Values and rights

The feelings we have towards pollution, conservation and so on, are dictated by the deeper values we have for such things, especially in regard to motivations that can either be intrinsic (e.g. valuing a resource for its own sake) or instrumental (e.g. valuing something because of the economic benefit it holds). In one of the earliest studies on the value of nature, Clark (1973) described blue whales by suggesting that the dead animal can only be valued according to its contribution in oil and blubber. The value of the species to science, aesthetics and so on cannot be estimated. Present-time accounting is akin to barbarism because it does not take into account the value of the resource to future generations, who would no doubt be eternally grateful to the present generation for preserving it. But for that matter, if we were truly ecologically savvy, we might suggest that attaching any sort of value, economic or aesthetic, to an entity is purely anthropocentric. Who cares what the animal means to humanity? What is important is only its role as a member its own species and its broader role within ecology.

The valuation of species has been discussed more recently by Myers *et al.* (2000) through their analysis of the world's leading areas of exceptional concentrations of endemic species, termed 'hotspots', which have been experiencing heightened levels of habitat loss (habitat loss being the main reason for extinctions) (see also Myers, 1988). Areas qualify as a hotspot if they contain 0.5% or 1500 of the world's 300,000 plant species as endemics, as well as habitat threat criteria. The authors identified 25 hotspots (e.g. Tropical Andes, Mesoamerica, Caribbean, Brazil's Atlantic Forest and so on), which contain the remaining habitats of 44% of all species of endemic plants, confined to an area of 2.1 million km^2 or 1.4% of the earth's land surface. These plants formally occupied an area of 17.4 million km^2, representing 11.8% of the surface of the earth. While the value of these regions is unquestioned on the basis of the plant variability alone, they also contain the richest number of animal species as well.

The value of these lands is such that, if conservation measures are not intensified soon, it is inevitable that they will be lost forever. Myers and his colleagues note that the amount of money that conservation agencies raise annually, about $40 million, is not nearly enough. They feel that $500 million will be needed annually, which equates to $20 million per hotspot, and that $300 billion annually is required to protect all biodiversity worldwide. The need for innovative strategies to maintain biodiversity is therefore imminent, and has prompted some of the world's leading authorities on ecology and biodiversity to consider ecotourism as a vehicle by which to conserve these resources (Wilson, 2000). In this regard, researchers continue to work on ideas that explore how best to use ecotourism for biodiversity conservation. One of these is the ecotourium concept, which is designed to maintain biodiversity while at the same time sustain communities who look to these areas for financial well-being. Essential in this model is the notion that tourists themselves need to play a more active role in the maintenance of park flora and fauna and that the money they spend be more effectively used to these ends (Fennell & Weaver, 2005).

Much of the discussion on species valuation has been framed in the context of intrinsic vs. instrumental values, as above, which has been attributed to the move from community rights and obligations to individual rights and needs (Mayr, 1988). Instrumental reason is both utilitarian and irrational, as noted earlier, because it is not based on the recognition of complexity and diversity, but rather on linearity of thought and action (e.g. farming monocultures or fishing a single species of fish to near extinction). Once resources like fish have diminished we move on to another because it is the most efficient way to sustain families and economies. Decisions are made using certain forms of technology, in space and time, based on efficiency and growth only, and with little regard to common sense and a well-balanced holistic approach tempered by ethics. This is a form of predestination, where we feel locked into systems of inevitabilities that remove our sense of responsibility and accountability (Saul, 2001). The problem with this inevitability is that our behaviour becomes increasingly irresponsible and antisocial, so that we can do what we wish without any consequences, at least none that are not shielded by our arrogance. Our behaviour is conditioned by socially institutionalised sets of norms that erode autonomy and free will.

Instrumental reason can be found almost everywhere. It has been detected in the document *Our Common Future* (WCED, 1987), which is our widely regarded blueprint for ethical development. In a provocative investigation of the concept of sustainable development, Mies (1997) feels that the document advocates what many view as an intensification of problems separating developing and developed economies. She cites the following passage from the Brundtland Report:

> If large parts of the developing world are to avert economic, social, and environmental catastrophes, it is essential that global economic growth be revitalized . . . this means more rapid economic growth in both industrial and developing countries, free market access for products of developing countries, lower interest rates, greater technology transfer and significantly larger capital flows, both concessional and commercial. (Mies, 1977: 12)

This has compelled Mies to draw the following conclusion:

> This insistence on further, more rapid, economic growth, both in the industrial and the poor countries, is evidence of the fact that the authors are obviously not ready to see the connection between growth on one side and impoverishment on the other, between progress and regression, between overdevelopment and underdevelopment. They are still wedded to the linear, evolutionist philosophy of unlimited resources, unlimited progress, and an unlimited earth, to an economic paradigm of 'catching up development'. (Mies, 1977: 12)

The vision painted by *Our Common Future* is one where there is the opportunity for LDCs to reach the same standard of living as the rest of the world.

There is an image of what they want to be, but it remains just that. In any game there are always winners and losers. Mies' conclusion is that, because the market system carries a philosophy that pushes the most important questions outside the circle of morality through instrumental reason, it cannot be considered a moral economy. Suggestions to the contrary fall short because we must first address the most significant, underlying issue.

Similar reasoning has plagued the tourism industry for years, even in attempts, such as the Brundtland Report, to control and plan effectively. For example, one of the first government policies for international tourism was established by Spain's Franco regime in the 1959 *Plan Nacional de Estabilization* (Holden, 2003). The plan, based on instrumental reason, emphasised growth and modernisation at the expense of the environment. Forty years later, tourism numbers had declined because of the perceived loss of quality in the natural environment. But even though there are countless examples of the same phenomenon, we have trouble grasping the concept of environmental checks and balances in tourism.

On the island of Zakynthos, Greece, Lang (2002) observes, environmentalists and tourism operators are at odds over the threatened loggerhead turtle population. Once prevalent in the area, development and tourism have driven the turtle to a few secluded beaches in the Greek archipelago. The beaches remained tranquil enough for the turtles until, in 1984, an international airport was built on Zakynthos, disrupting the nesting area and contributing to a decline in turtle numbers. This is directly attributed to the fact that the nesting season (July and August) corresponds to the high season in tourism, which brings with it the noise and lights of the hotels at night, litter, motorboats and umbrellas that pierce the eggs in their buried nests, as well as cars and motorbikes that roam the beach strip. Because the population of turtles is so threatened, a number of local and international organisations have become involved with the aim of protecting it. The result has been a series of measures that are designed to limit construction, keep people off the beaches at night and reschedule flights during daylight hours only.

These measures, however, are contested by local business people, who now resent the turtles, environmentalists and the government for passing laws that compromise their livelihood. As Lang notes:

> One shouting match over an abandoned nest between a landowner and biologists went as follows, 'Do you know this turtle belongs to me? I own this land, and the turtle does me harm. It harms my interests. I don't care that you are crazy about turtles; I don't like them'. (Lang, 2002: 3)

Furthermore, Lang reports that the Mayor of one of the municipalities was quoted as saying that the environmentalists were anarchists and that 'since man appeared on the earth he has destroyed whatever got in his way' (Lang, 2002: 3). Citing a British newspaper, Lang noted that volunteers who came to Zakynthos faced abuse, assault and threats of rape, with one biologist ending up in hospital after being attacked by angry businessmen.

This case study is yet another that underscores the value sets of two very distinct populations. Those who stand behind the preservation of the loggerhead turtle have perhaps a more holistic view of the situation, but they stand to lose nothing *personally*, it may be argued. On the other hand, businessmen see the intervention of the measures to control tourism as a direct loss to their economic viability (in reference to the three distinct value sets discussed earlier).

We can sympathise with the differing values of those who are involved in other industries when conflicts between sectors occur. But what about differing values of those in tourism alone? The same traditional conflicts often occur that are spawned by self-interest and profit. For example, Masterton (1992) found that almost all tour operators in his study agreed that the abuse of the planet is a bad thing, in theory. In practice, however, most operators did not want to discuss their environmental responsibilities, and one went so far as to say: 'I have to make a living, and if people want to go to polluted, or overcrowded, or disgustingly commercial tourist traps, then they're going to go. So why shouldn't I get the commission for the trip' (Masterton, 1992: 18). Masterton notes that most of the leading tour operators to Hawaii, the Caribbean and the Greek Islands who were contacted for her article on environmental ethics and tourism refused to discuss their advertised responsibility towards environmentally responsible behaviour. What tourism operators have discovered, as well as service providers in other sectors, is that these days anything green has tremendous marketing potential, to the point of seeing market forces replacing legislation in developing changes in attitudes with respect to the environment.

What is just as interesting (or rather problematic) about these issues over values in tourism is their persistence. In discussing the life and loves of the artist John Ruskin, who was a particularly keen admirer of nature and of different places, De Botton (2002) comments that Ruskin despised tourists for their failure to slow down and appreciate the true nature of the places they were visiting. In 1864, Ruskin was quoted as saying to a group of wealthy industrialists on tour in Manchester that:

> Your one conception of pleasure is to drive in railroad carriages. You have put a railroad bridge over the fall of Schaffenhausen. You have tunnelled the cliffs of Lucerne by Tell's chapel; you have destroyed the Clarens shore of the Lake of Geneva; there is not a quiet valley in England that you have not filled with bellowing fire nor any foreign city in which the spread of your presence is not marked by a consuming white leprosy of new hotels. The Alps themselves you look upon as soaped poles in a

> bear-garden, which you set yourself to climb, and slide down again, with 'shrieks of delight'. (De Botton, 2002: 223)

With great ease we have been able to justify tourism travel on the basis of cost-benefit calculations and numbers of participants (i.e. it must be right if we are all doing it). But is it right? This has been discussed by Aronsson (2000) in his report on energy impacts and costs from tourism in Sweden. He reveals that 95% of all energy requirements for tourism are from transport. And while international air transport represents only 8% of long-distance trips, it accounts for about 55% of energy requirements. Even more telling are the results for long-haul travel (over 3000 km), which represents only 10% of the overall number of journeys, but accounts for as much as 60% of the energy consumption for the travel of tourists. In examining the specific emissions from different types of transport, Aronsson reports that there is an increase in hydrocarbons when comparing cars and planes; cars emit the highest levels of carbon monoxide; and aircraft have the highest emissions of nitrous oxide and carbon dioxides, along with the highest energy and environmental costs. It should be noted that trains have by far the lowest emissions in comparison to the other modes of travel.

Thus, all forms of transportation are not created equal, with certain types more detrimental than others not only on the basis of fuel consumption but also in the support systems that are required to sustain them. Just to drive the point home, Orr (1999) offers the following comparisons:

> A bicycle, for example, moving at 20 miles per hour requires only the energy of the bicyclist. An automobile moving at 55 miles per hour for one hour will burn 2 gallons of gasoline. A transatlantic flight between New York and London on a 747 moving at 550 miles per hour for six hours will burn 100 gallons [approximately 400 litres] of jet fuel per passenger. ... A bicycle requires a relatively simple support infrastructure. An airline system, in contrast, requires a huge infrastructure including airports, roads, construction, manufacturing and repair facilities, air-traffic control systems, mines, wells, refineries, banks, and the consumer industries that sell all the paraphernalia of travel. (Orr, 1999: 149–150)

Although we try to implement a system of social ostracism at the point of purchase for certain types of vehicles, for example, cars, through posted information on fuel consumption per km, it does not seem to have much effect in places like Canada and the USA, where fuel costs have not approached the prices charged in Europe. There still remains a low financial cost to the rental of a sport utility for a vacation, as well as a low social cost. In fact there appears to be more of a social gain in these places by virtue of the choice for the larger vehicle, thus enhancing one's status by upscaling. (The innate sense that we have for detecting cheaters – see Chapter 2 – does not appear to apply to the rental or purchase market in North America. If it did, there would likely be more social ostracism, which might otherwise influence buyers to rent and purchase smaller vehicles.)

Reverence for the 'other'

Boorstin (1985) holds that, well before people thought about conquering mountains, these landscapes were revered by the people who lived in their shadows. The Nepalese, for example, idolised the mountains as the setting of their gods and also as their concept of the universe. During the 20th century this sense of reverence was eroded, as mountains were viewed through Western eyes more as an impediment to man's movement, colonisation and conquest of nature. The first to climb the Matterhorn, Edward Whymper, Boorstin notes, wrote that the mountains were 'an affront to man's conquest of nature'. This sort of reaction to nature was recognised by Sir Edmund Hillary after he climbed Everest in 1953, when he, like Einstein, knew he had taken a massive step forward in further distancing humans and nature. To his credit, Hillary worked hard in the latter part of his life to make sense of his accomplishment in a broader socioecological context. But the legacy of domination of mountains has become so ingrained that the words 'conquer' and 'conquest' became synonymous with humankind's craving for commerce and personal pride. Service providers now give clients an opportunity to scale Everest if they can afford the ticket. The most impressive natural assets of the planet have thus been reduced to trophies – extrinsically sought-after and ticked off a list without any reverence for their majesty and biophysical significance.

In writing on E.O. Wilson's (1984) biophilia (defined as 'the urge to affiliate with other forms of life', p. 85), Orr (1993) suggests that economic utopia, or the present predisposition towards more economic growth, wealth, televisions, lawyers, freeways and shopping malls, is something we have tried and does not work for the long-term sustainability of the planet. Biophilia, which is founded on a love for all life forms, is a concept rooted squarely in the realm of reverence, which is not a new concept, but merely one that needs to be consistently revisited.

Upon being asked where he was from, the Greek philosopher Diogenes referred to himself as a 'citizen of the world' (following from Fennell, 2004). That is, his thoughts and concerns were not solely confined to his current position, but transcended this to include the entire world. His perspectives on this topic led to the term *kosmopolites* (meaning 'world citizen'), suggesting that each of us dwells in two realms: the place of our birth and the global community. In more contemporary times, one's sense of space, place and scale has been considered in detail by the geographer J.K. Wright (1966), in one of his many geo-isms. Among these, Wright described the concepts of geopiety and georeligion. Geopiety refers to the emotional bond or awareness (piety) that people have towards space (geo), the latter of which is so central to the discipline of geography. In referencing the work of Tuan (1976), Singh *et al.* (2003) have suggested that geopiety is a religious concept that combines ecology and territoriality through attitude, beliefs and values. Indeed, Tuan notes that feelings of belonging transcend both culture and religion. And, as Singh *et al.* observe, 'it is the "rooted-ness" of values arising from such soulful attachments to place that creates communities' (Singh *et al.*, 2003: 8). From the spatial context, geopiety appears to be more closely linked with one's local environment and the territorial aspects that go along with this association. In a wider context, geopiety is a microcosm of georeligion, which refers to one's emotions about the earth in general. Typically, emotions associated with the concepts of geopiety and georeligion include love, reverence, affection, pity and compassion.

From Diogenes and Wright we have come to understand some basic spiritual and philosophical principles that are so vital to the stewardship of the planet today. The local–cosmopolitan perspective of Diogenes accentuates the importance of acting locally, but thinking globally, which is an underlying theme in environmentalism. And from Wright, the love and compassion for place, at multiple scales, underscores the importance of valuing the places where we live, work and play. For tourism, the interplay between place and emotion is critical if tourists are to be able to transfer their love of place beyond the scope of generating regions to those destinations that receive tourists. This means that the same love that we supposedly have for our own houses, streets and communities must be transferable to the places we visit as tourists.

Perhaps in a less spatially oriented fashion, the same 'geopietic' argument has held true in our regard for plants, animals and natural areas for years. In the English-speaking world this 'nature-piety' has its roots in England, where the worship of nature corresponded with the collapse of Christianity during the 1720s (Clark, 1969). As Clark suggests, however, faith was not completely eliminated at this time, but was to enter Western Europe *through* nature. In the beginning it would come from the minor poets, provincial painters and gardens, but later would take hold in the writings and art of some of the most famous minds of the time. Foremost in this regard was Rousseau, who was successful in forcing people to look at the landscape in a romantic and existential fashion. For example, mountains were valued spiritually and recreationally as more than just a nuisance to communication and development.

Perhaps even more significant though were his perspectives on what was to become existential philosophy, about the same time that David Hume had written on the topic. As described by Clark, Rousseau, when immersed in the natural world, 'lost all consciousness of an independent self, all painful memories of the past or anxieties about the future, everything except the sense of being' (1969: 274–275). Existence was thus a function of a series of events that were perceived through the senses; and these events were tied to the natural world first and later to people. The idea that virtue was intricately tied to the natural world ('natural man') stood in stark contrast to the rest of industrialised Europe. Rousseau's philosophy could not have been more unconventional.

In his book on *Regaining Compassion*, Birch (1993) writes that, in our materialistic world, science has dominated our existence to the point of divorcing us from the things that should matter most: feelings, affection and love. We are reduced to cogs in a machine, not unlike the materialistic world of Descartes and Newton, who saw the world as such. Unhappily we have lost the ability to relate to one another because we cannot make affective connections. His analogy is that we are rather like railway cars in our connections; the cars are connected but one merely pulls the other. If our relationships were based on compassion we would have connections that were intrinsically oriented rather than instrumental. This means that we often treat people as means to an end instead of ends in and of themselves. As such, we may cultivate relationships only because we see how these individuals may further our agendas, treating people as objects rather than as subjects. People, just like other sentient beings, are ends themselves because they all have value. And this can be said of the natural world too.

Reverence for animals

Social contract theory has been criticised for its insensitivity to nonhuman rights. Animals, this theory propounds, are incapable of entering into a contract (Olen & Barry, 1989) and therefore exist outside the rules that are allocated to only those who can understand them. Theorists have followed through with this in suggesting that, because animals do not have duties, they are not moral agents and, as such, could not have rights (Feinberg, 1985). Descartes, for example, felt that non-human animals did not have rights because, like robots, they were incapable of sensations like pain. However, in an oft-quoted passage on animals and pain, Jeremy Bentham suggested that 'the question is not, Can they *reason*? nor, can they *talk*? but, Can they *suffer*' (as cited in Harrison, 1997 and Fennell, 2000a). The same question is still being asked today and underlies our inability to understand how animals feel and communicate pain. Evidence points to the fact that animals react the same way that people do when they are hurt because of their biochemistry and brain construction: they cry out, they attempt to escape the situation, they examine the affected part, then withdraw and rest (Masson, 1999; Masson & McCarthy, 1996).

Today, it is acceptable to view animals as having rights, despite the fact that they cannot claim those rights (Regan, 1983). After all, babies cannot claim their rights, but rely on others to do so for them. In this, we have a duty not only to these animal populations but also to human future generations to conserve the biodiversity on the basis of rights. But these rights are also subject to conditions. This was the case in Greece, where cats and dogs were to have their right to life snuffed out in preparation for the 2004 Olympics. Athens is reported to have 50,000 stray dogs alone, which create an unsightly mess on the streets of the city (Smith, 2002). And since the country refuses to neuter animals (the reasoning lies in the feeling that all animals should have the right to a sex life), they were left with the development of a mass campaign of poisoning via a lethal cocktail of pesticides. Indeed the poisoning of these animals has taken place for some time, and enraged tourists so much that the Greek foreign ministry is constantly inundated with letters.

The mixed emotions on attitudes towards animals have led to the development of a spectrum of theories on their use and treatment by humans (Benjamin, 1985). The first set of theories, 'no obligation', establishes that there are no restrictions on what humans may do to other animals. The second set, 'indirect obligation', maintains that ethical restrictions are justifiable if they can be derived from direct obligation to people. For Thomas Aquinas, it was not acceptable to kill another's ox, only because you do injury to another man and his property, but not because of the act of killing the ox. The final set of theories, 'direct obligation', suggests that ethical restrictions on the use of animals are justified for the sake of the animals themselves, from the intrinsic point of view.

Fennell has explored many of these animal ethics theories in the context of tourism in a book on tourism, animal and ethics (Fennell, 2012a), as well as in a range of articles published in the journal *Tourism Recreation Research*. These theories include animal rights (Fennell, 2012b), utilitarianism (2012c), ecocentrism (2013a), ecofeminism (Yudina & Fennell, 2013), and welfare (Fennell, 2013b). Work from other scholars has strengthened this area in tourism ethics research. Shani and Pizam (2008) developed a framework for animal attractions from the perspective of three main theories: environmental ethics (ecocentrism), animal welfare and animal rights. Articles that focus on specific theories include utilitarianism and marine wildlife (Dobson, 2011); animal rights and animal welfare in reference to the UK dolphin industry (Hughes, 2001); ecocentrism and Australian dingoes (Burns *et al.*, 2011); and Buddhist compassion and the Tiger Temple, Thailand (Cohen, 2013). Bertella (2013) illustrates that promotional pictures of animals can be depicted according to three perspectives: utilitarianism, animal rights, and ecofeminism. Often these pictures represent what Bertella observes as a compromise between what is attractive to tourists and what is thought to be authentic local culture, with many demonstrating blatant disregard for animals – animals as passive and secondary in tourism experiences (Reis & Shelton, 2011). In other work, Bertella (2016) has shown that animals like whales are appreciated for their aesthetics, while sled dogs have more instrumental value in bringing tourists into contact with nature. Fish on the other hand, are viewed as commodities to be exploited, suggesting a strong anthropocentric value and attitude towards these animals.

Instrumental value and neoliberalism are at the heart of Duffy and Moore's (2010) work on elephant-back tourism in Thailand and Botswana. Elephants in these places are said by the authors to be re-regulated as they transition into new types of engagement with humans. In Botswana, for example, new ways of using elephants is conducive to the principles of neoliberalism because of the completely different way in which elephants are valued (elephant-back tourism). In Thailand, elephants have been used in labour for thousands of years so the transition is not so marked. The major change is a shift from the public sector (logging) to the private sector (tourism).

To be sure, however, the terrain of animal ethics is not confined to conventional theoretical approaches. New applications of theories serve to illustrate the breadth of the field. For example, Nussbaum (2006) argues that the rightness or wrongness of how we treat animals should be determined on the basis of how this treatment allows or obstructs their flourishing, in the Aristotelean sense. Every sentient being should have the opportunity to experience a life of dignity that corresponds to the way in which these species ought to live. For Diamond (2001) the injustices that animals endure at the hands of humans should be contextualised through our sense of shared vulnerability. Decisions on how we should treat animals therefore should not boil down to cost-benefit calculations, as in utilitarianism, or other metrics, but rather on kindness, compassion, love and pity. This rhetoric has been applied to polar bear ecotourism in Churchill, Manitoba, Canada – bears as performing spectacles – which Yudina and Grimwood (2016) argue are embedded in normative power imbalances with a dominant tourism industry that emphasise instrumentalism and anthropocentrism. The way out of this control and dominance is to view polar bears as stakeholders and subjects that should be recipients of our care, kindness and compassion.

In tourism there is the expectation that certain travel types (although this is an issue that should not be particular to certain types of tourism only) will have deeper sensitivities to animals than others. Ecotourism is one of these and can be loosely defined as a form of travel that: (1) focuses on the natural history of a region; (2) is sustainably oriented; (3) is small scale, affording local benefits; and (4) is learning-based (Fennell, 2015b). As arguably the world's greenest form of tourism, those who participate in it should be sensitive to the interests of animals as a first priority. However, there have been cases reported where students and scientists entering rainforest regions have done so with the main purpose of taking plants in an effort to patent biodiversity. Such biopiracy, under the guise of ecotourism, fails to take into consideration the rights of plant and animal species, but also the livelihood of local populations who have developed extensive knowledge over certain plants and animals in which the biopirates are interested (Pera & McLaren, 1999) (Box 7.2).

Intention is critical in understanding the motives of agents (Griffin, 1997). A person who is cruel, for example, intends on making another suffer gratuitously. If I do something to hurt you, but at the same time save you greater pain (by replacing a dislocated shoulder), I am still cruel. If, however, a doctor tries to help replace the shoulder and by accident breaks your clavicle, he or

Box 7.2 The myth of ecotourism

Writing on the newly developed ecotourism industry in Rurrenabaque, Bolivia, Ceasar observes that the authenticity and regulation of activities in the region are virtually non-existent. Ecotourism exploded in the area because of the adventures of the Israeli tourist Yossi Ghinsberg who got lost on a trip to the region only to emerge out of the forest three weeks later, and to later write a best-selling novel on the experience. Since then, the region has played host to over 12,000 tourists, mostly of the back-packer variety, but with little capacity to absorb the influx. Ceaser notes that guides often catch snakes and alligators and keep them in bags, only to be released at the most opportune times for tourists. The sewage system is unfit for tourism; most effluent is simply discharged into the river. Even if efforts are made to recycle, the recycled materials which have been taken to the major centres are later dumped into the river anyway. Companies are developing 'ecolodges' but doing little if anything to control impacts, such as waste. The problem lies in the fact that Bolivia has no environmental standards for hotel development so 'eco-hotel' is a self-imposed title. In order to remain competitive, tour companies have begun to under-cut each other, with an associated loss of service and regard for the welfare of people and natural capital.

Source: Ceasar (1999).

she is clumsy but not cruel. The centrality of the argument, according to Griffin, lies in the intentions of the helper. So cruel intentions do more harm than the same amount of pain generated through altruistic intentions. In the former case, the sufferer is merely a subject of another's power. Consequently, moral rules need to mesh with the motives of human agents. A failure to recognise the customs of the people in a different culture by behaving contrary to the practices of these people has an element of intention. One could argue that a failure to know these customs has prevented the proper conduct of the tourist. As such, there is no intentional motivation to cause uneasiness. If, however, the tourist chooses not to act according to custom then there is a deliberate intention to disrespect the local ways of life.

Like the example above on ecotourism and biopiracy, there is an element of intention that perhaps should disqualify certain types of activities as being ethical. Such was the case with a recent debate on billfishing (fish like marlin and sailfish) as ecotourism in a 2000 edition of the *Journal of Sustainable Tourism*. While some scholars have argued that consumptive forms of outdoor recreation like fishing can be ecotourism (see Holland *et al.*, 1998), others argue that fishing of any kind is not ecotourism (Fennell, 2000a). It is said that the *intention* to entrap the animal, which is not the same as the intentions of ecotourists, which should be geared towards minimum disturbance and impact in all cases; the *pain and stress* that result from catching the animal; the aspect of *consumptiveness*, where catch-and-release practices still may be viewed as consumptive along a continuum; and the *extrinsic* nature of the participation in the event, which is based on a different set of values related to the resource, may be reflective of what Curry (2011) has categorised as an anthropocentric (light green) or ecocentric (deep green) standpoint, rather than a focus on the interests of individual animals (animal liberation represented as mid-green environmental ethics). In the case of billfishing, despite the angler's best intentions to minimise stress on the animal, one can only do so to a point, after which he or she must cease to pursue and capture the animal. The argument follows that the treatment of animals cannot be based on an acknowledgement of healthy populations in the biological sense (ecocentrism) alone, but rather that respect must be shown to the individuals comprising these populations (animal liberation). This is an example that is more than just a matter of animal liberation. The other side of the discussion is the principle behind upholding the integrity of forms of tourism (and personal conduct) as they have been theorised, especially when the propensity to ascribe the ecotourism label to something that is not, comes at the hands of instrumental reason (Butcher, 2003; Holland *et al.*, 2000).

A more draconian interpretation of intention, however, can be found in the work of Dower (1989), who writes that, in order to have a more comprehensive ecoethics, we must take responsibility for the unintended consequences of our actions or inactions. Although it is easy to view morality, in part, as the avoidance of intentionally harming, stealing, assaulting and so on, in the real world our actions or inactions oftentimes have consequences for others despite any efforts to prevent them.

This entails taking responsibility for the unintended consequences of actions, which, in turn, requires a willingness not to take risks, because, as Dower notes, it is 'almost true by definition that if one takes a course of action which has a certain risk, one does not intend what is risked' (1989: 19). He uses the example of nuclear power, where risks from radiation levels and accidents arguably outweigh any benefits that might come about from the industry.

Environmental Ethics

The ontological argument for the preservation of nature is not a metaphysical one, as is the case in attempts to prove the existence of God, but rather exists on aesthetic and ethical grounds (Hargrove, 1993). Ethics for and about the natural world is also not an applied ethic similar to biomedical or business ethics, because it constitutes an incipient paradigm shift in moral philosophy by demanding a focus on the other, as noted above (Callicott, 1984). This section will more formally outline the meaning of environmental ethics in presenting an alternative world view alongside the application of business ethics in tourism.

Most scholars are in agreement that environmental ethics as a discipline came about through the social movement that inspired the world's first 'Earth Day' on 21 March, 1970. A few very good essays emerged in the literature to help launch this field of study, including Aldo Leopold's *A Sand County Almanac* (1949/1966), Rachel Carson's *Silent Spring* (1962/1987), Lynn White Jr's 'The historical roots of our ecological crisis' (1971) and Garrett Hardin's 'The tragedy of the commons' (1968). At the heart of the movement is the recognition of two opposing paradigms. The first (anthropocentrism) is more of a dominant world view, while the second (biocentrism) is more of a minority, harmonious world view. The anthropocentric paradigm posits that nature can only be conceived from the perspective of human values. Humankind, therefore, determines the form and function of nature within human society. Conversely, the biocentric philosophy considers that all things in the biosphere have the right to exist equally, i.e. they are all equal in intrinsic value (Devall & Sessions, 1985). These paradigms have been articulated in the work of Wellington *et al.* (1997), who have identified three distinct stages in the development of environmental ethics:

Phase 1. This phase was marked by an anthropocentric view of nature and usually took the form of an application of economic analysis to environmental problems. Issues that were analysed included those relating to industrial/commercial activities and human well-being. The role of ethics was rather simplistic, and progressed along the lines of applying an ethical principle to a real world situation and developing a solution to the problem. This 'engineering' approach to applied ethics evoked a good deal of criticism.

Phase 2. A new understanding of ecological relationships led to the belief that beings other than humans could and should have moral standing. This type of thinking worked as a catalyst for a whole host of new theories, including ecofeminism, deep ecology, animal liberation and ecocentrism. Distinctions were made between instrumental value (that which had value as a means to an end) and intrinsic value (that which had value in and of itself).

Phase 3. This stage was distinguished by the development of a more pragmatic, postmodern approach to understanding and solving ethical dilemmas. The authors argued that there has been a shift away from simply championing theoretical ethics or applied ethics, since real life situations are often far more complex. Not only is there the realisation that experts outside the field of environmental ethics have the ability to confront ethical dilemmas (philosophers joining non-philosophers), but also that facts alone, independent of public policy, are often not sufficient to solve society's ills.

Table 7.1 Contrasting paradigms

Dominant world view	*Deep ecology*
Dominance over nature	Harmony with nature
Natural environment as a resource for humans	All nature has intrinsic worth/biospecies equality
Material/economic growth for a growing human population	Elegantly simple material needs (material goals serving the larger goal of self-realisation)
Belief in ample resource reserves	Earth 'supplies' limited
High technological progress and solutions	Appropriate technology; non-dominating science
Consumerism	Doing with enough/recycling
National/centralized community	Minority tradition/bioregion

Source: Devall and Sessions, 1985.

In the early 1970s, the dichotomy between these two opposing positions was effectively articulated by Arne Naess Drengson and Inoue (1995), who coined the term 'deep ecology' through his ecosophy (*eco*, meaning 'earth', and *sophia*, meaning 'wisdom': earth wisdom). The essence of deep ecology is to probe beyond the limited information afforded to us through science, into the realm of the religious and philosophical (Devall & Sessions, 1985; see also Drengson & Inoue, 1995). Deep ecology is a representation of basic intuitions and experiences of the individual in the context of a broader holistic ecological consciousness. In this regard, deep ecology is the antithesis of the dominant world view of today (Table 7.1). What resonates through the deep ecology philosophy is the importance of ethics, harmony and equity. As such, it is intricately tied to the basic assumptions surrounding human behaviour (i.e. good vs. bad), the reasons for these behaviours, their effects, and how we may institute change.

A book that had a similar message during the same time, and is of particular importance to our discussion here, was Taylor's classic, *Respect for Nature: A Theory of Environmental Ethics* (1986), which provided a theoretical argument for the importance of respecting the natural world at a time when environmental ethics began to emerge. Our duties, obligations and responsibilities regarding the natural world must be governed by ethical principles. Taylor noted the importance of following an ethical ideal or ethical spirit. In order to realise the vision of a 'best possible world', stakeholder groups must be motivated to provide the needed will-power to implement shared measures for the benefit of all. This is particularly relevant to tourism studies, where there are virtually no underlying ethical principles – in a theoretical context – that might act in guiding a comprehensive vision for the importance of human values in tourism decision-making.

This is akin to Singer's (1981; see also Miller, 1991) circle of morality, which continues to expand outward from the family to the tribe, to the

Table 7.2 Definitions of environmental ethics

Author	*Definition*
Regan (1981: 34)	The need to 'postulate inherent value in nature'
Taylor (1986: 3)	That which is 'concerned with the moral relations that hold between humans and the natural world'
Dower (1989: 11)	'A set of principles, values or norms relating to the ways in which we interact with our environment'
Miller (1991)	The widening of the categories of ethical and moral reflection to include the natural world
Ehrlich (2000: 319)	That which analyses the 'ethical problems created by the population-resource-environment predicament'.

race and to all of humanity, and which will need to continue to expand outward further to all of the natural world. Most problems have resulted from atomistic and anti-ecological thinking. Indeed, modern moral philosophy has been attacked for its focus on the individual (moral atomism), which thus makes the contributions of ecoethics, i.e. the focus on a more holistic approach to morality and nature, that much more important (Johnson, 1984).

The momentum behind the research on human–human and human–environment relationships has culminated in a number of definitions of environmental ethics or ecoethics (see Table 7.2). Underlying these definitions is the aspect of individual responsibility for doing one's part to ensure the maintenance and sustainability of the earth's resources. As illustrated in Chapter 2, if there was an innate function within culture or the brain that was responsible for acting ethically towards the natural world, the development of an area of ethics to level the playing field, so to speak, would not be needed. Humanity has thus shown that long-term planning of resources was not highly valued (Ehrlich, 2000).

But although prevalent, ecoethics has suffered because of the tendency of its proponents to focus on what Miller (1991) has seen as a romanticised gender-neutral relativism in attempts to address our vast range of problems. In doing so, there is the belief that all can be well by discussing all things at once, and under the same philosophy. Its strength, it can be argued, lies in its recognition as a metaphor and the identification of value sets that are non-anthropocentric in their origin. These have been articulated well in the work of Merchant, who identified three different sets of values in reference to environmental ethics (Boxes 7.3 to 7.5), which, although lengthy, link to the previous discussion in this chapter, which briefly touched on light green, mid-green and deep green ethics as established by Curry (2011) in reference to billfishing as a form of ecotourism or not.

Critics, such as Narveson (1997), however, say that it is illogical to protect a tree for the sake of the tree, because trees do not have morals, interests

Box 7.3 The self-interest or egocentric approach to environmental ethics

Assumptions about

The nature of human beings. People are fundamentally different from all other creatures on Earth over which they exercise control.

Social causation. People are masters of their own destiny; they can choose their goals and learn to do whatever is necessary to achieve them.

Human society. The world provides unlimited opportunity for humans. It is proper for humans to control and have 'domination' over the rest of nature.

Constraints on human behaviour. Human history is a record of never-ending progress. For every problem there is always a solution. Progress need never cease.

Responsibilities and duties

The goal is the maximisation of individual self-interest: What is good for the individual will ultimately benefit society as a whole. Within the Judeo-Christian tradition, appeals to the authority of the Wholly Other coupled with the primacy of the individualistic salvation ethic minimise the need for social responsibility.

The metaphysical base

The Mechanistic Paradigm:

(1) Matter is composed of atomic parts and is to be so understood and viewed.
(2) The whole is thus equal to the sum of its parts.
(3) Causation is a matter of external action on inert/inactive parts.
(4) Quantitative change is more important than qualitative change.
(5) Dualistic: Separation of mind/body; matter/spirit.

Exponents

Plato, mainstream Christianity, Descartes, Hegel, George Berkeley, Hobbes, Locke, Adam Smith, Malthus, Garrett Hardin.

End goals

The knowledge of idealised 'truth'; of 'form' or 'idea' (Plato); of 'innate ideas' (Descartes); of 'faith' rather than 'works' (Christianity). 'Conception' (generalising principles prior to experience) as the way to knowledge.

Source: Merchant, 1990, adapted from Miller, 1991.

Box 7.4 The homocentric approach to environmental ethics

Assumptions about

The nature of human beings. Humans have a cultural heritage in addition to and distinct from their genetic inheritance and are thus qualitatively different from all other animal species.

Social causation. The primary determinants of human affairs are social and cultural (including science and technology) rather than individual.

Human society. Social and cultural environments are the crucial context for human affairs. The biophysical environment is essentially irrelevant.

Constraints on human behaviour. Culture is a matter of cumulative advance; thus technological and social progress can continue indefinitely. All social problems are ultimately soluble.

Responsibilities and duties

Greatest good for the largest number of people. Social justice rather than individual progress is the key value. Sense of responsibility to and for other people important. In religion, some sense of responsibility toward nature, of human stewardship over the creation.

The metaphysical base

A combination of mechanism and organicism depending on the particular approach. Philosophically, materialism and positivism fit here. Major subcategories would include most utilitarianism, consequentialsim.

Exponents

J.S. Mill and Jeremy Bentham (utlitarian theorists); Barry Commomer (socialist ecologist); Karl Marx and Mao Tse-tung (political theory); Randers and Meadows (limits-to-growth theorists).

End goals

Maximisation of utility and the good of the greater number. Through the study of the material, and the social conditions of society, we can discover the truth about human responsibility; sense perception within the historical process can alone enable us to discover the 'really real'.

Source: Merchant, 1990, adapted from Miller, 1991.

Box 7.5 The ecocentric approach to environmental ethics

Assumptions about

The nature of human beings. While humans have exceptional characteristics (culture, communication skills, technology), they remain one among many other species and are interdependently involved in the global ecosystem.

Social causation. Human affairs are influenced not only by social and cultural factors, but also by intricate linkages of cause, effect and feedback in the web of nature. Purposive human actions can have unintended consequences.

Human society. Humans live in and are dependent on a finite biophysical environment that imposes potent physical and biological restraints on human affairs.

Constraints on human behaviour. Although human inventiveness and power may seem to exempt us from the constraints of biospheric limitations, ecological laws will always provide the primary context for human and other animal life.

Responsibilities and Duties

A belief that all living and nonliving things have value. Duty is to the whole environment. Sometimes it includes a rational, scientific, empiricist belief system based on the laws of ecology; sometimes a religious and/or mystical approach to the wonder of the natural world.

The metaphysical base

The organic or holistic paradigm:

(1) Everything is connected to everything else.
(2) The whole is far greater than the sum of its parts.
(3) The world is active and alive; internal causation.
(4) The primacy of change and ongoing process is always affirmed.
(5) Nondualistic unity of mind/body, matter/spirit, people/nature.

Exponents

Taoism; Buddhism; Native American philosophy; Thoreau; Gary Snyder; Theodore Roszak and 'right brain' analysis; Aldo Leopold; Rachel Carson; Fritjof Capra; political ecology; deep ecology; ecofeminism

End goals

Unity, stability, diversity, self-sustaining systems, harmony and balance in nature and within competitive biological systems, survival of all organisms, democratic social systems, 'less is more', soft energy systems, self-sustaining resource systems, sustainable development.

Source: Merchant, 1990, adapted from Miller, 1991.

or desires. It is not the responsibility of government to enact legislation to protect natural areas or specific species, because of the belief that governments have sold out to bizarre special interests. Narveson says that the supply of natural resources to feed our desires is actually increasing as the population increases, and has been for hundreds of years. He would rather see the human mind as the central resource because it thinks up ways to use our world – more brains (people), more innovation. In focusing on resources as finite we do not consider what the human brain could develop as a substitute. Any scarcity that exists, such as of food, is not because there is a lack of the resource, but rather because of poor politics, poor management and poor technology.

The extent to which the environmental platform has taken off is reflected in the degree to which we consider environmental affairs on a day-to-day basis. Although it could be used more effectively by politicians and other decision makers, we see the tangible outcomes of the environmental crusade in the many products and policies that are geared towards minimising our impact on the natural world. An internet search of environmental ethics located some 50 institutes and centres directly involved in this area (Fennell, 2000b). The Center for Environmental Policy at the University of North Texas (evolving from a parent not-for-profit organisation developed in 1980) is involved in research and the dissemination of knowledge. This not-for-profit centre has as its mandate: (1) the publication of the journal *Environmental Ethics*; (2) the reprinting of important environmental books; (3) the training and promotion of research in environmental ethics through workshops and conferences; and (4) graduate education in environmental ethics.

The discussion on ethics and environment is particularly relevant to tourism studies, although there are virtually no underlying ethical principles – in a theoretical context – that might act in guiding a comprehensive vision for the importance of human values in tourism decision-making (Holden, 2005; Wearing & Wearing, 2014). Researchers have, however, explored a number of different avenues of research connecting tourism studies to the natural world with a burgeoning link to ethics.

A long standing focus in outdoor recreation, and now tourism, is the use of the New Environmental Paradigm scale to investigate environmental attitudes (Dunlap & Van Liere, 1978). Luo and Deng (2008) used this scale to measure the motivations of nature-based tourists and found, consistent with other studies, that differences exist between groups harbouring separate motivational perspectives. Nature-based tourists more supportive to limits of growth and who hold concerns over ecocrises, are inclined to want to be closer to nature, to learn about nature, and to escape urban environments. By contrast, those nature-based tourists motivated by skill development and who seek to experience new things, tend to have attitudes towards human domination of nature.

Similar results have been found in reference to core values among different types of tourists who seek nature as the basis of their touristic experience. Perkins and Brown (2012) discovered that biocentric values, i.e. the intrinsic worth of nature, are positively associated with ecotourism, pro-environmental attitudes, and environmental protection. Those holding egoistic values, in comparison, were found to be less interested in nature-based tourism, more interested in hedonistic-type activities, and were less supportive of the conservation and protection of the environment. Fairweather *et al.* (2005) investigated awareness of ecolabels among a sample of tourists to Christchurch, New Zealand. Sixty-one percent of the visitors in the study expressed biocentric values, and they believed that ecolabels were needed in New Zealand, they would choose accommodation with ecolabels, and they were concerned about the natural world in the places in which they travelled.

The lack of attention paid to the environment in tourism studies has promoted Reis and Shelton (2011) to argue for a more complete exploration of the different meanings of 'nature' constructed by societies, and how these are important in dictating various tourism practices. Examples of these different approaches include embodiment (Franklin, 2001), interagentivity (Latour, 2014), and indigenous perspectives (Higgins-Desboilles, 2009). In reference to Franklin (2001), there is the argument that activities like fishing allows the participant to have a more fully sensual or embodied experience, over ecotourism with its focus on simply the visual gaze. This theme was explored in detail in the work of Waitt and Cook (2007). Based on an investigation of the embodied experiences of kayakers in Thailand, the authors were forced to conclude that kayaking ecotourists protected their human subjectivities by remaining separate from a more sensual connection with the natural world. In this way:

> Non-human entities are prioritized as objects to be viewed. The non-human world remains understood as aesthetic despite the smells of mangroves and bat urine, being bitten by mosquitoes, squealed at by monkeys, grazed by rocks and burnt by sun, In part the responses may reflect an anxiety to speak of this sensual knowledge in tourist places valued for their alleged pristine qualities. (Waitt & Cook, 2007: 547)

Reinterpretation and representation of nature has also been explored by Grimwood (2014) in reference to canoe tourists and Inuit residents in the region of Baker Lake, Nunavut, Canada. Tourists were found to harbour perceptions of the Arctic and its people through the lens of colonialism and modernity; people who are passive recipients of tourism and powerless to control the pace and scope of change. A just and sustainable tourism in the Arctic – a new way to speak about the Arctic in Grimwood's words – requires cooperation between stakeholders, respect, inclusiveness,

Box 7.6 Beyond Leave No Trace

The importance of continuing to explore the different meanings of nature has been captured in work by Simon and Alagona (2009) on the Leave No Trace philosophy used so liberally in USA and other countries for the ethical management of wilderness areas. The traditional Leave No Trace philosophy is based on seven principles. These include:

- Travel and camp on durable surfaces.
- Plan ahead and prepare.
- Be considerate of other visitors.
- Respect wildlife.
- Minimize campfire impacts.
- Leave what you find.
- Dispose of waste properly.

The authors argue that a new framework is needed to recapture wilderness recreation as a more 'collaborative, participatory, productive, democratic, and radical form of political action' (2009: 17). Simon and Alagona recommend a new set of principles to be more reflective of changing values and priorities in society. These are:

- Educate yourself and others about the places you visit.
- Purchase only the equipment and clothing you need.
- Take care of the equipment and clothing you have.
- Make conscientious food, equipment, and clothing consumption choices.
- Minimize waste production
- Reduce energy consumption.
- Get involved by conserving and restoring the places you visit.

connection, and participation, and the rejection of universal and divisive expressions of value. Research on exploring the different meanings of nature is vital in moving the environmental ethics agenda forward. It is encouraging to see similar thinking in the more applied side of tourism and outdoor recreation (Box 7.6).

Conclusion

This chapter sought to shed light on the eroding relationship that is found to exist between humans and the natural world. Although we are cognisant of the importance of the ecological services that the natural world performs for us, we have been decidedly ambivalent in factoring these into our policy and financial calculations. We simply do not wish to acknowledge the inherent worth of the natural world, for its own sake, for fear that it will either reduce business or increase costs.

This form of behaviour towards the natural world, however, is not new. The section on 'The Myth of Stewardship' illustrated that humans have used and abused natural resources for millennia. This is in sharp contrast to the belief on the part of some that all traditional societies always practised an ecological ethic of care. At the heart of the issue are the values that we hold concerning the natural world. These values have been widely discussed in the ecological justice literature, where issues from pollution and marginalised communities in the most developed countries to logging in the LDCs continue to plague policy development. In this chapter, the discussion on rights was resumed with a focus on the rights of animals, acknowledging that rights is one of many different theoretical and applied perspectives that take into consideration the interests of animals. Ecotourism provided a natural link to this discussion, as issues related to intention, stress, consumptiveness and extrinsic motivation have eroded the basic purity of this form of tourism. A framework based on reverence was suggested as a manner by which to secure a more ethical foundation in ecotourism. Finally, environmental ethics was discussed for the purpose of identifying a template from which to aid in the development of tourism ethics.

8 Broad-based Concepts and Issues in Tourism

Most forward-looking people have their heads turned sideways
Harold Innis, n.d.

Introduction

This chapter examines a number of broad-based ethical concepts and issues in the tourism field that have moral relevance. The chapter begins with a discussion on common pool resources, conservation and community development, and property rights. The behavioural complexities of the commons problem are discussed using the social trap concept. The chapter further discusses the concept of governance from informal and formal standpoints, with a focus on interactive governance. Issues surrounding accreditation, best practice benchmarking and the precautionary principle in tourism are discussed as policy mechanisms that have implications at local and global scales.

Common Pool Resources

Concern over the distribution of resources, especially common pool resources, in the face of increasing demand has been around for some time, as outlined in the following 16th century anonymous poem (Butler *et al.*, 1992: 2):

But now the sport is marred,
And wot ye why?
Fishes decrease,
For fishers multiply.

Although the example above is one regarding sport, the commercial fishery has manifested itself in the present day as the tragedy of the commons, where short-term reinforcement of certain behaviours overwhelms any attempts to counter such behaviour for the long-term well-being of society. Societies find themselves on a road that is potentially lethal, but are unable to step away from such a pattern of behaviour (e.g. the use of certain technologies, which, although harmful, continue to be used because of the benefits to individuals and to society as a whole). The concept was made popular in the work of Garrett Hardin (1968), who was building on the work of Gordon (1954) regarding common property resources (fishing stocks) and an economic theory of natural resource utilisation. This, in turn, was based on the work of Baranoff (1918) on bionomics or bioeconomics; of Huntsman (1944), who was one of the first to argue that fisheries depletion was an economic problem; and of Burkenroad (1953).

Gordon's paper is fascinating as it synthesises a number of themes emergent in the literature before and during the time of his research. One of his most important conclusions was that the management of common pool resources, such as fisheries, is infinitely difficult because such resources yield no economic rent, and hence no legal entitlement to a section of the seascape. Since rent is not appropriated by anyone, fishermen are free to ply the waters wherever they wish. The result is a pattern of competition that can become so intense as to diminish the resource completely, as we have seen too frequently in recent years. This goes against the historical view of the fisheries resource as noted by Thomas Huxley, who wrote, in 1883, that restrictive measures on the resources of the sea were a waste of time and effort as these resources were inexhaustible.

> The cod fishery, the herring fishery, the pilchard fishery, the mackerel fishery and probably all the great sea fisheries, are inexhaustible: that is to say that nothing that we do seriously affects the number of fish. Any attempt to regulate these fisheries seems consequently, from the nature of the case, to be useless. (as cited in Gordon, 1954: 126)

It also goes against the law of diminishing returns, which is a theory that explains equilibrium in the market. Gordon argued that certain environmental conditions are necessary in order to prevent the common pool resource from being decimated by excessive exploitation. As he says, everybody's property is nobody's property. So the oil left underground is valueless to the driller because somebody else may legally take it. This has been documented by other researchers, who note that, because no one legally owns the space, and so cannot lay claim to benefits from husbanding the resource, self-interest takes precedence over conservation (Towsend & Wilson, 1990). Conventional wisdom suggests that property rights may lead to the orderly exploitation and conservation of the resource base, through either privatisation or nationalisation.

But nationalisation, which is a function of the regime in place and the power and intentions of its actors (who tend to divert power and especially money to themselves) and the cultural practices of its citizenry, does not always work. Such was the case in Africa during the 1960s and 1970s, where wildlife was nationalised over concern that poachers would decimate these resources. Because wildlife is a major currency of the African people in many areas, the citizenry were placed in a position of competition with federally owned wildlife and with little incentive to use their wildlife for wealth or trade. In South Africa, subsistence hunters were refused rights in the face of a diminishing resource, creating tension between those with access and those without. In recent years, the removal of wildlife resource rights from rural peasants has given way to an emerging policy of redistribution of resource rights through the philosophies and practices of ecotourism (Dieke, 2001). Here, it is argued, decision-making, control and benefits are greatly sought after by the economically marginalised from such a history of injustice. But this new-found freedom and accessibility to resources has understandably contextualised ecotourism as something of a pious hope in the face of the immediate social and economic issues that characterise marginalised populations.

This has motivated decision makers to champion a postmodernist philosophy of community development and conservation at the expense of other more traditional conservation forces based on science. The argument is that, although community-based conservation (CBC) has advantages (e.g. distribution of funds evenly in the community and sustainability after the monetary support is removed), it is activated in place of traditional conservation science, which to some has worked for years (Attwell & Cotterill, 2000). The ethical concerns over such policies and practices are such that CBC is presented to the community as a panacea for development issues through the empowerment of the community and all of its citizens. But these initiatives often ignore larger issues that will ultimately make community-based conservation meaningless in the long run. These include increased population growth, non-traditional means of hunting (e.g. motor vehicles and high-powered guns), the use of modern veterinary medicines to increase stock sizes and the desire for wealth to buy any number of commercial goods.

This is supported by other authors who feel that the argument against the traditional approach to conservation in Africa is ill-informed. Traditional forms of conservation are not designed to separate local communities from their sources of food at the expense of pure protection of flora and fauna. One has only to examine the past in concluding that the stereotype of all indigenous people as land stewards is contentious (Spinage, 1998), as noted in Chapter 7. Attempts to maintain the traditional approach are often fruitless because the large international organisations, such as the WWF, would not dare embrace the fortress mentality of two decades ago because it is far less humanistic. This is in spite of the view of some who continue to ask how long parks can exist while being surrounded by hungry populations (Ghimire & Pimbert, 1997).

Central to the issue of access and resource depletion is private property ownership, which Ridley (1998) feels is the prime reason for most environmental problems in the LDCs (as noted by Gordon, above). When people fail to have an opportunity to own land and become prosperous, especially when so much of this occurs at many different scales and locations, people have little hope and even less regard for common property (i.e. no rental value). Diamond (1993), for example, writes that, in New Guinea, over-harvesting is avoided by villagers only in cases where the species is owned by the individual. This study is supported by other researchers writing on indigenous knowledge and conservation, who note that areas conserved as open access will quickly be degraded (Gadgil *et al.*, 1993). Property rights therefore need to be defined. But even more important than this is the need to take responsibility for this property, which includes limiting access to the property only to those with a stake in it, and enforcing rules of appropriate conduct. This is generally supported by scholars even though there have been many small and traditional systems put into place that have been successful in the management of common property (see Ostrom, 1990; Pinkerton, 1989). The Valencia's 'Tribunal de las Aguas' is a 1000-year-old local court that meets weekly to administer the use and costs of the region's water source. Such models work because they are built around a small stable population in which penalties can be administered easily across generations. However, these systems break down through the arrival of outside influences (e.g. cheaters such as transnational corporations, as in our discussion of reciprocal altruism), where such groups can easily escape penalties of overuse. Well-managed resources have subsequently turned into unmanaged common resources, which are open for the taking in such cases (Chichilnisky, 1996).

Ellis (2001) writes that there have been various attempts to provide a solution to the commons problem. These include: (1) the imposition of taxes; (2) imposing limits on the physical quantity of the resource harvested or extracted; (3) issuing tradable quotas; (4) limiting the number of firms exploiting the resource; (5) creating private property rights for the resource; and (6) promoting collusive welfare maximisation. He notes that the first of these three require the intervention of a benevolent regulatory agency, which maintains complete information on the resource and its users. This includes the level at which to set taxes, knowledge about the resource for the purpose of imposing harvest restrictions, or the number of quotas that are allowable in order to sustain the resource and those who extract it. A restriction on the number of firms can work as well, Ellis notes, but there are often complications surrounding the number at which to maximise extraction and the under-utilisation of the resource in the face of excessive production costs. The creation of private property rights is appealing because of the sense of ownership and stewardship towards the resource. But as Ellis observes, this approach is difficult because of resources such as fisheries and oil, where it is unclear if property rights can be fairly established. The final example of an attempt to solve the commons problem, promoting collusive solutions, has been shown to be successful through game theory (Chapter 2), when negotiation costs are limited and there are few players.

Ellis's solution to the problem is an alternative approach termed 'Common Pool Equity', which he posits as a new property right that 'may be presented for a share in the profits of any firm that extracts from the common pool resource. The implication of this will be that profits per share will . . . be equalized across all firms in the industry' (Ellis, 2001: 141). This approach is likened to a situation in which all firms are allowed to be owners with equal shares of each firm in the industry, each of which would be induced to internalise costs. Ellis suggests that, as a multi-plant monopoly, the distortionary incentives (i.e. working for oneself in pure competition against competitors) that stem from the commons problem would be eliminated.

From the perspective of tourism, Holden (2005) writes that there still has not been a moral shift in human behaviour to settle the issue of the overuse of common pool resources because of the absence of an intrinsic value of nature. Both industry and consumers have not been able to move beyond the instrumental, intrinsic use of nature, and Holden feels that such a shift towards the intrinsic must be driven by the market. Common pool resource issues in tourism are also challenged by the inability for disparate activities to evolve at the same time. Vail and Hultkrantz (2000) suggest that activities like snowmobiling are not only inconsistent with traditional property rights and common pool theory, but it has effects on cross-country skiing, reindeer herding, alpine fish stocks, commercial forestry and air quality. Because of these problems, the authors note the following four challenges for sustainable nature tourism within the context of a common resource:

- Keeping demand pressure within capacity limits at prime sites and peak times;
- Balancing tourism and non-recreational activities in multifunction ecosystems;
- Controlling cumulative, irreversible landscape transformation; and
- Strengthening landowners' incentives to invest in conservation and value-added tourism. (Vail & Hultkrantz, 2000: 240)

Social traps

The commons problem, referred to as one of several types of 'social dilemma' by theorists such as Ostrom (2003), was examined by Platt (1973) in the 1970s in an effort to explain the behavioural complexities and responses that individuals and groups were faced with in regard to systems and choices. His social trap theory posits that the choices open to an individual depend on the system to which one is part. If these choices are unacceptable or limited, they may require the individual to act on decisions that are self-defeating for the person or for society as a whole. Individual behaviours (i.e. resource consumption) are rewarding in the short run (to the individual), and lead to negative outcomes for the collective – the commons – in the long run (Edney, 1980). Hardin's 'Tragedy of the Commons' (he used the example of grazing sheep on a public grassland) provides an example of a social trap because, by saving money on the free grass, farmers are able to increase their stock in the face of a diminishing resource until the resource is destroyed entirely (collective demand exceeding supply). The trap, according to Platt (1973), is that the individual continues to work to his own advantage (grazing his stock), even though collectively the resource is damaged to the point where he can no longer benefit from it.

Platt identifies three social trap scenarios: one-person traps, missing hero traps and collective traps. One-person traps include situations such as cigarette smoking, where there is a short-term benefit for the individual (social and biochemical reinforcement) but long-term costs (lung cancer). The missing hero trap includes situations where someone is required to act for the betterment of the group. Platt uses the example of a mattress on the highway that has created a lengthy traffic jam. Those in the back of the line cannot help because they have no idea what the problem is, and will not help when they reach the mattress because of the time already lost. Once past the mattress there is no incentive to stop and remove the mattress. In such cases, Platt suggests we need a hero who will sacrifice their time and effort for the benefit of the group. He notes that different settings, along with their social groups, create the impetus for goodwill in individuals unless there is too much difficulty or danger. Collective traps include the 'Tragedy of the Commons', which is described in detail in the following Box 8.1.

Box 8.1 The Tragedy of the Commons

The Tragedy of the Commons, a Prisoner's Dilemma game but at the level of society rather than two individuals, occurs when there is a failure of the market to provide an optimal and efficient allocation of resources. The failure takes place as a result of the imposition of external costs on those who extract the resource. The competition built into this scenario ultimately leads to the over-exploitation of the commons. When there is competition for common pool resources such as fisheries and forests, rivalries create conflicts that ultimately undermine sustainable development efforts (Briassoulis, 2002). If people are informed about an issue and are rational, they should end up better off than if they acted irrationally. The Tragedy of the Commons provides a negative answer to this dilemma. Using the following example we can show why (adapted from Sober, 1988, as follows). Let us say that in a population of 100 freshwater angling operators, there is opportunity to purchase a device for $100 that limits the amount of oil and gas entering the water system. At this stage it is a matter of individual choice, not law, which mimics the voluntary codes of ethics presently in use in many marine jurisdictions. You as an operator must ask yourself what the benefits are from the purchase of this device. If no one else purchases a device it would not be beneficial for you to purchase one because the positive impact on the water system would be negligible, meaning that the gain is so trivial that you might as well save your money. However, if everyone purchases one the gain to the system is exceptional. Having said this, you may also wish not to purchase the device yourself if everyone else has done so, because you are assured that the environment has been taken care of (99 out of 100 operators purchasing the device – all except you). Your preferences, with 4 as best and 1 as worst, are as follows:

		State of the world	
		Everyone else buys one	Nobody else buys one
Your action	You buy	3	1
	You do not buy	4	2

The rational act, by the scores above, is not to buy the device, that is an action which dominates the alternative, because no matter what everyone else does, you are better off not buying (4 > 3 and 2 > 1). But what is available to you (not to buy) is also available to the other 99 operators, and they may rationally decide not to buy the device as well. The result is that no one has purchased the device, with everyone only able to attain 2 points, as outlined above. But the freshwater system could have been better off (3 points) if they all collectively decided to purchase the device. As the reader may detect, the model above is limited because it does not account for intermediary ranges. For example, 65% or 80% of operators may have elected to buy, which would have broadened the scale of the game. But it does show the basic premise, which is this: 'the rational action for each individual to choose is known in advance to make all the players worse off than they would have been if they had all chosen the irrational action' (Sober, 1988: 86). This is the paradoxical dilemma that underlies the Tragedy of the Commons.

Source: Briassoulis (2002); Sober (1988).

In reference to tourism, the Galapagos Islands is a good example of the social trap phenomenon, whereby the decisions by policy makers to increase visitation yearly undermine initial attempts to control the impacts of tourism in this sensitive region. The lure of tourists and associated tourist spending is too powerful. Another example is the cruise industry. As no one owns the oceans, and hence cannot regulate activities that take place in these regions, cruise companies are free to pollute their waters. According to Platt (1973), morality and greed are not the central problem in this phenomenon, but rather the arrangement in time of costs and rewards of those involved – another example of instrumental reason.

Platt's work, however, has been heavily criticised by Edney (1980) for its reliance on behaviouralism, i.e. it is focused on rewards and punishments for engaging in certain behaviours. In this light, Edney notes that there is no reliance whatsoever on feelings, interpretations, morality, altruism and so on, beyond the cost-benefit calculation, which posits that we are all basically alike in our cognitive processing. In Edney's words, the social trap phenomenon 'essentially treats humans as neutral entities which engage in self-satisfying behaviors that sometimes lead to crises... . The formulation thus places reward over reason in human choices in the commons' (Edney, 1980: 134).

In response to the behaviouralistic shortcomings of Platt's work, Edney set out to explore a number of other theories that he felt could be used to examine problems of resource allocation within society. He underscores the philosophy of Hobbes in suggesting that a person's basic drives are built on self-interest, egoism and competition. People in an open-market system are ruled by competition and these aggressive tendencies are thought by some to be biologically driven (see Ardrey, 1970). Similarly, Dawkins' (1999) selfish gene theory posits that social behaviours are a function of biological selection. Selfishness is thought to be necessary and basic to one's survival. Conversely, universal love and welfare of the group 'are concepts which simply do not make sense' (Dawkins, 1999: 2). The influence of competition and selfishness are so invasive, according to Dawkins, that associations of mutual benefit within the community are often unstable because stakeholders continually strive to get more out of the association than they put in. Sharing and openness are concepts that have been described as critical to the success of community development. Most important to Edney, however, is a focus on values such as equity, freedom of choice and social power, as the fulcrum from which to discuss commons problems. This is proposed instead of a system based on rationality, which conflicts with other forms of rationality and the semantics over what is sensible rather than what is good. Trust also forms the cornerstone of Edney's argument, in suggesting that cooperation is possible and thus of long-term benefit to all members of the commons (see

Ostrom, 2003 for an excellent breakdown of the relationship between trust and cooperation, in the context of reciprocity).

In this regard, and citing the work of Hobbes (1651/1957), who said: 'And Covenants, without the sword, are but words, and of no strength to secure a man at all', researchers have found that, despite previous research, enforceable agreements (covenants) can be made between parties without some external enforcer (sword) as an intervention (Ostrom *et al.*, 1992). In a series of laboratory experiments that replicated common pool resources, Ostrom and colleagues found that communication by itself can be an effective agent in inducing players to show environmental restraint; even more so than punishment. The implications of the management of common pool resources are that those involved in governance of smaller-scale common pool resources should not assume that the stakeholders involved in the resource are caught in an inescapable Tragedy of the Commons, as noted previously. Stakeholders can arrive at jointly derived strategies if provided with an arena for dialogue and if they have sufficient information, leading the authors to conclude that self-governance is possible among such groups.

In related research, Costanza (1987) has suggested that escaping social traps is a matter of following one of four different methods: (1) *education* can be used to warn people of the effects of certain actions (e.g. warnings on cigarette packages), although it involves a great deal of time and there are inherent problems in trying to educate many people for the same purpose; (2) the *superordinate authority* option includes legal systems, government and religions, which may forbid or regulate actions that are or will be socially inappropriate. The problem with this method is that it must be rigidly enforced and monitored and success based on religion necessitates the inclusion of a series of like-minded individuals. Traps may be changed to a trade-off through the imposition of (3) *compensatory fees*. The most common option would be to tax those individuals or organisations where there is any consumption or impact that goes above the optimum level of resource efficiency (e.g. pollution taxes). The final method (4) *insurance*, although not discussed in detail, can be used to hold industry accountable to a certain level or standard of conduct/action. In concluding, Costanza writes that we need to be able to calculate the long-term social and environmental costs of our actions and charge these to those who are able to reap short-term gains. Uncertainties should also be considered. If there are unknowns, worst case/impact scenarios should be forecasted and assumed with costs absorbed by those responsible for the impacts – not the public.

Governance

The importance of the link between political rule and the natural world can be seen in the following myth, which describes how the king of the

Persians evaluated the work of those who had been given the responsibility to govern:

> He would simply observe the condition of the land and the forests within particular jurisdictions. If the land and the forests were well cared for, he automatically rewarded his governors. If the land was ill-tended and restoration efforts to repair damage delayed, the king replaced the caretakers. He evaluated the overall governing ability of his subordinates by their care for the natural world. The political principle is clear and as applicable today as it was in the mythical past. Those who genuinely care for the Earth, who are sensitive to the effects of human impacts on the environment, can be trusted to govern well generally. (Miller, 1991: 10)

The ethic of care emerges as the most important virtue here in efforts to balance human agency and the environment. Success in governance was thus intricately tied to a heightened level of reverence and respect for nature. This is a simple criterion on which to base one's ability to rule (although the methods of ensuring a healthy natural world may have been more complicated), and it serves nicely as a contrast to the meaning of governance today. In contemporary times, governance has been defined as a 'set of regulation mechanisms in a sphere of activity, which function effectively even though they are not endowed with formal authority' (Rosenau, 1992: 5). More specifically, Glasbergen (1998: 2) illustrates that governance refers to 'opportunities for goal-oriented and deliberate intervention in society', and entails three elements: (1) subjects and objects in a specific relation, for example public/private, state/market, government/society; (2) the notion of an intent to induce change; and (3) a conception of the social context in which the intent must be realised, including principles and objective notions. Governance is a more general term than government and includes the necessary consensus, acquiescence or consent, to carry out programmes where there are many groups involved, especially in areas where the state cannot play a lead role (Hewitt de Alcantara, 1998). Policy is perhaps the most tangible outcome of governance structures that provide the needed framework from which to articulate the goal orientation as outlined by Glasbergen above. In tourism, for example, there are four distinct forms of policy that have been used (Parker, 1999, based on the work of Anderson, 1994):

Regulatory. As the easiest to understand, this form involves the use of direct and often forceful restrictions on the actions of individuals, groups and businesses. For the latter category, this includes, for example, the issuance of licences to service providers for the purpose of controlling the number of operators in any one region. Other regulatory policy includes directives on land-use planning, site planning and architectural design. Tourists can be regulated through bans on certain activities or through carrying capacity limits. Whole societies can be regulated through system-wide resource protection laws that are designed to, for example, regulate land in fragile coastal zones.

Distributive. With its link to justice, this form of policy is designed to distribute benefits to specific groups. It is promotional in nature and thus contrived to stimulate tourism activities within the destination. The promotional policy can focus on local resources, or it can seek to attract external investment. This can take the form of exemptions from property taxes, exemption from import duties of materials and a reduction of income taxes.

Self-regulatory. The focus in this form of policy is on the regulation of groups, like regulatory policy, but the key difference is that such policies are pursued and implemented by a group for its own advancement. This may come in the face of governmental financial burdens, which prevent the development of policy for the safeguarding of people and or resources. An example is the Antarctic tour operators' code of ethics, which is designed to provide a measure of control over the use of resources in this region. This form of policy often takes the lead in the development of codes of ethics for operators, tourists, local people or communities.

Redistributive. Involves governmental programmes that are designed to shift wealth from one social group (those who have resources) to others (those who do not). Shifting resources may be based on geography, ethnicity or class. Ecotourism is said to have redistributive qualities in that it encourages tourism to areas that are economically depressed or marginalised, affording citizens in these areas an opportunity to generate income. The problem, as cited by many authors, is that redistributed benefits are too meagre or become tied up in the hands of relatively few at the expense of others.

The importance of regulation within society is such that it has compelled some theorists to note that it has become a fact of life (Rolston, 1988). In tourism, most regulations seem to be generated by issues of safety, health or justice. The speed at which knowledge flows into the development of regulations and the willingness for the industry to respond to these directives is proportionately related to the significance of the impact or dysfunction. Hotel fires, shipwrecks and higher hygienic standards are all examples of situations that will demand immediate learning within the industry (Hjalager, 2002).

However, some theorists contend that, although formal governance structures like policy (and regulations) are important, we have had a tendency to place too much emphasis on these at the expense of informal governance concepts such as values, love, ethics and cooperation. In this regard, a more recent incarnation of the work of Edney, above, can be found in the research of Millar and Yoon (2000), who argue that how resources are used and shared is ultimately based on moral choice rather than solely institutional structure. Citing the work of Redclift (1997) and Ostrom (1990), they suggest that sustainable development is as much about the institutions we construct – the rules and roles inherent in these – as it is about the future condition of the environment. It follows that institutions are just as much about roles (repeatable patterns in social relations, according to these authors) as they are about rules (the structure and flow of information, the range of decision-making, sanctions to enforce behaviours). Roles are thus important as a basis from which to articulate responsibility in the transformation of rules. What often constrains the full potential of institutions is the tendency to dictate the will of the collective from the vantage point of rules only. Millar and Yoon observe that the will that lies in institutions devoted to sustainable development and resource management must lie in a shared morality based on love. As a sentiment, love is one of the most entrenched narratives in society because it is the foundation for trust, knowledge of the other, respect, responsibility and personal commitment, making it intimately tied to goodness. Love is said by the authors to have the ability to cut across and bind all of the various roles within the firm, while acting as a foundation for the development of formal and informal rules (see Chapter 3 in reference to Kierkegaard's ethic of care based on love).

The tragedy of decision-making in resource management, according to Millar and Yoon, is that it creates winners and losers. These decisions underscore certain values and virtues. Whether the most threatened or vulnerable of these is preserved is a function of the instrumental forces at hand within the organisation. As discussed previously in this book, we have long been obsessed with instrumental reason because it is the fountainhead of systems, and we feel most comfortable living within these systems. So this has meant that we have grown to feel uncomfortable with our own special qualities and strengths, especially if these work against the grain of the corporation or organisation (Saul, 2001).

Given the progression of research and practice on governance, some theorists argue that interactive forms of governance have become more useful recently in environmental policy development because they are founded on consensus. This means that policy can proceed further if there is cooperation between different stakeholder groups, including public, private, NGOs and others. Consensus is reflected in the cooperative management model of governance, where change lies in the communication between groups, and where results are laid down in voluntary agreements among the involved parties. In allowing communities to garner greater access and control over resources, co-management is seen as a realistic tool (see Plummer & Fitzgibbon, 2004). However, in cases where there is agreement on the value of a cooperative approach to management, such as fisheries, the problem lies in the institutional frameworks for change that have been unsystematic and marred by institutional constraints (Noble, 2000). At the centre of the issue lies a paternalistic, government-led approach to fisheries management, at least in Canada, which is decidedly biological in its makeup. This is a key impediment to management change as it excludes industry from being a responsible partner (Lane & Stephenson, 2000). In earlier work, Pomeroy and Berkes (1997) suggested that efforts to support co-management structures should be centred on a philosophy of decentralisation, which they characterise as the 'systematic and rational dispersal of power, authority and responsibility from the central government to lower or local level institutions' (p. 469; see also Ryan, 2002). In an effort to overcome this problem, these authors propose the consideration of a number of key principles to facilitate institutional arrangements for effective co-management strategies for fisheries. These principles and others have been articulated in the work of Noble (2000):

Interactive organisations. Conflict resolution can be ameliorated by the development of advisory boards, and by more streamlined membership.

Local control. Includes local ownership and control along with consideration of decentralisation of power and delegation.

Community support. This takes the form of support from many vantage points and collaboration.

Planned process. Based on the satisfaction of goals, the long term, knowledge and adaptive capacities.

Substantive diversity. Goals must take into consideration equity, economic development, sustainability.

Holism. Involves inclusiveness of different groups and integration of interests.

In drawing this section to a close, it is important to note that, based on the previous discussion, policies are not human, but rather are the manifestation of human thought and the potential for human action. Although we may generate a myriad of different consulting reports, governmental policies and procedures, these only come alive, in a metaphorical context, when they are activated by people. People, not reports, are the most important part of the equation here, yet it is the latter that too often become elevated to positions of superiority over the people they serve. In this, the people who activate policy reports have a tendency to alienate others by taking sides over issues like development and the environment. In referencing the South African Truth and Reconciliation Commission, which enables countless victims of torture and the families of those murdered by officials to be heard, Saul (2001) illustrates that, in order to move past the horrors of this situation, the protagonists and antagonists had to meet in such a forum in order to create a shared memory for the purpose of establishing an ethical relationship. But as Saul notes, this was a case of reconciliation, not closure. Nothing, especially something as horrible as this, is forgotten. Policies and technical reports have no memory. But the people who must live by them do.

The instrumental reason behind the 'policy first over people' can be demonstrated in many of tourism's most important meetings (e.g. the 2002 Ecotourism Summit in Quebec City). The apparent need to have such meetings controlled by governing bodies and NGOs, those entities that are the biggest and have the most prestige and clout, at the expense of independent free speech and scholarship, emphasises the policy fetish. These groups have in common the same specific organisational agenda that becomes threatened by thought that takes place outside the organisational vacuum. But the nature of this vacuum creates an inescapable capsule, which prevents and compromises their ability to strike a balance between what they want and what is really needed. Here they have banished their imagination and instead chosen to go down the path of linear and short-term self-interest. The free thinker, who is not bound by dogmatic ties to Kuhnian paradigms, is as dangerous as a fox in the hen house to the organisation with an agenda based on ideology (Saul, 2001). The Quebec City meeting, which was quite restrictive in its programme and organisation, provides ample evidence of how instrumental reason can both flatten and narrow our options.

In stepping out of this black hole of utilitarian determinism, governance must be open to reason and an ethics based on due regard for the other, together. If reason alone forms the basis of our decision-making it becomes what Cottingham (1998) likened to a 'committee of ethics', because of the inability to move beyond a perspective based on inefficiencies and allocations, to one which is based on the development of close personal relationships, which are intrinsically meaningful. Conversely, because we often do strange things in the name of love, it too can be an agent of biased judgement. However, if grounded in reason, love can become a form of logic and provide the needed context in structuring relationships between identities (see Chapter 2).

Accreditation

In the last 30 years there has been increasing recognition that codes of ethics, accreditation, best practice, benchmarking and auditing are important in raising professionalism, standards and quality in the tourism industry. This is done by helping businesses to become more financially successful through the identification of areas of improvement, improving sustainable environmental practices and enhancing client experiences (Issaverdis, 2001). The range of different environmental management strategies has been examined by Mihalic (2000), who identifies four of these in ascending order of importance: (1) the least effective includes voluntary environmental codes of ethics that are said to be effective because they raise the level of understanding in the region as well as creating political support for specific issues (the topic of the next chapter); (2) uncertified or self-declared labels or brands are thought to be more beneficial than codes of ethics, but do not have the marketing value that the larger logos have; (3) more effective still are the environmental management branding schemes that are allocated on the basis of competitions for excellence in environmental practice or certified environmental good practice; and (4) the most effective category includes the group of accredited schemes that are granted by third parties on the basis of specific criteria. This last category includes Blue Flag, an organisation concerned with the water quality of European beaches and characterised by an independent, non-profit status; the use of experts in adjudication; the approval of awards on the basis of a European jury; and the allocation of these awards on a year-to-year basis (Mihalic, 2000).

Proponents of accreditation argue that it is necessary as a form of regulation for the industry, especially for products such as ecotourism, where consistency around programme offerings is essential. They argue that it is time for a universally accepted framework for accreditation in order to offset the aforementioned problems. But the issue most vexing in ecotourism, where there are about 100 certification and ecolabelling programmes, is overlap, lack of uniformity, and consumer and industry confusion regarding which companies are able to declare themselves as sustainable, green or eco-friendly (Font, 2002; Honey & Rome, 2001). This was identified in 2000 by the WWF who published a report on tourism certification arguing that programmes like Green Globe 21 have failed to establish clear brand recognition, the result of which is the undermining of the potential to bring about sustainable tourism. This resulted because of Green Globe's choice to award logos to companies who had environmental management systems, despite the fact that there was no requirement for these to achieve a certain level of performance. The WWF argued that these accredited service providers were process-oriented rather than performance-oriented.

This is true in Australia, where a number of different accreditation programmes have been assembled, all of which have identifiable problems. Such problems include: (1) limited ability of tourism industry associations to operate their accreditation schemes if they are not self-funded; (2) the ease of getting accredited; and (3) the decreasing strength of accreditation programmes (a watering-down effect). It has been argued that a replacement of the current scheme in Australia is needed in favour of a single industry-wide programme, which could be more easily handled (Harris & Jago, 2001). Other authors suggest that the future of accreditation programmes can only be secured in the hands of international brands, where there will be less confusion for tourists (Kahlenborn & Dominé, 2001).

These issues were tackled at the Mohonk Conference (Mohonk Mountain House, New Paltz, New York, 17–19 November, 2000), which acted as a formal mechanism to articulate general principles for ecotourism or sustainable tourism certification programmes. These programmes are characterised by membership fees, a logo or brand, voluntary enrolment, auditing or other forms of assessment, and compliance to the regulations that are built into the programme. The overall framework set out by the participants of the Mohonk Conference was as follows (Chester, 2002). The programme should:

- State clear objectives.
- Be participatory, multisector and multistaker.
- Provide tangible benefits to tourism service providers and a means for tourists to choose wisely.
- Provide benefits to local communities.
- Set minimum standards while encouraging and rewarding best practice.
- There needs to be a process to withdraw certification in the event of non-compliance.
- Control logos for appropriate use, expiration dates and loss of certification.
- Include provisions for technical assistance.
- Be designed such that there is motivation for continued improvement of the scheme and the products certified.

What makes these accreditation programmes ethical is their focus on trying to bring about sustainable tourism through the regulations that they hold to be true. For example, the goals of Green Globe 21 (2002) are to lower costs through reduced energy consumption, improve competitiveness in the market by becoming more 'green', conserve the ecosystems, improve the quality of life in host communities, reduce greenhouse gases, improve community relations between the industry and local people, and target marketing to customers who care about the environment. But as Font (2002) has noted, this organisation has been cited for being too profit-motivated, for certifying improvement not performance and for permitting companies who are committed to the principles but not certified to use the logo.

Ecolabels place the responsibility for improving the environment, and the systems responsible for this, in the hands of the consumer. This means that there are issues of equity and efficiency, as described by Buckley:

> People who buy low-impact ecolabeled tourism products, however, are incurring a private cost shared between a small number of individuals. In so doing they generate a public benefit shared among a much larger

> number of individuals, including themselves and other tourists who buy non-ecolabeled products in the same area, as well as the area's residents and host community.
>
> This is not a pure market mechanism of an individual choosing to incur a cost to acquire a benefit. The benefit to the buyer of ecolabled products is diluted.... The dilution occurs because the total improvement in environmental quality is less than it would be if everyone had contributed towards its cost. The more tourists who choose not to buy ecolabeled products, the smaller the benefit gained by those who do. Hence, tourists who choose not to buy such a product are gaining benefit without contributing to its cost. But those who favor this purchase enjoy only part of the benefits they could expect to experience without this dilution factor, with the residual enjoyed by people who have not paid. (Buckley, 2002: 198)

This suggests that there is an element of altruism required on the part of tourists who must pay a premium to experience the ecolabelled product. This can be a major barrier because, although many are willing to pay their fair share for a common benefit, as discussed earlier in this chapter, there are decidedly fewer who will pay more than their fair share. This means that the number of tourists who pay extra for ecolabelled products will be smaller than those who would pay for such products if the costs were borne by all other tourists. Buckley says that, despite this factor, we gain an indication of the strength of environmental concern among tourists due to the high level of response to products that carry ecolabels. This is true, however, only for those in the most developed nations. It is also true for those customers who are unaware of environmental issues, and who are not willing to pay extra costs for ecolabelled programmes (Font & Tribe, 2001; see also Romp, 1997).

The issue that emerges from this discussion is the extent to which one can regulate morality through schemes of this nature, especially when morality comes at a price (charging fees for accreditation). It may also be that the organisation has different motivations for their involvement in accreditation, which may run contrary to the ideals of ecotourism or sustainable tourism (see the section on policy, above). Again, while organisations can have a moral platform, it is the individuals who are very much removed from these groups who are responsible for implementing these measures. But it is not only the tourist who has the responsibility of improving the environment through his or her purchases, but also the moral actions that are nested within the accreditation process, as the foregoing discussion suggests. The following problem/solution construct, although not of a tourism evolution, has the same implications for tourism audits that are not founded on the basis of ethical conduct (Box 8.2).

Box 8.2 Certification

Problem

ISO 14000 is a voluntary standard for environmental management systems. A company can declare itself in conformity with the standard, but third-party certification of conformity is available and is generally regarded more highly by purchasers. The company seeking third-party certification contracts a 'registrar' who sends an audit team of three trained professionals to review the company's management system. The audit usually takes three days. On the third day of an audit of a facility, one member of the team reveals to the lead auditor that he has worked for the facility recently as a consultant. Is this an ethical problem? What should the lead auditor do?

Solution

It is an ethical problem for an auditor to have a relationship with the company being audited. It is considered to be a conflict of interest in that the auditor may wish to please the company with a good report so as to enhance his/her chances of working with the company in the future. The lead auditor must:

- call the ISO registrar and inform the registrar of what has happened;
- remove the compromised member of the team;
- void all the work done by the removed team member;
- request the company to permit an extension of the audit; and
- redo all the work done by the removed team member.

In the event the redone work confirms the findings of the disqualified auditor, the auditor should apprise the auditor of this fact in writing. While the removed auditor clearly exhibited poor judgement in accepting the appointment to the team, his/her reputation should not remain compromised in the absence of incriminating evidence.

Source: Beim, 1998: 258–259, 268–269.

Best Practice and Benchmarking

Best practice originated in business theory and corporate planning, with clear implications for the service sector, as well as for the natural resources sector (Pigram, 1996). It is a term that is used to describe how well a company can perform in time, space and in view of resource availability. Wight (2001) illustrates that one single best practice does not exist in tourism because the tourism industry differs on the basis of geography, politics, vision, culture, environment, technology and so on. But she says that best practices have been shown to obtain superior results, which often means more profit.

Best practice benchmarking (BPB) is a technique used by firms to become as good or better than other companies in many elements of their operations. Wight suggests that 'profitability and development comes from a clear understanding of how their operation is doing, not just against its performance in the previous year, but against the best they can measure' (2001: 153). The 'best' therefore may be the best in the country or the best in the world, and their success should act as a motivator for the firm to be better. Wight offers the following benefits of BPB in tourism, which enable service providers to:

- benefit from the practices of others so they themselves do not have to reinvent the wheel (i.e. in reference to costs and research);
- re-examine their current processes and operations, which often leads to improvement;
- accelerate change through revealing gaps in operations;
- view different ideas to better the current operation;
- make implementation easier and faster;
- better understand markets and competitors;
- develop a stronger reputation;
- gain faster awareness of important trends and innovations, and avoid the costs of making their own mistakes.

We gain an appreciation of how best practice could help in developing sustainable tourism enterprises in the work of Williams (1999), who writes that there are three stages in this process. The first is the identification of sustainable tourism principles; the second is the translation of these into practice; and the third is the development and implementation of environmental auditing systems, which force service providers to examine how active they have been in implementing such systems. For Williams, this may include a cocktail of approaches, including regulatory systems, conservation initiatives, innovative approaches to planning and design, carrying capacity strategies and facility operations, which, for example, reduce waste. These are the tangible mechanisms by which to operationalise ethics through tourism operations. And they are important because we as society deem them to be important; they give us greater market share or they allow us to provide consistency across a jurisdiction.

As noted by Pigram (1996), industrial countries have been promoting best practice environmental management since the adoption of Agenda 21 at the Earth Summit in Rio de Janeiro in 1992. Like the aforementioned authors, Pigram notes that best practice, in the context of environmental governance and management, offers a framework for the realisation of environmental goals, such as cleaner technologies, recycling and conservation of resources. Organisations must be committed to environmental excellence; make environmental responsibility a concern for all levels of the firm; develop environmental action plans; adopt audits and indicators; acquire skill sets related to the natural environment; and support staff who are involved in processes of change for excellence.

The Tourism Operators' Initiative for Sustainable Tourism Development (2002), supported by UNEP, UNESCO and WTO, is an international group dedicated to assisting tour operators from around the world to develop best practices for sustainable tourism. Objective 1 of their mandate highlights the importance of best practice for attaining their goals:

(1.1) Facilitating the exchange of best practices adopted to implement the principles for sustainable tourism and the Initiative's Statement of Commitment.
(1.2) Developing new management tools and adapting existing ones, specific to the industry (EMS, ISO 14001, etc.) for tour operators' own operations, and for sustainable management practices in the supply chain.
(1.3) Providing a platform for dialogue with other partners.
(1.4) Involving tour operators in the relevant programmes of work of UNEP, UNESCO, WTO.
(1.5) Assessing progress made on a regular basis.
(1.6) Facilitating partnerships between members to address issues with a common voice. (Tourism Operators' Initiative, 2002)

The commitment of the Tourism Operators' Initiative to sustainability is emphasised in the pursuit of best practices that use natural resources responsibly, reduce and prevent pollution and waste, conserve biodiversity, conserve cultural heritage, involve local communities and use local products and skills.

The main drawbacks of best practice are that the organisation must be in the programme for a prolonged period of time before being recognised with an award, and only after receiving the award can the company begin to act to improve performance (Font & Tribe, 2001). While it is debatable whether these are drawbacks, depending on the perspective one maintains, perhaps more central in terms of what is ethical or not is what is viewed as best. Is a company that is identified as a 'best practice' unit a more ethical form of enterprise than others? This is a question that we need to consider closely. Is it like saying that laws are all ethical because they have been passed by the legal arm of a government. Mechanisms such as best practice and certification should not only be about efficiency and development, but also about citizenship, fairness, justice, altruism and so on. The extent to which best practice for environmental management or sustainability may be realised too often rests on the profit margin.

The Precautionary Principle

One of the challenges for industry is the assessment of risk relative to: (1) the knowledge of the citizenry regarding the product; and (2) the ethics behind selling a product that may have a deleterious impact on the well-being of humans or other beings. At what level, therefore, is the product acceptable? Should we adopt a zero-risk set of standards in marketing our product? Do we establish risk on the basis of benefits or costs to us? Is it appropriate to expose certain segments of the population to higher levels of risk than other segments?

The precautionary principle was conceived in Germany (*Vorsorgeprinzip*, meaning precautionary principle) during the 1970s for the purpose of exercising foresight in matters of environmental policy and resource protection (Boehmer-Christiansen, 1994). It was introduced internationally in 1984 at the First International Conference on Protection of the North Sea (Tickner & Raffensberger, 1998). Since then, the principle has been extended into national and international environmental policy by more than 40 countries and is now affirmed in many international treaties and laws (e.g. the 1990 Bergen Declaration; the 1992 Rio Declaration; the 1992 Maastricht Treaty on European Union; the Convention on International Trade in Endangered Species of Wild Flora and Fauna and in 1992 at the Earth Summit in Rio) (Dickson, 1999; Ellis, 2000; Freestone & Hey, 1996; Rogers *et al.*, 1997; Tapper, 2001).

The precautionary principle has been defined as 'a culturally framed concept that takes its cue from changing social conceptions about the appropriate roles of science, economics, ethics, politics and the law in pro-active environmental protection and management' (O'Riordan & Cameron, 1994: 12) and contains the following core elements (VanderZwaag, 1994: 7):

- a willingness to take action (or no action) in advance of formal scientific proof;
- cost effectiveness of action, that is, some consideration of proportionality of costs;
- providing ecological margins of error;
- intrinsic value of non-human entities;
- a shift in the onus of proof to those who propose change;
- concern with future generations;
- paying for ecological debts through strict/absolute liability regimes.

The NGO Steering Committee Tourism Caucus (1999: 2) has observed, in reference to the health impacts of tourism, that 'All environmental initiatives must include a health component of exposed populations taking into account the environment-health precautionary approach as there are clear casual links between the degradation of the environment and disease'. This group goes on to suggest that government intervention is needed to provide safe drinking water, sanitation and waste disposal in efforts to contain diseases such as malaria and dengue. It is essential that the use of DDT and other hazardous chemicals be eliminated in efforts to eradicate adverse health effects on humans and the environment.

Although the precautionary principle makes intuitive sense, it has been hampered by the notion that it stands in the way of economic productivity because of issues related to uncertainty, risk, cost-benefit analysis and science (Fennell & Ebert, 2004). In regard to cost-benefit analysis, regulators must not use financial methods only to gauge health and environment issues. The more utilitarian or instrumental groups, such as industry, push harder for cost-benefit analysis in the precautionary principle only when there is reputable scientific evidence pointing to serious or irreversible damage. Other groups argue that precaution should be used when there is 'reasonable suspicion' that a product may cause harm. Science becomes the most important agent in determining what is acceptable and what is not. But although we may place unlimited stock in its ability to be objective and pave the way to truth, the simple fact is that science does not always produce data or establish cause in a timely manner in order to protect human health or the environment (CBD, 2001). A case in point is the cod fishery in Canada which was managed from a form of precaution based on science alone. More recently, the federal government has adopted a policy hinged on science-based risk management that is guided by judgement based on values and priorities (Government of Canada, 2001b).

In tourism, despite the few examples that mention the precautionary principle, there is little research available. One could argue that principles such as low impact, sustainability, local control, responsibility, non-consumption, and so on, are implicit applications of the precautionary principle. In reality, however, what is needed is a more explicit understanding of how it applies in a tourism context, especially in reference to many of the following questions (Fennell & Ebert, 2004):

- What in the community will be sacrificed for this tourism development enterprise?
- What are the anticipated direct and indirect social, economic and ecological impacts?
- Who inside and outside the community has been consulted, over what period of time?
- Who will be compensated for loss (included here is the environment) and how? Is the possibility for loss built into the proposal? How?

- What legal or policy directives exist on precaution in the region as they apply to the project?
- What is the political and industry receptivity to the precautionary principle?
- How does the precautionary principle interface with other regulatory or non-regulatory initiatives on tourism and development in the region?
- How can precaution be built into these existing structures or vice versa?
- What are the opportunities to implement the precautionary principle across different spatial, political and economic scales?
- How can the precautionary principle act as an agent to catalyse a better relationship between natural and social scientists, different types of science and multiple stakeholder groups?
- How can the precautionary principle ensure that these groups are allowed to have more control through equal decision-making power as regards the influences on their communities?
- How is it that the scientific data that drives decision-making might be more accessible to the public?
- Have stakeholders taken into consideration the concept of reversibility? Developments should not so damage the integrity of the natural environment that it cannot, over time, be reinstated to its original condition.
- Who has the knowledge to effectively plan with the interests of the community in mind?

The precautionary principle has led to a backlash in industry because it accentuates the process of pulling back the reigns on unfettered growth. As such, while no caution is dangerous, too much caution may be equally counterproductive. With the rise of technology, humanity has substituted the artificial world of science and technology for the real world – or that which lies outside of the human influences, which Heidegger refers to as 'Being'. The precautionary principle should be a mechanism that attempts to keep us from going too far down the road of technology, science and inauthenticity. As Heidegger has noted, only when humanity has learned to be 'calm', i.e. when people no longer attempt to be master of nature through the aid of technology, will they have a chance to be the guardian of 'Being'. The precautionary principle affords us the opportunity to step back and re-evaluate the state of being not just from the perspective of growth, but also in full view of the costs that may accrue in the present as well as in the future.

Conclusion

The issues surrounding common pool resources or social dilemmas surface regularly in ecological economics and resource management. The same, however, cannot be said of tourism, despite the connection between such resources and the tourism industry. In this chapter, the history behind common pool resources was discussed along with the concept of social traps and the Tragedy of the Commons (in reference to the Prisoner's Dilemma). Selected tourism works are included in the discourse that relate to the aforementioned concepts (again, from what appears to be a limited pool), along with examples within places that have been affected by these resource issues. Governance was also included as a broad-based concept and issue, again in reference to resource economics and management, with implications for the tourism field. The pros and cons of accreditation were discussed in the chapter, with particular attention paid to those schemes that have yielded the most success and controversy. This treatment was followed by a review of best practice and some relatively new knowledge on the precautionary principle and tourism. These, it is argued, are policy- and governance-based strategies that are felt to provide a link to the resource issues identified previously in the chapter, which may thus provide some leadership on how to tackle some of the most emergent problems in the tourism industry.

9 Codes of Ethics

Tourists are like lichens: they are hearty as hell and you find them in the damndest places. While there are about 20,000 species of lichens, thankfully there are only just over 100 of tourists

Adapted from a biological source, details unknown

Introduction

In the previous chapter, accreditation was examined as one of the most important strategies to manage the impacts of tourism (Mihalic, 2000). Codes of ethics was said to be one of the least sophisticated of these strategies; however, although not sophisticated, codes of ethics demands a chapter of its own because of the sheer weight of its influence in tourism. As such, ethics in tourism has traditionally taken the form of codes of ethics or codes of conduct developed by industry, government, NGOs and researchers. Although it may be argued that these instruments have not been established from a theoretical foundation, they have provided a significant degree of industry guidance. This chapter examines the many different sides of codes of ethics, including pros and cons from a multidisciplinary perspective. An effort is made to contextualise codes within a more broadly based ethical continuum. This is accompanied by a code development process section, as well as an illustration of a comprehensive ethical programme, including the code of ethics as just one aspect. The chapter concludes with a brief overview of the World Tourism Organization's Global Code of Ethics for Tourism.

What Are Codes of Ethics?

One of the most recognisable codes of ethics was devised by the Greek Hippocrates, a physician in the 5th century BC from the island of Cos, Greece, who pledged the medical profession to the preservation of life and the service and well-being of humankind (the Oath of Hippocrates or Hippocratic Oath). Although it is said that Hippocrates is the author of this oath, theorists note that it was most likely written about 100 years after his time (Veatch, 2003). The oath is divided into two parts: an oath of initiation and the code section of the oath, the latter of which includes directives on dietetics, pharmacology and surgery. In reference to pharmacology, the oath states that:

> I will never give a deadly drug to anybody if asked for it, nor will I make a suggestion to this effect. Similarly I will not give to a woman an abortive remedy. In purity and holiness I will guard my life and my art. (Veatch, 2003: 193)

During the 1790s, Thomas Percival gained notoriety by developing a formal code of ethics for his hospital in Manchester, England, in response to the typhus and typhoid epidemic of the time. This had a significant effect on the organisational structure of the hospital, including the roles and responsibilities of nurses, physicians and various levels of administration.

Other fields have much less of a history in regard to codes of ethics. In public administration, for example, the first code was developed by the International City Managers' Association in 1924 (Pugh, 1991). In recreation, the Country Code, as closer to our discussion here, was developed by the Countryside Commission in the UK in 1951 (formally the National Park Commission). The purpose of such was to mandate how those who spend time in the countryside should act in appropriately respecting the natural environment. The code is shown in Box 9.1.

Box 9.1 The Country Code

Leave livestock, crops and machinery alone.
Take your litter home.
Help to keep all water clean.
Protect wildlife, plants and trees.
Take special care on country roads.
Make no unnecessary noise.
Enjoy the countryside and respect its life and work.
Guard against all risk of fire.
Fasten all gates.
Keep your dogs under close control.
Keep to public paths across farmland.
Use gates and stiles to cross fences, hedges and walls.

Source: Mason and Mowforth (1996).

The Country Code has recently gone through another incarnation, now called the Countryside Code, developed by the Countryside Agency and the

Countryside Council of Wales. This code was launched on 12 July, 2004, as part of the process to help prepare people for the introduction of the public's new right of access to the countryside (Countryside Rights of Way Act of 2000). This new code, formulated through significant stakeholder involvement, is built around values such as respect, protection and enjoyment, the result of which is an increase not in the number of guidelines, but rather in the detail built into each (see https://www.gov.uk/government/publications/the-countryside-code). Two main sections are included in the Countryside Code, one for the public and the other for managers:

For the public:
Be safe – plan ahead and follow any signs
Leave gates and property as you find them
Protect plants and animals, and take your litter home
Keep dogs under close control
Consider other people

For land managers:
Know your rights, responsibilities and liabilities
Make it easy for visitors to act responsibly
Identify possible threats to visitors' safety

Ethics, conduct or practice?

Inconsistency in the terminology of codes in tourism research is an extension of the dialogue in other fields. Codes have been designed as codes of conduct, codes of practice or codes of ethics, and characterised by content and focus. Codes of ethics are said to be different from codes of conduct and codes of practice, based on the belief that the former are more philosophical and value-based, while the latter are more applicable and specific to actual practice in a local situations (Scace *et al.*, 1992). An example of a guideline falling within a code of ethics would be 'respect for the frailty of the earth', whereas a code of practice guideline would likely be oriented more towards acceptable business practice with reference to the organisation's 'commitment' to the customer.

In support of this, Bayles (1981) observes that codes of ethics are based on principles that prescribe a sense of responsibility, rather than precise conduct, and include actions formulated from a set of principles that incorporate prevailing values to serve as moral yardsticks for everyday life. More specifically, Bayles suggests that codes are meant to translate more formal philosophical theories of ethics into guidelines that can be applied to day-to-day decision-making. Codes usually mandate behaviour above that of the law, encouraging higher ideals of normative convention, and are the result of years of influence and modifications in many fields, including science, medicine, government, business and the legal profession (Backoff & Martin, 1991). In each of these professions, codes of ethics typically originated from public outcry over irresponsible practices and consumer deception. As a result, governments have influenced the development and revisions of codes to foster public confidence and responsible practices.

The British Columbia Ministry of Development, Industry and Trade (1991: 2–1) defines a code of ethics as 'a set of guiding principles which govern the behaviour of the target group in pursuing their activity of interest'. From an industry perspective, they function as 'messages through which corporations hope to shape employee behaviour and effect change through explicit statements of desired behaviour' (Stevens, 1994: 64). In tourism, such codes have gone beyond the realm of business to ensure that local people, government and tourists abide by predetermined guidelines. There has been a proliferation of such codes over the past few years from a variety of organisations, including government, NGOs and industry, many of which can be found in the work by Mason and Mowforth (1995), and the United Nations Environment Programme (UNEP) on Industry and the Environment (1995).

The Government of Canada (1998), in association with Industry Canada, has developed a comprehensive guide on voluntary codes of ethics with reference to their development and use. The document makes little distinction between codes of conduct, codes of practice, voluntary initiatives, guidelines or non-regulatory agreements. This document notes that codes are usually developed in response to pressure from consumers or competitors, or for fear of regulations or trade sanctions. Conversely, and in a review of codes of ethics for tourism in the Arctic, it was found that, although codes of ethics are usually part of a strategy to regulate tourism, there needs to be a clear distinction between regulations, codes of conduct and guidelines (Mason, 1997) (see also Mason & Mowforth, 1996; Stonehouse, 1997). Regulations, by contrast, have some legal status behind them.

Citing the work of UNEP, Mason (1997) illustrates that codes of ethics have five main objectives: (1) to serve as a catalyst for dialogue between government and other tourism bodies; (2) to create awareness in government and the industry of the need for sound environmental management; (3) to heighten awareness among tourists of the need for appropriate behaviour; (4) to make host populations aware of the need for environmental protection; and (5) to encourage cooperation between government agencies, host communities, industry and NGOs. These objectives are said to rest on a number of key values. Some of these values as identified in the literature include justice, integrity, competence and utility, in illustrating that the tourism industry must: (1) recognise that its basis is a limited resource, the environment, and that sustainable economic development requires limits to growth; (2) realise that it is community-based and that greater consideration must be given to the sociocultural costs of tourism development; and (3) recognise that it is service-oriented, and that it must ethically treat employees as well as its customers (Payne & Dimanche, 1996: 997).

Genot (1995) underscores the practical reasons for the tourism industry to adopt codes of conduct citing, among other factors, a sound environment, which means good business, meeting consumer demand, unifying industry efforts and image, and assuring product quality. Genot feels that the following principles are at the core of any code of ethics: environmental commitment, responsibility, integrated planning, environmentally sound management, cooperation between decision makers and public awareness. Many of these principles are elaborated upon in other publications devoted specifically to the ethical conduct of operators and other members of the tourism industry (see Dowling, 1992). Genot further suggests that codes of ethics will be more successful if they are action-oriented and positive in their tone. Also, implementation is a key factor for those who develop such codes, and, as these are public statements of ethics rather than legal ones, it is essential for service providers to buy into their meaning and significance.

One of the first countries to make a concerted effort to link tourism and codes of ethics was Canada. The reason stated in Canada's National Round Table on the Environment and Economy, a response to the Brundtland Report of 1987 (supported by the Tourism Industry Association of Canada (TIAC), and the International Institute for Peace through Tourism at their 1991 meeting), for the use of codes of ethics in Canada's tourism industry, include (TIAC, 1991):

- enhancing Canada's image as a destination and its ability to compete in international markets;
- ensuring Canada's capacity to provide quality tourism products and services in both the short and long term;
- attracting tourists who are increasingly seeking environmentally responsible tourism experiences;
- providing a source of motivation, and team spirit, for staff at all levels;
- improving the quality of life within host communities;
- reducing costs through more efficient practices for energy conservation, water conservation and waste reduction.

TIAC's codes of ethics for tourists and the industry are outlined in Boxes 9.2 and 9.3. The importance of the code for service providers is reflected in the more comprehensive set of guidelines that has been specified for this group in comparison to tourists, perhaps setting an expectation for heightened ethical awareness and responsibility on the part of the former over the latter.

Box 9.2 Code of ethics for tourists: Sustainable tourism

(1) Enjoy our diverse natural and cultural heritage and help us to protect and preserve it.
(2) Assist us in our conservation efforts through the efficient use of resources including energy and water.
(3) Experience the friendliness of our people and the welcoming spirit of our communities. Help us to preserve these attributes by respecting our traditions, customs, and local regulations.
(4) Avoid activities which threaten wildlife or plant populations, or which may be potentially damaging to our natural environment.
(5) Select tourism products and services which demonstrate social, cultural and environmental sensitivity.

We wish you a pleasurable and rewarding visit and look forward to welcoming you again soon.

Source: TIAC (1991).

Box 9.3 Code of ethics for the industry: Sustainable tourism

The Canadian Tourism Industry recognizes that the long-term sustainability of tourism in Canada depends on delivering a high quality product and a continuing welcoming spirit among our employees and within our host communities. It depends as well on the wise use and conservation of our natural resources; the protection and enhancement of our environment; and the preservation of our cultural, historic and aesthetic resources. Accordingly, in our policies, plans, decisions and actions, we will:

(1) Commit to excellence in the quality of tourism and hospitality experiences provided to our clients through a motivated and caring staff.
(2) Encourage an appreciation of, and respect for, our natural, cultural and aesthetic heritage among our clients, staff, and stakeholders, and within our communities.
(3) Respect the values and aspirations of our host communities and strive to provide services and facilities in a manner which contributes to community identity, pride, aesthetics and the quality of life of residents.
(4) Strive to achieve tourism development in a manner which harmonizes economic objectives with the protection and enhancement of our natural, cultural and aesthetic heritage.

(5) Be efficient in the use of all natural resources, manage waste in an environmentally responsible manner, and strive to eliminate or minimize pollution in all its forms.
(6) Cooperate with our colleagues within the tourism industry and other industries, towards the goal of sustainable development and an improved quality of life for all Canadians.
(7) Support tourists in their quest for a greater understanding and appreciation of nature and their neighbours in the global village. Work with and through national and international organizations in helping to build a better world through tourism.

Source: TIAC (1991).

What the preceding discussion illustrates is that there is not universal acceptance regarding the nomenclature behind codes of ethics. Regardless of the type or message, we can perhaps agree that codes of ethics should be value-laden and therefore be established on the basis of a solid moral platform, whether that be for behaviour that transcends law, or behaviour that is specific to a particular industry situation. For the purposes of this discussion, codes of ethics, practice and conduct will be treated synonymously and referred to as codes of ethics.

Tourism Studies on Codes of Ethics

In the last three decades the importance of applied ethics in the service industry has grown tremendously in response to the consuming public's growing suspicion and perceived empowerment. As a reaction to the heightened concern and awareness of ethics in business contexts, codes of ethics have been developed to provide moral guidance to organisational members and consumers of the services provided, as noted above. Tourism as a point of departure to foster ecological awareness has an extremely important vested interest in having ethically sound business and educational practices. As a result, the application and implementation of codes of ethics has direct significance to the tourism industry (Fennell & Malloy, 2007). Results of market research suggest that 45% of tourists are more likely to book a holiday with a tour company that has a written code of ethics that guarantees that the operator minimises environmental impacts and maximises wages and working conditions. As many as 55% of the sample noted that they would be willing to pay 5% more for holidays that guarantee the abovementioned conditions (Tearfund, 2000a).

In a study of 10 major whale-watching operators in Johnstone Strait, British Columbia, Canada, Gjerdalen and Williams (2000) investigated service providers' views concerning the utility of the code in encouraging appropriate behaviour, as well as the factors that influenced operators to comply with the code. The authors discovered that whale-watching operators felt that direct legal sanctions were best for encouraging compliance of a range of different sanctions, followed by exclusion from the whale-watching network and publicising the names of non-compliers. Of a list of normative sanctions, 'personal guilt' was found to be most severe, followed by personal disruption and shame/embarrassment. Respondents also felt the need to comply with the code because they felt it was fair and appropriate for managing the situation. Although respondents said the rewards for complying exceeded the costs, they were not at all in agreement that, by following the code, business had increased. Despite this, there was a feeling among whale-watching operators that the code offered optimism about the future of their business; that there was a shared sense of conservation among the operators; and that it is an effective community-based, self-regulating management tool for the region.

As front-line workers in the travel and tourism industry, travel agents need to engender a significant level of trust if they are to be thought of as knowledgeable professionals (Dunfee & Black, 1996). In order to secure this trust it is essential that they operate on the basis of standards developed from within the industry, especially as the relationship between buyers and sellers places salespeople in morally precarious positions (Wotruba, 1990). By subscribing to standards that come from within, travel agents will have an easier time respecting these standards and internalising them. Professionalism, therefore, can partly be demonstrated through a significant degree of self-regulation. This is an important consideration because it places the onus on the individual or service to abide by a set of shared beliefs and values. Those who choose not to abide by these standards will have breached the travel agents' social contract. In so doing they may be fairly characterised, within the bounds of the establishment, as unethical.

But it is not just specific aspects of the travel trade that have an interest in codes of ethics, but also individual countries. Both Japan and Germany have developed codes of ethics that need to be adhered to by the countries who desire to attract Japanese and German tourists (Fleckenstein & Huebsch, 1999). Implemented by the Japan Association of Travel Agents, Japan's code says that if a tour is unable to deliver what it promises, then a percentage of the price of the tour must be returned. Similarly, the German Federal Book of Travel Law is said to leave little room for negotiation under the same conditions as specified in Japan.

Primary research in the area of tourism codes of ethics has been largely reliant on content analysis methodology. In a content analysis of 42 codes of ethics (Stevens, 1997), 26 of which were from hotels and the other 16 from management companies, it was found that 96% of codes did not discuss civic and community affairs, 88% did not discuss environmental affairs, while 96% did not discuss guest safety. Where more emphasis was placed, relative to other categories, was on competitor relations, payments or political contributions and personal character. Management codes were found to place a similarly low level of emphasis on environmental issues (6%) and civic and community affairs (12%), with a main conclusion from Stevens that these companies would be wise to focus on environment and community in their codes because of the close tie with these aspects of the industry.

Other researchers have analysed codes of ethics using content analysis on the basis of who guidelines were developed for, who guidelines were developed by, the type of tourism they were directed towards, the orientation of the code, the mood of the message and the main focus of the guideline (Malloy & Fennell, 1998a). These variables were juxtaposed with two philosophies of ethics, deontology and teleology, and whether the codes were relevant to a local condition ('local' meaning local, regional and national scales) or cosmopolitan (meaning a universal code).

Deontological codes included those that specified certain duties (e.g. be aware of the periphery of a rookery or seal colony and remain outside it. Follow the instructions given by your leaders). Conversely, teleological codes were those that specifically illustrated a consequence from their performance or non-performance (e.g. for the statement 'Do not enter buildings at the research stations unless invited to do so. Remember that scientific research is going on, and *any intrusion could affect the scientists' data*', the consequence of the action is highlighted in italics). In this research, the authors discovered that approximately 77% of all guidelines were deontological in nature. This led the authors to conclude that the deontic nature of the statement failed to provide the decision maker with the rationale for abiding by a particular code, with the implicit assumption that an explanation of consequence (i.e. the teleological premise) is unnecessary to the tourist or other stakeholder. Other conclusions illustrate that most of the codes were developed by associations (NGOs) and for tourists; many of the codes were ecologically based, rather than socially or economically based; about 85% were stated using positive language and the focus of such codes was on people or the resource base. The authors recommend that future research in the area of codes of ethics should explore the deeper philosophical meaning of codes, advocate more of the teleological perspective in their development (i.e. the consequences of one's action if the code is not adhered to) and better understand the extent to which codes actually affect or change behaviour. To this we can add that codes should be simple and clearly understandable if we are to expect to educate and affect behaviour (Tribe *et al.*, 2000). Although this latter comment is simple and straightforward (i.e. educate and affect behaviour), it is an important one. For this reason the teleological or consequential aspects of codes is thought to provide a value-added dimension that is lacking in deontologically based codes, as noted above.

The WWF (n.d.) Code of Conduct for Arctic Tourists provides an excellent example of the teleological component that can be built into a code of ethics. The second guideline asks that tourists 'Support the preservation of wilderness and biodiversity'. This guideline is supported by the following information, which addresses the 'how' and 'why' questions that are essential in getting the message across to tourists, but which are so often neglected in the construction of codes of ethics:

- Learn about efforts to conserve Arctic wildlife and habitat, and support them by, for example, giving money, doing volunteer work, educating others on conservation or lobbying governments and business.
- The large undisturbed wilderness areas of the Arctic are a unique environmental resource. Oppose developments that fragment these areas or that may disrupt wildlife populations and ecosystems.
- Visit parks and nature reserves. Visitor demand and tourists' expenditures support existing protected areas and can lead to the protection of additional nature areas.

A content analysis methodology was also employed by Garrod and Fennell (2004) in their work on whale-watching codes of ethics used around the world. The authors discovered that, although whale-watching is growing at an alarming rate, it has done so from a patchy regulatory framework, which has raised concerns over the sustainability of the industry. One of the main concerns was the lack of a universal code of ethics for this activity. This is thought to be important in minimising the various regionally based effects on different cetacean populations. Indeed, this goes hand-in-hand with what appears to be a dearth of research on the basic biology of many species of cetacean, as well as the impacts that are borne out of the whale-watching experience itself. The authors concluded by observing that, although many theorists argue that the voluntary approach represents the best way forward for the regulation of whale-watching, it must progress by maximising the benefits of the voluntary approach while avoiding its many pitfalls. Perhaps the best means of achieving this end would be through shared knowledge between social and natural scientists, which may ultimately contribute to a universal code of conduct based on sound scientific advice.

In concluding this section, perhaps one of the most useful applications of codes of ethics is one that has rarely been explored. McGregor (1997) notes that aboriginal environmental ethics constitute an important aspect of traditional knowledge. This was recognised in the Brundtland Report, which underscored the importance of learning from such cultures in managing complex ecosystems (WCED, 1987). An area of research that seems to be untapped concerns the benefits that could accrue through the combination of traditional and Western codes of ethics. This coproduction of knowledge as it applies in time and space could serve to help build partnerships with aboriginal people, and help to preserve traditional knowledge, as well as benefit the people and ecology of these places. This could include an analysis of virtues such as justice, integrity, competence and utility, among others, in structuring effective codes (Fleckenstein & Huebsch, 1999: 140–141). Some of these are reflected in the Gwaii Haanas Watchman Program, as discussed in Box 9.4.

Box 9.4 The Gwaii Haanas Watchman Program

The Gwaii Haanas Watchman Program was developed in order to facilitate the protection of natural and cultural heritage in the Queen Charlotte Islands, British Columbia, Canada. Home to the Haida Gwaii, an indigenous tribe on the west coast of Canada, the region is both a Canadian national park and UNESCO World Heritage Site containing magnificent cultural artefacts (e.g. totem poles) as well as biodiversity, which includes refugia from the Pleistocene era. The designation of this region as such has spawned a thriving tourism industry, which is thought essential in stabilising and diversifying the economic base of the region. However, in generating revenues, there has also been significant growth in visitors, and associated impacts, including the vandalism of Haida village sites. In an effort to control the impacts, the Haida nation and Skidegate Council initiated the Watchman Program, which includes the placing of Haida individuals at various posts spread throughout the region. The main responsibilities of the Watchmen are to guard the natural and cultural heritage as well as educate visitors by sharing their knowledge and personal experiences. The Program is an important part of the management plan of the park, as well as the co-management strategies that have been established by the park authorities and the indigenous people.

Parks Canada has developed a handbook in association with the Watchmen, which includes a voluntary code of ethics that was developed by tour operators, charter boat owners, guides and the Haida Watchmen, outlining proper conduct in relation to the park and its cultural and natural attributes. As noted by Hoese, the code is essential in stressing the importance of 'general etiquette; behaviour towards wildlife and at archaeological, cultural and historic sites; visitor safety; and practices regarding food gathering, garbage and camping'. Those tour operators who choose not to abide by the code of ethics are shunned by other operators, or refused information in the network.

From over 20 years of involvement in the Watchman Program, there have been many lessons learned. The most important successes include:

(1) As guardians, the Watchmen have ensured the cultural and ecological integrity of the area.
(2) The Program has encouraged the involvement of Haida elders and youth, allowing for a cross-cultural experience which benefits indigenous people and visitors.

(3) The economy has benefitted through employment and business opportunities.

The challenges include:

(1) Because of the high level of visitation, quotas have had to be established, tour operators need to be licensed, and first-time visitors need to attend a pre-trip orientation.
(2) Some island residents feel the growing number of tourists takes away from their quality of life.
(3) Monitoring visitor impacts has been difficult, without good baseline data, with the need for more comparative data in the future.

Source: Adapted from Hoese (1999: see p. 105 in reference to the direct quote).

Other Studies

An intriguing study outside the literature, but which has appeal to tourism studies, is the work of Fry (2000), who developed a typology of cross-cultural conflict resolution strategies that can be used in different capacities within different regions of the world (and in reference to the preceding section on traditional knowledge). The value of the study lies in its ability to articulate a series of strategies that can be used apart from or in addition to codes of ethics for appropriate behaviour in many non-Western contexts. He lists nine of these strategies:

Avoidance. The temporary or permanent avoidance of interaction with a disputant. For example, this takes place in Finnish, Fijian, East Indian and Mexican cultures.
Tolerance. The issue under question is simply ignored and the relationship with the offending party is sustained (e.g. Bolivia).
Self-help/coercion. A unilateral action to handle a grievance that may ultimately lead to cycles of violence manifested in feuds.
Negotiation. Involves handling a dispute through joint decisions by disputants and results in mutually agreeable outcomes.
Friendly peace-making. A third-party role that separates or distracts adversaries. The third party stays independent of the issue.
Mediation. Neutral third parties assist antagonists in arriving at an agreement. Mediators do not make rulings and do not have the authority to impose an agreement.
Arbitration. A third party renders a decision but cannot enforce it.

Adjudication. A judge offers a recommendation and is empowered to enforce it.
Repressive peace-making. As the most authoritative method, the third-party settlement role is defined by governments who interject for the purpose of imposing peace on warring indigenous peoples.

In these strategies we gain an appreciation of the breadth of cultural reconciliation rituals that can take place in order to restore the peace. These include such acts as gift-giving (e.g. liquor, kava, whales' teeth), payment (e.g. pigs, horses), sharing food or drink, physical contact (e.g. kissing, shaking hands), appeasement postures (e.g. bowing one's head, bowing and looking down), expressions of apology, remorse or contrition, and the participation of other persons besides the disputants in reconciliation. This study led Fry to observe that written codes and court systems are not the only way to maintain social order. And there might be something in these strategies for tourists who find themselves in situations that demand clarity and compromise, which could be articulated in codes of ethics. The problem with these, however, like voluntary codes of ethics, is that they are difficult to enforce and parties can always decide to walk away if the outcome is not in their favour. Notwithstanding, it would be both interesting and helpful to explore more examples of these within tourism research, especially for forms of tourism, such as adventure tourism, cultural tourism and ecotourism, which draw tourists into many remote regions.

Although cultural studies disciplines have a great deal to offer in the discussion of codes of ethics, it appears that the majority of work that is salient to tourism has been undertaken in business. This has resulted, as noted previously in the book, because of the recognition of the unethical nature of businesses, which has thus necessitated the development of a many-pronged approach to ethics through academic journals, university programmes and a focus on corporate codes of ethics. For example, research suggests that, during the 1960s, polls indicated that 42% of Americans felt that 'most businessmen will do anything – honest or not – for a buck'; in the 1980s, 56% of respondents felt that businesses were 'fair' or 'poor' at contributing to the communities in which they were located; and, in the 1990s, only approximately 20% felt that business people had 'high' or 'very high' ethical standards (Stevens, 1997; see also Baumhart, 1968; Lipset & Schneider, 1987; McAneny, 1992, respectively).

Despite the volume of these transgressions, theorists assure us that the value of business ethics lies not only in the benefits to consumers, but also to the entire business community, which relies on trust and confidence (Bowie, 1978). In striving to attain an ethical business community in this regard, codes of ethics are seen as essential for all involved. If only one firm elected to subscribe to such a code, it would not be to their competitive advantage because they would suffer at the hands of competitors who maintain a more relaxed policy on moral business practice (the Prisoner's Dilemma conundrum as reported in Chapters 2 and 8). If only one tourism company has invested in low-emission jet boats, they open themselves up to higher costs in the face of a competitive market. The approach for these companies is to convince a better educated and more environmentally conscious clientele that it is in their best interests to pay higher prices for this technology.

In other related research, Adams *et al.* (2001) investigated the effects of a code of ethics on the perceptions of ethical behaviour in organisations. These authors discovered that individuals in companies that had a code of ethics felt that the presence of a code had a more positive impact on what constitutes 'ethical behaviour' than individuals in companies that did not have one. More specifically, organisations with codes rated company support for ethical behaviour, levels of satisfaction regarding outcomes of ethical dilemmas, freedom and the ethical behaviour of their colleagues higher than companies that did not have such codes. On the question of improvement, Brookes (1991) notes that, in a survey of 84 CEOs in Canada, 92.9% said that their existing code of ethics could be better developed. In this regard, Brookes discovered that 59.9% of respondents said that improvement could take place through training programmes; 31% through compliance mechanisms of one form or another; 49.4% through measures of performance; and 51.2% through reports of performance.

There is also evidence to suggest that there are variances in the factors affecting ethical management practices between Western and Eastern countries. In research that sought to uncover ethical management practice differences between Sri Lanka and Australia, researchers found that, although 33.8% of Sri Lankan businesses had a written code of ethics as compared to 28.7% for Australia, 36.8% of Australian firms had a forum to discuss ethical issues, while only 26.3% of firms in Sri Lanka had the same. The authors further observed that 21.4% of firms in Australia recognised that their practices led to environmental impacts, while only 7.6% of firms in Sri Lanka said the same (in view of firm size and practices) (Batten *et al.*, 1999).

Pros and Cons of Codes of Ethics: A Critical Analysis

Some theorists have pointed to the fact that, although organisations use codes for the purpose of making the firm look good, the real aim is to make performance standards explicit for employees so that they cannot argue that no one actually told them how to behave (Brookes, 1991). While this has consequences for the organisation internally, inappropriate behaviour also has repercussions of lost confidence by the public. Still others feel that codes are unnecessary because of the corporate laws and regulations that currently exist in society. But in an era of public accountability, there appears to be an expectation by the legal field that codes of ethics be employed as a stop gap for ethical transgressions. For this reason firms have modified their codes two and three times in their efforts to decrease immoral acts.

One of the main reasons why codes of ethics fail in generating ethical behaviour is that they are often seen to have value in other areas. These areas include attracting employees, the organisation's public image, avoiding government regulation and boosting employee morale (Weaver, 1995). So while they may succeed in these other areas, there may be confusion on the ethical purpose of a code of ethics or how it should be worded to exact its ethical prime directive. In investigating whether the design of codes matters in aspects of justice perceptions and content recall, Weaver found that common alterations in the code of ethics are of little consequence for individual perceptions and beliefs. This, he notes, is in sharp contrast to non-empirical studies that have identified the importance of design factors in codes of ethics. A potential reason for the variability in perceptions and content recall can be found in the suggestion that codes of ethics have multiple functions within the organisation, and the leadership and context of the firm may greatly influence how employees interpret and respond to the code of ethics. This is supported by Wakefield (1976: 663), who notes that research has identified the following problems with codes of ethics: they fail to serve as effective supports of ethical behaviour; they serve diverse interests; they are vague and general; they tend to proscribe rather than prescribe; and they fail to answer specific questions on the behaviour of individuals and groups.

More broadly, Cooper (1987: 321) argues that the development of a code of ethics for public administration needs to include the following: (1) an understanding of ethical principles; (2) an identification of virtues that are supportive of those principles; and (3) analytical techniques that may be employed in specific situations to interpret the principles. He uses the concepts of practice, internal goods, external goods, and virtues to suggest that organisations need to do a better job at being ethical. These criteria provide the backdrop to an interesting application of internal and external goods to the practice of ethics in a public context. Internal goods include those that cannot be attained through purchase or acquisition (in any way) but rather must be developed through interaction with a practice and by submitting to its standards. These goods are important in that they lay the foundation for the development of ethical norms within the organisation; however, although such goods are essential as a foundation for the organisation, they are also good for the broader community. Conversely, external goods are those that do not contribute directly to the practice, and include such things as money, prestige, status and power, where there is competition and where there are winners and losers. These become the property of the individual, whereas internal goods become the property of the community of practice and the larger community. Although practices should be oriented more towards the internal goods and the relationships that evolve in securing them, Cooper illustrates that organisations are dominated by the pursuit of external goods for their survival and growth. The incorporation and consideration of virtues within the organisational setting are important and are required to cultivate attempts to achieve internal goods and at the same time to relegate external goods to a lesser position. As such, if an organisation has the obligation to pursue the public interest, internal goods necessary for the achievement of this might include the application of beneficence for the citizenry and justice. The virtues that might be employed in this regard could include benevolence, courage, rationality, fairmindedness and prudence.

The 1986 NASA flight is an example of where external goods were more important than internal goods. Senior officials in the Reagan Administration ordered the flight of the shuttle even though they knew that there were mechanical problems. Safety, redundancy (a standard for achieving excellence) and prudence, all internal goods, were apparently sacrificed so that the programme could ready the technology for commercial use in order to defray costs of operations. This led Cooper to conclude that the current trend towards the development of codes of ethics is hampered by a lack of focus on internal goods and virtues, which should be examined and cultivated and which need to form the basis of solid practices. Without these, members of the organisation are disconnected from the core of the practice, resulting in confusion, futility and self-deception. This is supported by others, who observe that codes of ethics are inadequate without a larger vision to give purpose and coherence (Millar & Yoon, 2000). The larger vision of the 'good' should be used to guide decision-making through vehicles such as codes of ethics.

Other theorists have argued that the modern institution, with its normative ethics structure, although designed to improve the moral standing of employees and the organisation, has taken away from the ability of the individual to take responsibility for his or her actions. Bauman (1993) refers to codes of ethics and other normative vehicles as the substitution of ethics for morality; morality being a concept that is rooted in broader issues and many other human qualities, as noted in Chapter 3. In this he says that:

> the modern mind is appalled by the prospect of 'deregulation' of human conduct, of living without a strict and comprehensive ethical code, of making a wager on human moral intuition and ability to negotiate the art and usages of living together – rather than seeking support of the law-like, depersonalized rules aided by coercive powers. (Bauman, 1993: 33)

Bauman recommends the disposal of imposed rules within the organisation or setting for the purpose of returning and placing responsibility for action squarely on the shoulders of the individual. The price we must pay for this is the loss of a sense of consistency and security. And intuitively this makes so much sense, because there is the feeling that we have lost the ability to be accountable. By taking a route to good behaviour, which is dictated by an external agent, it seems to defeat the purpose of a standard of conduct. This was examined by Kant, who felt that something is lost when we act in accordance with a rule rather than acting out of respect for a rule (see Chapter 3). In the case of the latter the issue is more a function of acting in order to escape punishment rather than from some internal drive and out of pure virtue. So virtue can be defined as the ability to act purely from some intrinsic drive to be good instead of some other extrinsic motivator.

Ironically, the codes of ethics employed by many tourism organisations for the betterment of the industry have been developed using instrumental reason. Holden (2003: 103) alludes to this in his review of the World Travel and Tourism Council's plan for environmental improvement, which establishes that a 'clean and healthy environment is essential for the furthering of tourism'. Note that it is not the intrinsic valuation of the environment for its own sake in this statement, one which is perhaps hinged on spiritual value or love, but rather a premise that positions the environment as an icon for furthering economic development through tourism. While some organisations such as NGOs believe in the intrinsic value of the natural world largely because of their mandates, private and public agencies may not.

This can be observed in the TIAC's (1991) code of ethics, where one can still sense an element of instrumental reason built into the rationale for applying these codes. For example, the first of these guidelines supports the competitive image clause, which prevails in the industry approach to business. This 'Canadian approach' to 'promoting' sustainable tourism is not necessarily the ethics that Holden had in mind in his discussion. This has prompted some tourism scholars to be sceptical as to the utility of codes of ethics, especially in areas such as ecotourism, where there are no answers to the confusion surrounding both ecotourism and sustainable tourism, but simply:

> a never-ending series of laughable codes of ethics: codes of ethics for travellers; codes of ethics for tourists, for government, and for tourism businesses. Codes for all – or, more likely, codeine for all. ... But who really believes these codes are effective? I am pretty wary of platitudinous phrases like 'we are monitoring progress'. Has there been any progress – indeed, has there been any monitoring? Perhaps I am missing it and the answer itself is actually in code. (Wheeller, 1994: 651)

Although not based on empirical research, it is difficult not to heed the words of Wheeller, and others in reference to codes of ethics. For example,

Castañeda (2012) argues that the UNWTO Global Code of Ethics must be viewed as a neoliberal manifesto directed towards unfettered development, and by extension, a directive towards increased conflict, because of its anthropocentric and development undercurrents. In other fields, such as business, there is a great deal more empirically based work on codes of ethics. For example, McDonald concludes that:

> given the empirical support of the role of corporate codes and policy for improving ethical perceptions and standards in organisations… . clearly, some form of normative structure in the form of policy and codes does have an impact on ethical attitudes and possibly behaviour. (McDonald, 1999: 145–146)

In tourism, there appears to be more support for codes of ethics by tourism researchers than not. As alluded to earlier, this comes as a result of the belief that codes of ethics: (1) are objective in that they have external transcendent values; (2) go beyond the individual in providing a basis from which the community can pass judgement in an organised and cohesive fashion; and (3) can promote courageous behaviour within the community (Chandler, 1983: 33–34). This has prompted theorists to conclude that the recent emergence of codes of ethics provides recognition that all forms of tourism have an impact on the resource base, and further that the industry should be accountable for mitigating these impacts, as noted above. Tourism, it is said, is not merely jumping on the bandwagon of ethics, but rather demonstrating the beginnings of a profound shift of attitudes on business and the environment (Williams, 1993b).

In an effort to examine the pros and cons of codes of ethics more closely, Box 9.5 is included, and based on two studies. Added together they provide a comprehensive breakdown of the main strengths and weaknesses of codes for tourism industry purposes.

Box 9.5 Pros and cons of codes of ethics

Good codes[a]

- Gain commitment of leader in the organisation.
- Gain commitment of front-line workers.
- Offer clear statement of objectives, expectations, obligation or rules.

Bad codes

- Are not backed by action.
- Lead to deceptive or misleading advertising.
- Bring bad publicity, leading to a loss of consumer trust.
- Slow/prevent implementation of laws.

• Are open and transparent. • Offer a continuous flow of information. • Offer a resolution system which is also transparent. • Outline meaningful benefits for participation. • State negative repercussions if principles are not complied with.	• Discourage competition and encourage collusion. • Hamper or prevent trade. • Create an uneven playing field. • May not be transparent or inclusive. • May not be enough to address non-compliance. • Can attract negative attention, for example, Tobacco Industry Voluntary packaging and Advertising Code.
Good codes[b]	Bad codes
• Provide a *common vocabulary* about what is right and wrong. • Offer a *framework for conflict resolution* and policy development. • *Clarify ethical issues* and help to resolve disagreement about moral dilemmas, decreasing unethical practices. • *Impose constraints* on and individual behaviour. • *Reduce uncertainty* regarding what is ethical or not. • *Suggest course of action* to very follow up on charges of unethical conduct. • *Facilitate improved cooperation* among parties, by enhancing ethical understanding of norms of action. • *Promote environmental awareness*, by sensitising the public to shared social and environmental values.	• Are little more than '*window-dressing*' and 'public relations gimmicks' designed to impress outsiders, but not to be taken seriously by practitioners. • Are *too abstract* – too broad, and difficult to apply in situations. • End up being *too vague*, and too weak in practical guidance. • *May be counterproductive*, if they formalise the status quo that they are attempting to change. • Are *difficult to enforce*, because they are often not covered by law. • Are *unnecessarily restrictive* on individual rights/freedom of choice. • *Unnecessarily complicate matters of management*, by introducing new rules and standards to be enforced. • *Are ineffective* in handling systematic corruption.

Sources: [a]Based on Government of Canada (1998); [b]Stefanovic (1997: 247).

If we are to better understand the pros and cons of codes of ethics in tourism, researchers will surely have to undertake more sophisticated studies. In this regard, the work of Farrell *et al.* (2002), reporting on a number of different strategies to gauge the effectiveness of codes of ethics, may prove useful. They found that empirical research on codes has concentrated on: (1) self-reporting by employees on their own individual behaviours, but which is subject to bias (see Weaver (1995) for a good description of this); (2) questionnaire methods, which seek to gather information from firms on the types of violations that have taken place, reasons for their occurrence, and the presence of organisational attributes, such as codes of ethics at the time of the transgressions (see Mitchell *et al.* (1996), for a good description of these processes); (3) the use of vignettes or hypothetical situations, which are used to gauge how the respondent would answer under certain conditions (see also Bersoff (1999), who nicely explains the importance and overall use of vignettes and hypothetical examples in this type of research. Here research has found that difficulty lies in demonstrating the link between reported responses to the items on these scales and actions under real conditions); and (4) testing differences in behaviours of two groups of companies: those that have a code of ethics vs. those that do not (this approach was used by Marnburg (2000), but focused on the attitudes of employees as a surrogate rather than actual behaviour). In view of the shortcomings of these methods, Farrell developed a method that employed direct observation of the behaviour of managers and employees from eight large Australian firms on the basis of several defined behavioural patterns recorded on a Likert scale. While this study itself failed to find associations between the consistency ratings of the firms and their distinct strategies in ethics, it does stand as a viable option that may be useful for overcoming the biases and challenges that are seen to exist in other methods.

Contextualising and Operationalising Codes of Ethics

This section illustrates three different processes related to the contextualisation and operationalisation of codes of ethics. The first process deals with codes of ethics as one of a few different levels of moral discourse. It follows from the work of Fox and DeMarco (1986), who maintain that there is a natural tendency to search for philosophically 'higher ground' in resolving issues, beyond the utility of codes of ethics. A comprehensive treatment of this need has been developed by Veatch (2003) in his work on bioethics, which can serve to illustrate the connection between different levels of moral discourse. The second section illustrates a formal process by which to develop codes of ethics; the last section articulates a comprehensive ethical programme, which may be used by any tourism or non-tourism organisation for the purpose of training.

Veatch writes that there are essentially four different levels of moral discourse in making ethical choices. The first of these he refers to as cases (casuistry). In situations where the ethically correct course is obvious, practitioners can rely on well-ingrained moral beliefs in their decision-making. This may come from other cases that were clinically similar, biblical stories or legal cases. The hallmark of this is that, if the features of the cases are similar, then cases should be treated the same – providing the origin outcomes were positive. For example, it may be that we can all agree in tourism that taking pictures of people without their permission is unethical. The ethically correct course is obvious. However, if we cannot agree on an appropriate type of intervention, because of the lack of guidance from the past, or because of the severity of the situation, Veatch says that decision makers must search for answers in the second level of ethical discourse: rules and rights (codes of ethics). Rules and rights that are grounded in a moral system, which he defines as 'an ultimate system of beliefs and norms about the rightness or wrongness of human conduct and character' (Veatch, 2003: 3), will be viewed as ethical, as opposed to those that are legal (and that may not necessarily be ethical). Veatch says that rules and rights often express the same moral duty from different perspectives. For example, the rule, 'Always get consent before surgery', expresses the health provider's point of view as the patient's right, which may be stated as: 'A patient has a right to consent before surgery.' In this way the rule and the right are viewed as reciprocal. When organisations gather together a series of rules or rights for the collective, these codified guidelines are referred to as codes of ethics. How serious these codes of ethics are is a function of how rigidly they are applied (see Figure 9.1).

If there are no exceptions to the rules or rights as stated in the code of ethics, this represents a form of legalism. The legitimacy of these legalised codes of ethics is very much tied to the methods of enforcement that are held within the jurisdiction. An infraction against these rules or rights would then lead to some form of punishment by law. At the other extreme it could be argued that each and every case or situation is such that it would be next to impossible to implement a rule or right for the purpose of deciding what one ought to do in the situation (referred to as antinomianism). As Veatch says, it is more likely in the use of codes of ethics that one of two intermediate positions (situationalism or rules of practice) would be employed. In the case of the former, moral rules are said to be 'guidelines' or 'rules of thumb' that could be used in each situation. On the other hand, rules of practice are specific guidelines or practices that are employed because they have been deemed morally obligatory, with exceptions made in only rare situations (see the preceding section on nomenclature of codes of ethics and codes of practice).

If rules and rights are unable to solve the problem, Veatch suggests that normative ethics, as a more complete analysis of the role of ethics, may need to be operationalised (the subject of Chapters 10 and 11). Here, the broad and basic norms of behaviour are considered in providing guidance. These, he notes, can

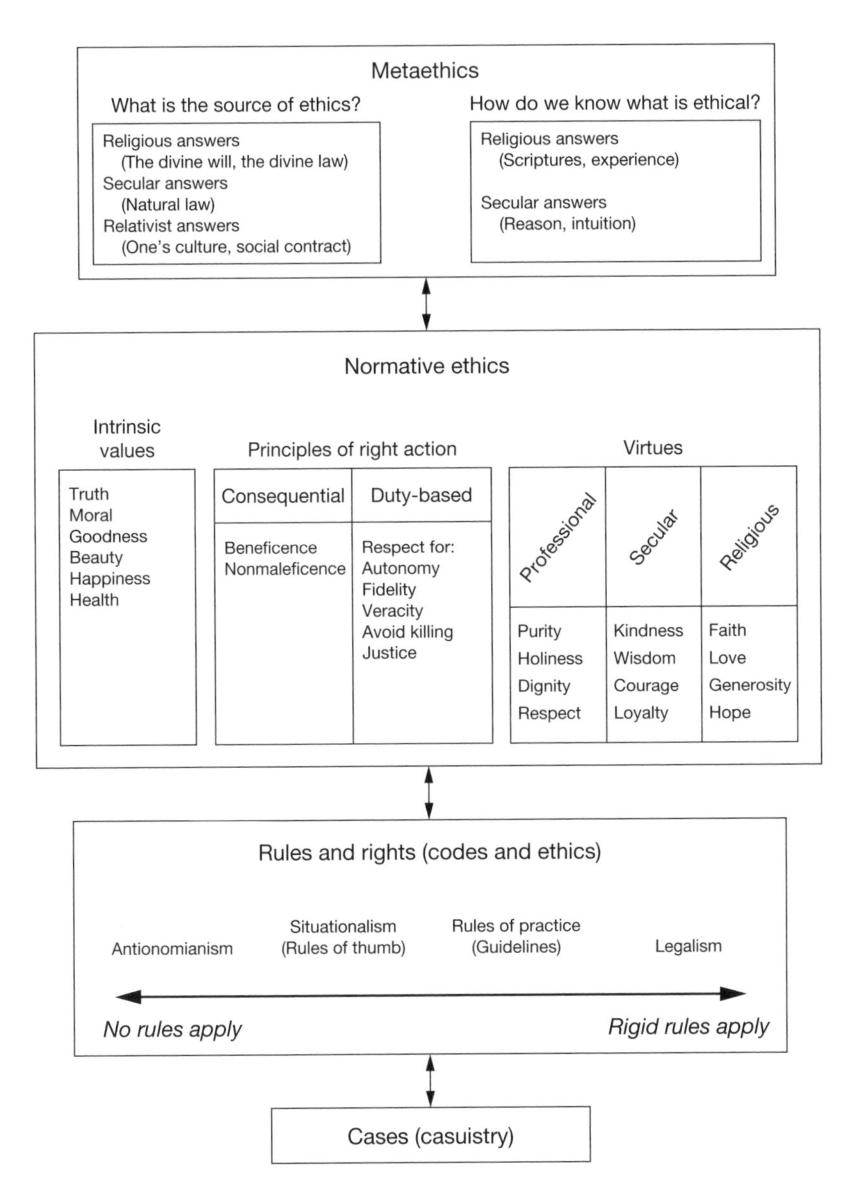

Figure 9.1 Levels of moral discourse
Source: Adapted from Veatch (2003).

be organised into three general categories: action theory, value theory and virtue theory. This includes consequence maximising principles as outlined in teleological approaches, and duty-based principles as noted in deontology.

At the individual level this might entail beneficence (producing good consequences) and non-maleficence (avoiding bad consequences) from the teleological position; and respect for autonomy, fidelity, veracity and the avoidance of killing (deontology). At the social level this may include social utility (what is best for the greatest number), as well as justice. What precedes the action, however, is the question: what kinds of consequences or actions are good or valuable? Values can provide the needed grounding for the purpose of employing the right prescriptions. Veatch's model identifies a number of intrinsic values, such as goodness, beauty, happiness and health, which provide guidance. The last normative theory, virtue, is to be understood as referring not to the character of action, but rather to the character of the people who engage in actions. The people charged with the responsibility of doing good need to maintain virtues such as benevolence (the will to do good). Veatch organises these virtues according to professional, secular and religious realms.

If, however, agents are unable to decide on the most appropriate values and virtues, or the principles of right action, the discourse must move to the level of meta-ethics. Veatch uses the example of one person giving priority to the principle of beneficience (acting so that good consequences result), while another thinks that autonomy should take precedence. Meta-ethics can help by allowing the differing parties to examine the basic meaning and justification of ethics – the ultimate grounding of ethics. This can be found in religion, natural law, social contract, reason, intuition, one's own culture or experience, or observation. But as the reader will no doubt realise, even at this fourth level there are vast levels of disagreement over the utility of one approach over another. Consequently, this model is not a one-way model, as outlined in the arrows that link the four levels, but is rather based on consensus that any of these levels may need to be consulted in reaching an ethically correct decision. As Veatch notes, it is less important where one starts in this model, as long as one is able to reach a reflective equilibrium through consideration of some or all of what these four levels offer.

Code development process

Previously in this chapter it was suggested that one of the main gaps in tourism research was the lack of empirical data for the purpose of discussing the pros and cons of codes of ethics. Another area where there appears to be a gap is in the process of how to go about constructing a code of ethics; that is, the methods that might be included in the planning, implementation and evaluation of effective codes of ethics for the tourism industry. The only document uncovered in the literature search for this section providing guidance for this is depicted in Figure 9.2. (A noteworthy exception to this is the UNEP (1995) *Environmental Code of Conduct for Tourism*, which provides examples of implementation, including publicity campaigns, publications, seminars and conferences, pilot projects, awards, education and training, and technical assistance, as well as some guidance on the monitoring and reporting process.) Accordingly, Figure 9.2 may be used as a template in its current form, or adapted to suit the particular needs of the organisation.

One of the strengths of this plan is the inclusion, at several stages, of the stakeholders who will be involved or affected by the code of ethics. It also highlights the importance of compliance in the implementation stage. In a review of the literature on compliance and non-compliance factors in codes of ethics, Gjerdalen and Williams (2000) state that individuals or groups tend to support codes because of:

Moral obligation. By making people aware of the consequences of their action (or inaction), they will often take responsibility for their action.
The need to belong. Operators who comply with codes experience a sense of belonging and security within the community.
Threat of punishment. The gravity of sanctions and consequences of ignoring codes often weighs heavily on people.
Shame and embarrassment. The threat of these often leads to compliance.
Social bonding. Individuals are less likely to break rules when there is the shared commitment to something, like a code, because of the threat of expulsion.

Conversely, they will not support codes because of: (1) lack of awareness or understanding of expected behaviour; (2) inappropriate environmental behaviour of operators and visitors, who may imitate what they see others do; (3) a lack of awareness regarding the consequences of violating a code; (4) codes or norms that are seen to be irrelevant or unrealistic; and (5) financial gain, protest, revenge or malice. Cole (2007) found that there were limits to the level of compliance that tourists demonstrated in reference to a code of conduct in Flores, Indonesia. Tourists admitted to the researcher responsible for developing the code that they were simply not prepared to change their behaviour. Furthermore, reminding tourists of their relative worth as compared to local people did not prompt them to change their bartering practices, i.e. they did not express empathy towards the villagers' poverty.

As a strategic process, the Government of Canada code in Figure 9.2 allows for review and evaluation of the document, which later provides the context for amending the plan to suit the needs of the group in question (a feedback loop). The importance of monitoring and evaluation is underscored by Mason and Mowforth (1995), who acknowledge that there are few if any who have provided the necessary template by which to accomplish this. By nature, codes of ethics are voluntary, so it may very well be that forms of persuasion will be unlikely to have the same effect as legalised schemes.

The Global Code of Ethics for Tourism

The Global Code of Ethics for Tourism (Code), developed by the World Tourism Organization, came about because of the perceived need to provide

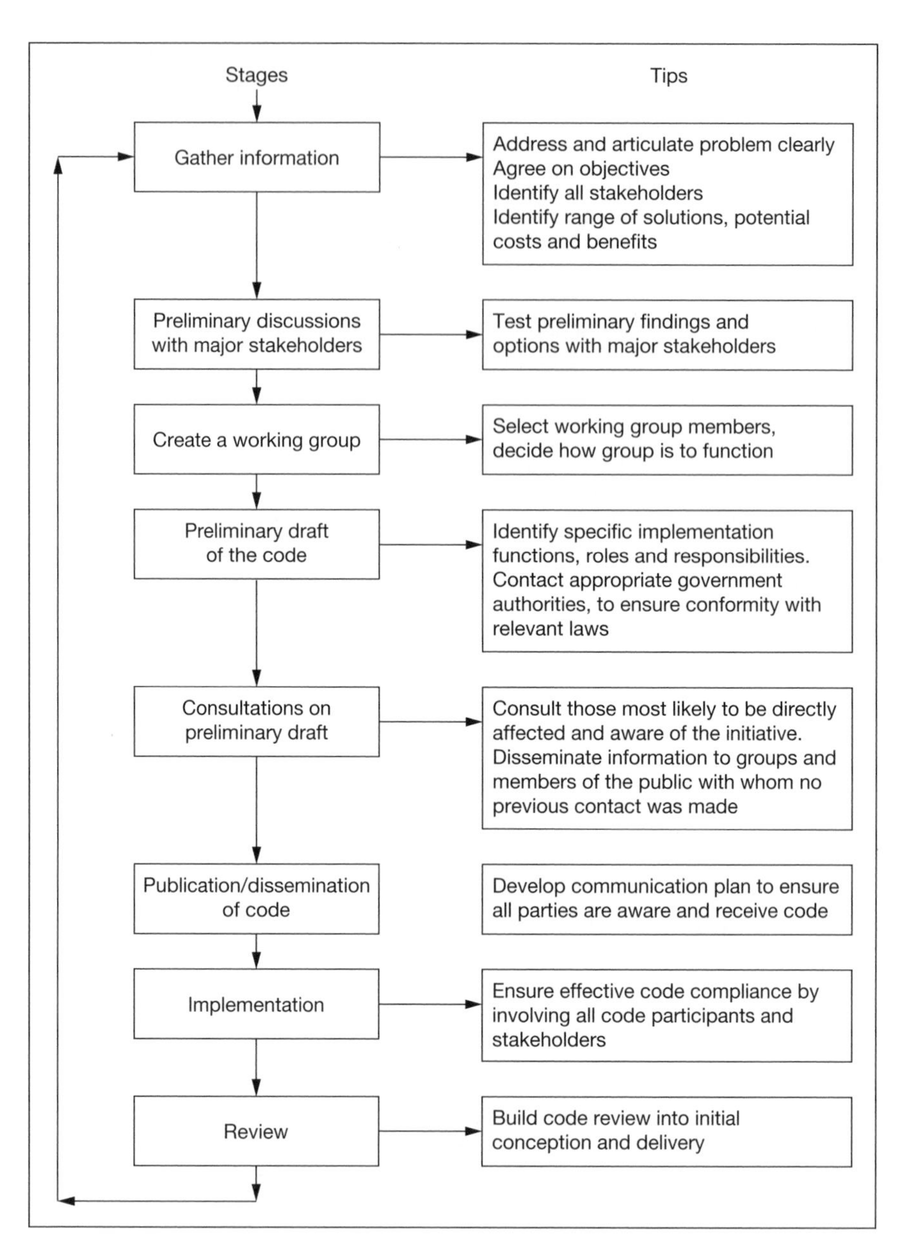

Figure 9.2 Code development process
Source: Government of Canada (1998).

a frame of reference that would govern tourism activity across borders. Approval of the Code came about through the 13th session of the general Assembly in Santiago, Chile, 27 September–1 October, 1999.

In a message by the Secretary-General of the WTO, Francesco Frangialli, it was outlined that the Code was reflective of the changing values of society at the end of the 20th century and because of the anticipated threefold increase in the volume of international tourism by 2019 (accessed 2 January, 2004 at http://ethics.unwto.org/content/global-code-ethics-tourism). Over the course of two years, from 1997 to 1999, the WTO requested input from all the various sectors of the tourism industry, which were vetted through a special committee for the preparation of the document. The code includes nine articles outlining appropriate behaviour for destinations, governments, service providers, developers, travel agents, workers and travellers http://ethics.unwto.org/content/global-code-ethics-tourism. The 10th article focuses on redressing grievances and thus takes a formal approach to enforcement. This will take place through a World Committee on Ethics, appointed as follows:

- six members and six alternate members designated by the WTO Regional Commissions, on the proposal of the State Members of WTO;
- a member and an alternate member designated by the autonomous territories that are Associate Members of WTO from among their members;
- four members and four alternate members elected by the WTO, General Assembly from among the Affiliate Members of WTO representing professionals or employees of the tourism industry, universities and non-governmental organisations, after conferring with the Committee of Affiliate Members;
- a chairman, who may be an eminent person not belonging to WTO, elected by the other members of the Committee, on the proposal of the Secretary-General of WTO (accessed 11 November, 1999 at http://ethics.unwto.org/content/global-code-ethics-tourism).

One of the challenges, which must surely have been the topic of debate in the development of this document as a lingua franca, was the application of these guidelines to such a large and diffuse group of entities that comprise tourism. In attempting to identify the magnitude of the problem in this, Fleckenstein and Huebsch note that businesses and organisations that hold membership in the tourism industry, include:

> lodging, tour companies, and attractions, . . . restaurants, entertainment complexes, casinos, shopping districts and malls, airlines, cruise ships, trains, buses, auto rentals, taxis, banks, foreign exchange sites, gasoline stations, medical clinics, hospitals, sporting activities, convention centers, meeting and conference centers, television, radio, and newspaper advertising, florists, wedding chapels, educational institutes, building trades, insurance companies, marketing agencies, churches, etc. (Fleckenstein & Huebsch, 1999: 138)

Compounding this is the fact that, because tourism is based on a service rather than a physical good, it does not lend itself well to control or standardisation. There is also the matter of the importance of issues in tourism codes of ethics. It has been the practice to include issues related to social and ecological impacts. But Fleckenstein and Huebsch (1999) note that the reality is that the main issues continue to be bribes, political issues, government intervention, customs clearance, transfer of funds and business practice differences. Furthermore, Fennell (2014) contends that the Code cannot be truly sustainable or responsible because it fails to take into consideration the welfare of the many millions of animals used in the tourism industry, because of its overriding anthropocentric tone. Until these issues are fully articulated in such codes, it can be argued, they (codes) will be unable to fully ameliorate the many problems that continue to plague the industry. Added to this is the failure of organisations, including the WTO, to cast the ethical net more widely, beyond just codes of ethics. This would include consideration of the theoretical framework as proposed by Veatch (see Figure 9.1). If codes of ethics are felt to be the least effective in addressing moral issues, as noted by some authors, then it is essential that normative ethics, meta-ethics and casuistry provide further guidance to the tourism industry.

Conclusion

Codes of ethics have garnered the most attention of all aspects of ethics in tourism. While they have been criticised for a number of shortfalls in regard to implementation and compliance, there are many who still feel that they have an important role to play in tourism by virtue of their contribution as a touchstone for guidance, reminding all employees and stakeholders to see the bigger picture (Guy, 1990). At the same time, it will be important to extend the range of study that has involved codes of ethics, as well as consider other aspects or vehicles of ethics in addressing the problems within the industry. For example, codes of ethics are seen to be essential as we embark on the new age of tourism in space. Given the recognition that space has a value beyond science and aesthetics, the development of the space environment for tourism is seen as a natural extension of our current business agenda. The need for a code of ethics for space tourism has been identified as an outgrowth of the need to regulate our activities, financial and otherwise, which has value in acting as a precursor to more formal policies and regulation (Williamson, 2003). Still, the adoption of a code of ethics for space travel, or other forms of travel, will be difficult for corporations who must balance profit in a climate of competition. Here, the question of the commons re-emerges in consideration of values, both organisational and industry-based, and resources in determining success.

10 Models and Methods of Moral Reasoning

What is morality in any given time or place? It is what the majority then and there happen to like, and immorality is what they dislike
Alfred North Whitehead, 1953

Introduction

This chapter presents an overview of a number of models and methods of ethical decision-making drawn from the business and psychosocial development literature. The intent is to briefly outline these and to revisit them in the following chapter, where they will be applied to different tourism scenarios. This chapter follows from the previous one, where there was recognition of a critical absence of theoretical knowledge that is deemed essential if we are to better understand the nature of the problems in tourism, and widen the universe of approaches we need to consider in overcoming these problems. The chapter concludes with a discussion on whistleblowing, and a framework of responses regarding whistleblowing, and the gains and losses inherent in these. This section is included not only because of the prevalence of whistleblowing in the current business literature, but also because of the framework for risk and response in regard to wrongdoing in the workplace.

Models of Moral Development and Reasoning

In a survey of the general managers of the largest lodging properties in the USA, Vallen and Casado (2000) sought to determine perceptions regarding ethical issues in the hospitality industry. They did this using the Josephson Institute of Ethics' (a body that provides training and workshops in ethics) 12 core ethical principles for moral decision-making. These are accountability, commitment to excellence, concern for others, fairness, honesty, integrity, law abidance, leadership, loyalty, promise-keeping, reputation and respect for others (in alphabetical order). Although their response rate was only 9%, yielding just 45 surveys, they discovered some interesting findings in their investigation. Only one of the 45 managers (in each case) said that 'respect for others', 'promise-keeping and trust', 'reputation and morale' and 'loyalty' were deemed most important. None of the respondents said that 'concern for others' was most important. When asked to rank which of the 12 principles they thought were breached most often in their work, the managers cited three, including 'accountability', 'commitment to excellence' and 'respect for others'. Qualitative comments collected by the researchers were also found to be salient. For example, some noted that 'Owners routinely lie and deceive employees'; 'A large number of associates are more concerned about "what is in it for me" versus dedication and loyalty for the long-term'; and 'Too many junior managers want the rewards without the accountability'.

The organisation is thus a setting that often places individuals at odds with their colleagues and challenges one's moral fortitude. Added to this is that much of the unethical behaviour that takes place within the organisation can be the result of pressure on the individual to act in ways that are external to his or her moral code (Grossman, 1988). There are many practical and empirical cases (almost weekly on the news) verifying this fact. In a study on organisational dependence and compliance, Wahn (1993) found that it is only when employees have choice in the face of the demands of bosses that they display heightened levels of moral fortitude when asked to do things deemed unethical. This choice can range from: (1) not to think about it; (2) go along and get along; (3) protest; (4) consciously object; (5) leave; (6) secretly blow the whistle; (7) secretly threaten to blow the whistle; (8) sabotage; and (9) negotiate and build consensus for a change in the unethical behaviour (Nielson, 1987).

The sheer fact that we have choice in our lives provides us with the opportunity to exercise our moral skills, which we must continue to acquire and nurture (Pinker, 2002). And although we can do this by educating ourselves, it helps to be immersed in a culture that supports efforts to be more ethical (see Box 10.1 on a code of conduct for ethical decision-making). Where the individual can help in this regard is through leadership within the organisation, through encouragement and through open dialogue with colleagues. This is corroborated by Saul (2001), who suggests that ethics is like a muscle that must be exercised each and every day in order to maintain optimal functioning – an analogy first used by Xenophon and much later by Thomas Jefferson.

If we choose not to use our ethical choices in work environments or in our daily lives, we open ourselves up to stress when our decisions or

Box 10.1 PricewaterhouseCooper's code of conduct for ethical decision-making

(1) *Recognise the event, decision, or issue*
- You are asked to do something that you think might be wrong.
- You are aware of potentially illegal or unethical conduct on the part of others at the firm or clients.
- You are trying to make a decision and are not sure of the ethical course of action.

(2) *Think before you act*
- Summarise and clarify your issue.
- Ask yourself, why the dilemma?
- Consider the options and consequences.
- Consider who may be affected.
- Consult with others.

(3) *Decide on a course of action*
- Determine your responsibility.
- Review all the relevant facts and information.
- Refer to applicable firm policies or professional standards.
- Assess the risks and how you could reduce them.
- Contemplate the best course of action.
- Consult with others – if you are uncomfortable or uncertain who to talk to, call the Ethics Helpline or write to the Ethics Mailbox.

(4) *Test your decision*
- Review the 'Ethics Questions to Consider' (see below) again.
- Apply the firm's values to your decision.
- Make sure you have considered firm policies, laws and professional standards.
- Consult with others – enlist their opinion of your planned action.

(5) *Proceed with confidence*

Ethics Questions to Consider

Is it legal?
Does it feel right?
How would it look in the newspaper?
Will it reflect negatively on you or the firm?
Who else could be impacted by this (others in the firm, clients, etc.)?
Would you be embarrassed if others knew you took this course of action?
Is there an alternative action that does not pose an ethical conflict?
Is it against firm or professional standards?

What would a reasonable person think?
Can you sleep at night?

Source: Adapted from Wolf (1990), in Hoffman *et al.* (2001).

non-decisions go against our ethical tendencies or dispositions, e.g. receiving more change back from the purchase of a ticket than you are entitled to (Ruse & Williams, 1986). The only way to remove the stress in such cases is to find a remedy. This usually means doing what your ethical nature tells you to do (e.g. giving the excess money back to the cashier). Moral decision-making may thus be viewed as a way to restore an ethical equilibrium in the body by arbitrating both cognitive and/or physiological tensions. Defined, ethical decision-making is 'the process of identifying a problem, generating alternatives, and choosing among them so that the alternatives selected maximise the most important ethical values while also achieving the intended goal' (Guy, 1990: 157). In this there are two important aspects involved, including moral reasoning and moral judgement. Moral reasoning can be defined as a conscious mental activity that consists of transforming given information about people in order to reach a moral judgement. This means that the process is intentional, effortful and controllable. Moral judgement, on the other hand, includes evaluations (as good vs. bad) of the actions or character of a person that are made with respect to a set of virtues held to be obligatory by a culture or subculture. These purposeful and intentional actions based on cultural norms are different from moral intuitions, which relate to the sudden appearance in consciousness of a moral judgement, including an affective valence (good–bad, like–dislike), without any conscious awareness of having gone through steps of searching, weighing evidence or inferring a conclusion (Haidt, 2001: 817–818).

The remaining part of this section focuses on the identification of a series of moral development and decision-making models that provide different options regarding how individuals may be able to navigate morally through a maze of often competing ethical challenges. Many of these are premised on the basis of the different objectivist and subjectivist theories which have been discussed earlier in the book. I make no attempt to favour one model over others, as it is assumed that one approach may be more useful under certain circumstances than others. The intent is to include a range of models in demonstrating: (1) how decision-making can take on many forms; and (2) that it often has a sound theoretical basis. In the following chapter I present a number of ethical situations in tourism and break these down using some of the models exhibited below.

Kohlberg's model of moral development

The study of moral development has been dominated by Kohlberg (1969, 1981, 1984), whose empirically-based hierarchical model of moral or immoral choices has generated considerable debate surrounding not only the methodology used, but also the interpretation of his basic philosophical assumptions, i.e. that autonomous justice is a *superior* position compared to that of relational benevolence or caring (see Elm & Nichols, 1993). Despite this debate, his work has been widely accepted by researchers in many disciplines as a foundation for moral thought and development (see, for example, MacLagan, 1996; Rest, 1986; Sridhar & Camburn, 1993; Trevino, 1986, 1990; Victor & Cullen, 1988).

Kohlberg describes moral development as consisting of six stages grouped into three levels of moral reasoning (Box 10.2). The first level, termed 'pre-conventional', reflects moral reasoning based upon a highly egocentric rationale. The decision maker's choices are to avoid punishment (stage 1) or to seek pleasure (stage 2) from external sources (e.g. revenue lost or gained from clients). The obligation to societal norms or laws or to global or ecological principles of justice are obfuscated or disregarded in favour of a hedonistic impulse.

The second or 'conventional' level describes reasoning based upon the desire of the decision maker to receive approval from significant others (stage 3) or from society in general (stage 4). The stage 3 individual would be more inclined to do what others (such as peers, co-workers, family or friends) feel as appropriate. Stage 4 individuals will operate based upon the influence of nomothetic norms such as organisational policy or societal laws (Malloy & Fennell, 1998b). It is important to note that, for both the pre-conventional and conventional levels, the locus of control for the decision maker is external.

The final and most advanced level, termed 'post-conventional', reflects a mode of reasoning that has progressed beyond the influence of external sources and is internally driven and self-regulating. Individuals at stage 5 base their reasoning on community-based notions of justice (i.e. social contract). In this regard, what is right is what the community has deemed to be right as opposed to what has been traditionally or externally imposed. An example of stage 5 reasoning can be observed in a statement by Martin Luther King, who suggested that the moral agent had an obligation to obey just laws and disobey unjust laws (i.e. segregation). Stage 6 moves beyond the locus of the community to a universal perspective. What is just is that which is just for all of humanity (and ecology). At this point in the hierarchy, the decision maker has developed a moral sense that extends beyond his or her own needs, and beyond the expectation of family or social mores. Moral reasoning is cosmopolitan in nature and is characterised by a deep sense of personal commitment. The psychological locus of control is internal and the philosophical correlate is existential (Gibbs, 1977).

The manner by which an individual progresses through these levels is a function of experience and education. In particular, it is the individual's recognition of the inadequacy of his or her moral perspective to resolve ethical dilemmas comprehensively that spawns a new and more advanced intellectual perspective. When the individual's cognitive schema is confronted or challenged through discussion, debate or from what they read or hear, a re-evaluation may occur and may result in a more complex level of moral reasoning (Kohlberg, 1969, 1981).

Box 10.2 Kohlberg's stages of moral development (*e.g. in running your business what is considered to be morally right*)

Pre-conventional level

Stage 1: Punishment and Obedience Orientation
(Will I be caught¿; sticking to rules to avoid physical punishment).

Stage 2: Instrumental Relativist Orientation
(What will I get out of it¿; little consideration given to social norms and ecological principles; right is an equal exchange, a fair deal).

Conventional level

Stage 3: 'Good Boy/Nice Girl' Orientation
(Living up to what is expected by peers and people close to you; people act to gain approval in society by adhering to social sanctions).

Stage 4: Law and Order Orientation
(Laws promote societal welfare; fulfilling duties and obligations of social system).

Post-conventional level

Stage 5: Social Contract Legalistic Orientation
(Societal standards through consensus apply; being aware that people hold a variety of values).

Stage 6: Universal Ethical Principle Orientation
(Ethical principles chosen regardless of society; when laws violate principles, acting in accordance with principles).

Source: Elm and Nichols (1993); Malloy and Fennell (1998b).

Generally, as Kohlberg comments, children up to the age of nine are associated with the pre-conventional level, while adolescents and most adults are tied to the conventional level. It is only relatively few individuals, less than 10%, who reach the post-conventional level of moral development.

Research on various aspects of the model has been supportive, as noted above. For example, Penn and Collier (1985) found that students were able to move to a higher stage of ethical development through ethical training. This is supported by Kohlberg (1984), who discovered that increased exposure and learning about ethical situations enables individuals to move through the stages more quickly. Nisan and Kohlberg (1982) conducted a longitudinal study and found that the six moral stages did not vary between cultures, contributing to his support for the model's universal application. One of the main criticisms of Kohlberg's model, however, is that it deals exclusively with how people reason about ethical situations and matters, not on their actions.

Gilligan and the ethic of care

Gilligan (1982) criticised Kohlberg's work in the belief that the model is based on the logic of fairness through a focus on justice rather than on other philosophical elements such as character and emotion. Kohlberg's studies, which formulated the basis of his model, Gilligan observes, were based on the response patterns of boys and men in his efforts to represent universal experiences of moral development. Gilligan argues that justice alone cannot frame morality and that a more comprehensive formulation of moral development ought to take into consideration the ethic of care (see Chapter 3, especially the work of Held, 2004; Robinson, 1997). So, while men are more strongly associated with a morality of rights, women, by contrast, are more firmly guided by a morality of care. Gilligan's model, like Kohlberg's, is based on successive levels of morality and described in the following passage:

> The first level is similar to Kohlberg's pre-conventional level as it is fundamentally self-centred and focused upon personal survival. Here, one cares only for oneself. This level is followed by the first transition in which the female begins to recognise that there are other individuals to whom she is responsible. In this stage, the female begins to break out of her egocentric perspective and starts to realize additional obligations and attachments to others. This transition is followed by the second level in which the female becomes altruistic in her concern for others at the expense of her own moral care. She is not only aware of her relations and responsibilities to others, she also believes that she must care for all of

> these individuals and becomes the selfless mother figure that Gilligan argues is a function of societal expectation as opposed to the truth (i.e. that altruism is not necessarily the same as goodness). This level leads to a second transition in which the woman recognizes that truth is not altruism and that she herself is in need of care and control. The final stage of Gilligan's model describes the morality of non-violence or caring. Here the woman has rejected the traditional criterion of feminine morality (i.e. self-abnegation and self-sacrifice) in favor of one in which care becomes the universal obligation where 'the worth of the self in relation to others, the claiming of the power to choose, and the acceptance of responsibility for choice' now becomes the paramount issue. (Malloy *et al.*, 2000: 83)

In support of this perspective, studies have found that men in fact moralize differently than women. Bersoff (1999) found that men can be more unethical because: (1) men have weaker socialisation pressures to obey rules; (2) men are bigger risk takers than women; and (3) women possess a stronger sense of empathy for the welfare of others (see also Walker (1984), who illustrates that, conversely, men and women differ very little in how they moralise).

Beyond the social psychology of gender and morals, Baron-Cohen (1995) found a biological link as to why males seem to be less empathetic than females. He discovered that there is a correlation between the level of foetal testosterone in boys and the amount of eye contact they make with mothers at 12 months old; the more testosterone, the less eye contact. As expected, female children, who had far less testosterone in their systems, made a much higher level of eye contact with their mothers (Lutchmaya *et al.*, 2002). In a further study, researchers gave 24-hour-old babies an opportunity to look at either a mechanical toy or a picture of a human face (Connellan *et al.*, 2000). It was found that baby boys preferred to look at the mobile and the girls the human face. The relationship between testosterone and empathy was discussed in a later study by Baron-Cohen (2002) on autistic children (children who are afflicted generally have trouble empathising with others), in which it was reported that autism is the masculinisation of the brain as a result of extreme levels of testosterone.

For tourism, this means that we can better understand the roles of and relationships between men and women in the tourism industry, not only for those who manage and administer it, but also for those who are working on the front lines or behind the scenes. It may also mean that male and female tourists and service providers have differences in opinion as to what constitutes acceptable practices and behaviours in travel destinations.

Trevino's person-specific interactionist model

Trevino's (1986) model postulates that ethical decision-making within an organisation is based on the interaction of cognitions, individual moderators and situational moderators, as illustrated in Figure 10.1. Trevino's model uses Kohlberg's stages of moral development in the cognition stage in providing a basis from which to examine the individual and situational factors that make his approach unique. The three main aspects of the model are explained below.

Cognitions

Individuals react to a specific ethical situation through cognitions that are a function of the person's stage of moral development, as defined by Kohlberg (see above), under the premise that each stage provides the basis from which to think about what is right or wrong in the context of the ethical dilemma (most adults in society operate at stages 3 or 4 according to research, as noted above). In general, Kohlberg thought that the reasons used to justify an ethical decision became more sophisticated with development. The longer someone serves in a capacity such as the work environment, thus gaining work experience, the more stimulus is created to further his or her level of moral development.

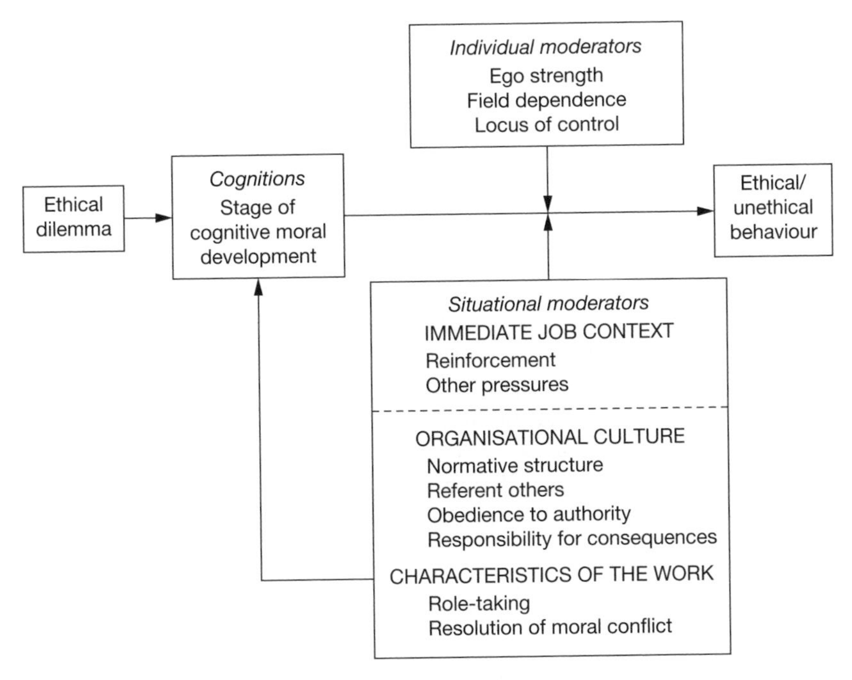

Figure 10.1 Interactions model of ethical decision-making in organisations
Source: Trevino (1986).

Individual moderators

Three individual moderators – ego strength, field dependence and locus of control – are suggested to influence the likelihood of a person to act on the cognitions regarding what is ethical or unethical, as noted above. Individuals who are high on ego strength are said to resist impulses and follow their intuitions regarding what is right or wrong. These individuals are more likely to act on what *they think* is right or wrong – than those who are low on this measure. Individuals deemed to be field dependent are those who more typically refer to the guidance of others in determining what is right or wrong. The opposite of this, field independent, includes individuals who function with more autonomy than their counterparts. Individuals who are at the conventional level of moral development, or lower, are said to rely on field independence in moderating the relationship between moral cognition and moral action. The last factor, locus of control, is a measure of the control that one exerts in his or her life. Externals are those who rely on other forces for guidance and are less likely to take personal responsibility for their actions, whereas internals believe that outcomes are a result of their own efforts or powers.

Situational moderators

Trevino identifies three categories of situational moderators. The first, immediate job context, makes reference to the fact that ethical behaviour is not just individualistic but subject to the pressures and interactions that take place at work. Reinforcement contingencies (rewards and punishments) and other external pressures, which include things such as personal costs (e.g. people under great time pressure), are said to have an impact on individual moral behaviour. The second category, organisational culture, also plays a role in determining ethical behaviour through the normative structure of the firm (the norms that guide behaviour), referent others (important and influential people in the firm), obedience to authority, responsibility for consequences, and the development of codes of ethics to guide behaviour. The third category, characteristics of the work, is illustrative of the notion that role taking and responsibility for the resolution of moral conflict are contributing factors to the continuation of moral development within the individual, and thus need to be encouraged.

Although extensive, Trevino offers the following propositions in further articulating the relationships between her three categories and individuals in the workplace:

(1) The majority of managers reason about work-related ethical dilemmas at the conventional level (stages 3 and 4).
(2) Managers at the principled moral reasoning level (stages 5 and 6) will exhibit significantly more consistency between moral judgement and moral action than those at lower stages.

(3) Managers' moral judgements in actual work-related decision situations will be lower (in the cognitive moral development stage) than their judgements in response to hypothetical dilemmas.
(4) Moral judgement development scores will be significantly higher for managers with higher levels of education than managers with lower levels of education.
(5) Participants (students or managers) in ethics training programmes based on cognitive moral development training strategies will exhibit significant pre-test to post-test increases in moral judgement development scores.
(6) Managers with high ego strength will exhibit more consistency between moral judgement and moral action than those with low ego strength.
(7) Field independent managers will exhibit more consistency between moral judgement and moral action than field dependent managers.
(8) Managers whose locus of control is internal will exhibit more consistency between moral judgement and moral action than managers whose locus of control is external.
(9) Conventional level (stages 3 and 4) managers will be most susceptible to situational influences on ethical/unethical behaviour.
(10) Principled (stages 5 and 6) managers will be more likely to resist, attempt to change, or select themselves out of unethical situations.
(11) In a culture that has a strong normative structure, there will be more agreement among organisational members about what is appropriate (ethical) or inappropriate (unethical) behaviour.
(12) In a weak culture, organisation members are more likely to rely on subculture norms for guidance regarding ethical/unethical behaviour.
(13) Managers' ethical/unethical behaviour will be influenced significantly by the behaviour of referent others.
(14) Managers' ethical behaviour will be influenced significantly by the demands of authority figures.
(15) Correspondence between moral judgement and action is significantly higher where the organisational culture encourages the individual manager to be aware of the consequences of his or her actions and to take responsibility for them.
(16) Codes of ethics will affect ethical/unethical behaviour significantly only if they are consistent with the organisational culture and are enforced.
(17) Managers' ethical/unethical behaviour will be influenced significantly by reinforcement contingencies.
(18) Managers' ethical behaviour will be influenced negatively by external pressures of time, scarce resources, competition or personal costs.

The model is important because it directly links the cognitive and behavioural approaches based on individual and situational factors in a comprehensive fashion. In this light, the emphasis is on the value of understanding how people search outside themselves for the required guidance in attempting to solve ethical dilemmas.

Martin's framework for ethical conduct

Martin's moral framework, although not comprehensive, is included here because of its inclusion of stakeholder groups in forging meaningful relationships between the organisation and other groups, rendering it applicable in a tourism context. Martin (1998) proposes that ethical conduct consists of three dimensions: (1) stakeholders; (2) areas of focus; and (3) a spectrum of conduct. Stakeholders include all of those individuals or groups that are affected by the company's actions (see Figure 10.2). These include customers, unions, government, employees, suppliers, financiers and the public – as well as the environment.

What needs to be explored through these relationships is the nature of the stakeholders' association with the company, as well as the rights of these groups in regard to the relationship with the firm. Martin argues that a company is more ethical if it gives consideration to a wider range of stakeholders. The types of issues that confront the company and the stakeholders (dimension 2) can be captured by four elements: the legal perspective, where

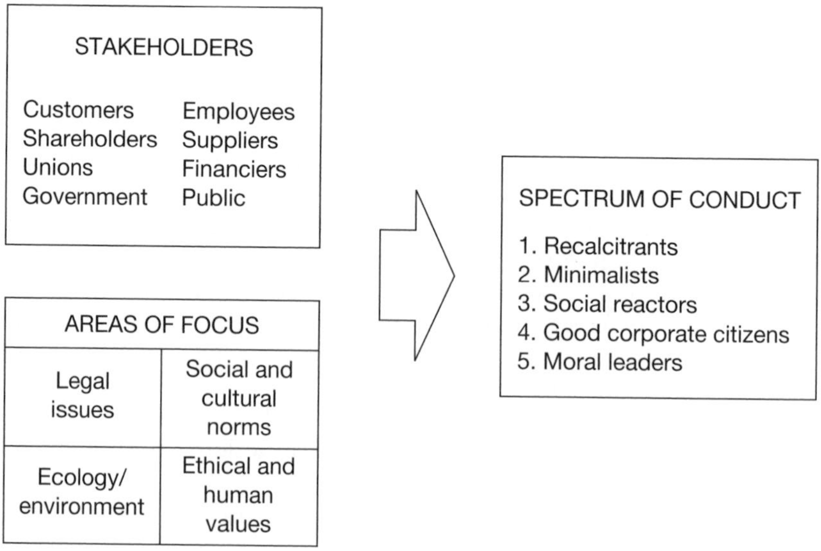

STAKEHOLDERS

Customers	Employees
Shareholders	Suppliers
Unions	Financiers
Government	Public

AREAS OF FOCUS	
Legal issues	Social and cultural norms
Ecology/ environment	Ethical and human values

SPECTRUM OF CONDUCT

1. Recalcitrants
2. Minimalists
3. Social reactors
4. Good corporate citizens
5. Moral leaders

Figure 10.2 Framework for ethical conduct
Source: Martin (1998).

companies should not operate outside the law; social and cultural norms, where ethical companies are said to be those that are sensitive to the social climate of their operating environment; the ecological perspective, which includes questions about the company's operations in light of the finite nature of the resources we use; and ethical and human values, which include the recognition of core values that might be used to ensure ethical conduct in the firm.

Dimension 3 of Martin's model includes the spectrum of conduct, which includes a range of behaviours related to minimum standards of conduct or higher ethical aspirations. In moving ethical actions beyond the level of behaviour that is 'acceptable', from that which is 'not acceptable', Martin proposes the following typology:

Recalcitrants. Companies that flout the law in their dealings, and which are deliberately dishonest.

Minimalists. Companies that comply with the law, but in a minimalist way, offering to do the bare minimum by law and no more.

Social reactors. Companies that respond to social criticism, in the appearance of acting in a socially responsible manner. Appearance is valued more than the need to be socially good.

Good corporate citizens. Companies that demonstrate the willingness to be socially responsible to many of the stakeholders, including the natural environment. They exhibit fairness, respect and social reciprocity.

Moral leaders. Companies that go beyond social responsibility to be proactive in promoting social good, through efforts to foster peace and human development.

Martin contends that the strength of his model lies in the fact that it emphasises that companies have different ethical strengths and weaknesses. While some companies place emphasis on being environmentally responsible, others may place too much effort into training and development at the expense of other areas. The assumption is that those who spread their efforts according to the stakeholders and the areas of focus will likely be the ones that sit at the higher end of the spectrum of ethical conduct. The 'spectrum of conduct' aspect of Martin's model has direct comparative implications with the work of Kohlberg, as discussed previously, in regard to a moral hierarchy.

Haidt's social intuitionist model

Julie and Mark are brother and sister. They are traveling together in France on summer vacation from college. One night they are staying alone in a cabin near the beach. They decide that it would be interesting

> and fun if they tried making love. At the very least it would be a new experience for each of them. Julie was already taking birth control pills, but Mark uses a condom too, just to be safe. They both enjoy making love, but decide not to do it again. They keep that night as a special secret, which makes them feel even closer to each other. What do you think about that? Was it OK for them to make love? (Haidt, 2001: 814)

Haidt notes that, upon hearing of such a situation, most would suggest that the act was immoral; however, connected with this claim is the inability to know why such an act is wrong. They suggest that inbreeding is a problem, but remember that Julie and Mark had two forms of birth control. Evaluators note that both brother and sister may be hurt emotionally, but also recall that both appeared to be unharmed and happy that they shared this experience. Evaluators of this scenario further suggest that although they cannot put their finger on the reason, they just know that it is wrong. This intuition strikes to the heart of the issue on moral reasoning as it suggests there are times when an individual senses that something is wrong, but cannot determine why. Conversely, the more rationalist approaches to moral reasoning are built from the belief that judgements may be determined a priori by processes of reasoning and reflection. In the case of Mark and Julie, if no condemning evidence is found, there is no reason to issue a condemnation.

Haidt's social intuitionist approach takes the stance that moral reasoning is a post hoc construction that takes place only after an agent makes a moral judgement, which in turn has been based on quick moral intuitions. We may not be able to understand why something is right or wrong initially (our moral intuition), but we make a judgement based on this intuition and later reason out why it may be right or wrong (moral reasoning). This approach is based on new cognitive research which has found that most of our cognitions take place outside consciousness; that people have difficulty telling us how they make judgement; that emotions are not as irrational as once thought; and that reasoning is not as reliable as once thought. The social intuitionist model is based on six links, which have been explored in the social and psychological literature (see Figure 10.3). These include: (1) *the intuitive judgement link*, where judgements appear in consciousness automatically as a result of moral intuitions; (2) *the post hoc reasoning link*, where moral reasoning is an effortful process that takes place after a moral judgement, in which a person searches for arguments that support the already made judgement; (3) *the reasoned persuasion link*, where moral reasoning is produced and sent forth verbally to justify one's already made judgement to others (i.e. we tell others why it is acceptable to act in one way or another); (4) *the social persuasion link*, which occurs when the moral judgement exerts a direct influence on others that may shape the judgement of these others (i.e. this influences others to act in the same way); (5) *the reasoned judgement link*, where

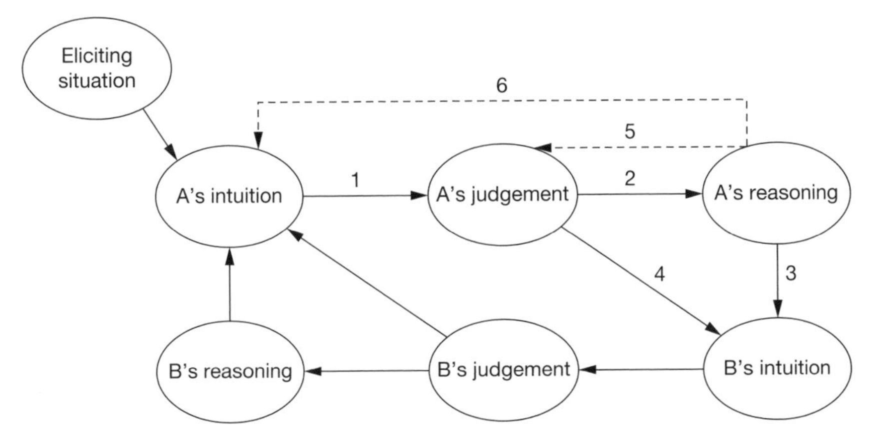

Figure 10.3 Social intuitionist model of moral judgement. The numbered links, drawn for Person A only, are (1) the intuitive judgement link, (2) the *post hoc* reasoning link, (3) the reasoned persuasion link and (4) the social persuasion link. Two additional links are hypothesised to occur less frequently: (5) the reasoned judgement link and (6) the private reflection link
Source: Haidt (2001).

people may reason their way to a judgement by force of logic and thus override the initial intuition (in these cases reasoning is truly causal and not the slave of the passions – such reasoning is said to be rare and occurs only when intuitions are weak); and (6) *the private reflection link*, where in thinking about the situation a person may spontaneously activate a new intuition that contradicts the initial judgement. This latter element can take place through role-playing or putting oneself in another's shoes.

Hunt and Vitell's general theory of marketing ethics

One of the most comprehensive models of ethical decision-making in the literature is one produced by Hunt and Vitell (1986). In their general theory of marketing ethics, the authors develop a descriptive approach, which attempts to explain the decision-making process that takes place for problems that have ethical content (see Figure 10.4). As illustrated at the 'Start' of Figure 10.4, individuals initially confront a problem that is viewed to have ethical content. The perception of this problem triggers the process depicted in the model. In analysing the nature of the problem, the agent first begins the task of perceiving different alternatives that might be used to ameliorate the problem (keeping in mind that it is difficult to envision a complete set of alternatives in the mind of the agent). In establishing alternatives, the individual considers two broad normative evaluations: deontology and teleology. In the case of the former, he or she evaluates the rightness or wrongness on the basis of pre-established rules, norms or values. From the teleological

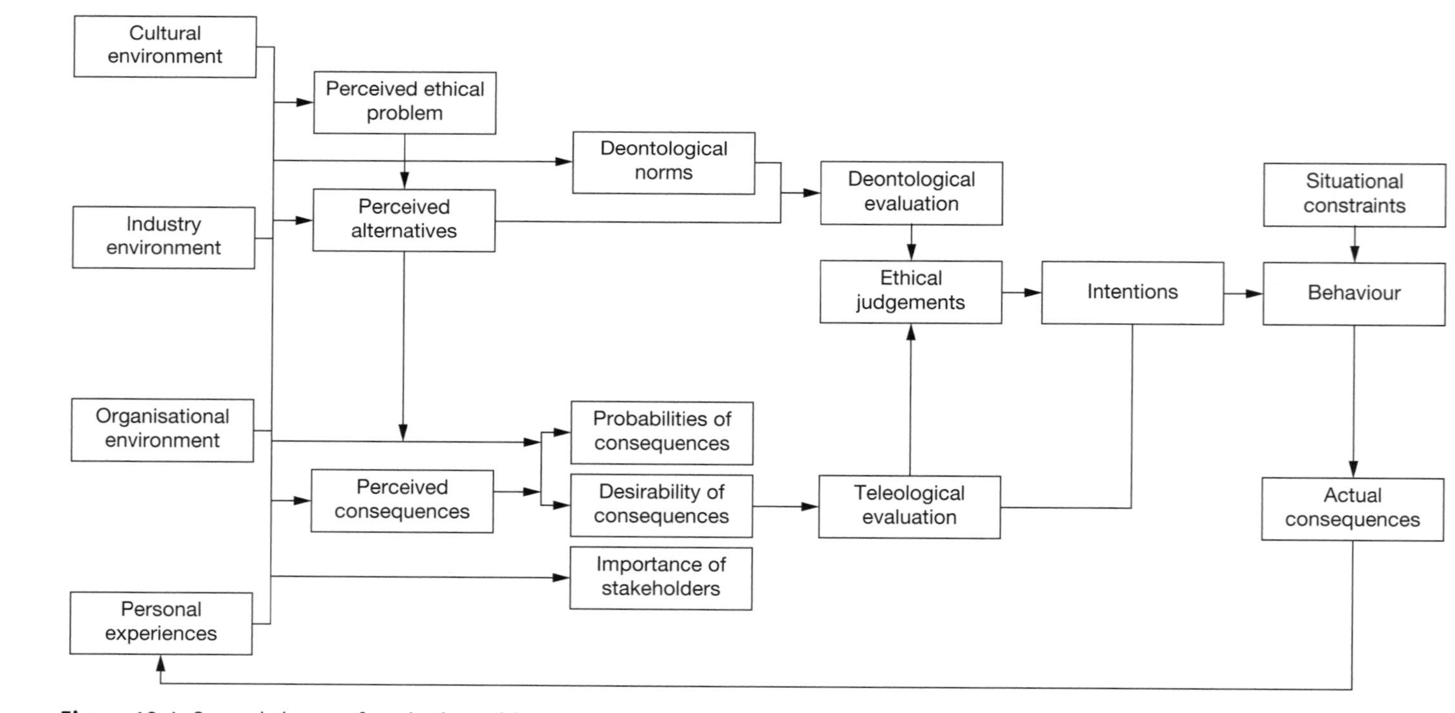

Figure 10.4 General theory of marketing ethics
Source: Hunt and Vitell (1986).

standpoint, however, evaluation is based on four constructs, including: (1) the perceived consequences of each alternative for stakeholder groups; (2) the probability that each consequence will occur to each group; (3) the desirability or undesirability of each consequence; and (4) the importance of each stakeholder group. Hunt and Vitell observe that the importance of other stakeholder groups will be different for different people. For example, travel agents must serve both clients and suppliers. While suppliers typically view the travel agent as working foremost for the supplier, it is in the best interest of the travel agent to mediate their time and interest between suppliers and clients. Hunt and Vitell note that the probability construct is essential in providing guidance for the calculation that one action contributes to the attainment of goals over another. The teleological aspect of the model will result in a sensitivity to the relative goodness or badness that may result from each of the alternatives perceived by the individual.

At the central part of the model are the ethical judgements derived from a combination of deontological and teleological evaluations, as supported in the literature (see Reidenbach & Robin, 1990). Consistent with the general research on behaviour, although an ethical judgement has been made, the individual will act only if there is an intention to do so (see Fishbein & Ajzen, 1975). Hunt and Vitell characterise this aspect of the model 'as the likelihood that any particular alternative be chosen' (1986: 9), with the belief that ethical judgement and intentions are better predictors of behaviour in circumstances where issues are central rather than peripheral. The authors note that ethical judgements will at times differ from intentions because of a link between teleological evaluations and intentions. In this manner, there may be situations where ego-induced altruism or empathy-induced altruism (see Chapter 2) push one to select less ethical alternatives because of the benefits to self or others who are close. In subsequent research, Mayo and Marks (1990) found support, first, for the notion that ethical judgements are determined by deontological and teleological means and, second, for the relationship between ethical judgements and intentions to adopt a particular alternative.

Situational constraints may also have an impact on behaviour on the basis of opportunity. When presented with a range of different opportunities, the individual may select a behaviour that is less or more ethical. The individual is able to learn about the consequences of the alternative selected through the 'actual consequences' aspect of the model. This may have a short-term effect or more of a long-term effect, depending on the nature of the alternative selected. This learning gets fed back into the 'personal experiences' aspect of the model, suggesting that it is possible for people to learn how to be ethical/unethical through conditioning. The final aspect of the model includes the four constructs of: (1) personal experiences; (2) organisational environment; (3) industry environment; and (4) the cultural environment. The latter three of these have an impact on all of the aspects of the general model, including the perception of alternatives, probabilities of consequences and deontological norms. While 'personal experiences' include the level of moral development of the individual (as noted by Kohlberg), industry and organisational environments may also have a significant impact on ethical decision-making. It may seem acceptable to behave in unethical ways if colleagues in the firm are doing so or counterparts in the industry have done so, especially in efforts to stay competitive. With respect to the cultural environment, Hunt and Vitell suggest that laws, rights, religion and so on, are important factors that also have an impact on ethical decision-making.

Schumann's moral principles framework

Schumann's (2001) moral principles framework for human resource management is built around five main ethical theories: utilitarianism, rights, distributive justice, the ethics of care and virtue theory. An abbreviated version of his work is outlined in Table 10.1, and is representative of a growing volume of work that tends to favour a combined or pluralistic approach to ethics, over just one or two theories.

The strength of the framework lies in the theoretical support that researchers have placed into the foregoing normative theories. Schumann also argues that, beyond theory by itself, these approaches continue to be used in day-to-day practice even though the people using them are unfamiliar with their theoretical basis. Finally, each of the five theories is unique and therefore provides the opportunity of looking at ethics in a different way. This diversity allows for the consideration of many different perspectives in arriving at morally correct decisions. Schumann further argues that ethical relativism should be rejected (as most theorists do, in his opinion) because people with differing opinions on a subject cannot be right; because it produces incoherent and inconsistent consequences; and because it contradicts itself and so renders itself logically false. The search for universal moral principles should thus take precedence over ethical relativism.

The final category, 'Resolve the conflict', provides an opportunity to assess whether or not the five theoretical principles come to the same conclusion or reach different conclusions. It also suggests that if one or perhaps two of the principles deviate from the others, the agent should decide which principles should take precedence over others. It may be decided that one principle is not more important than the others, but rather that it has special implications for the issue under scrutiny.

Malloy *et al.*'s comprehensive approach to ethical decision-making

Like the Schumann model above, Malloy *et al.* (2000) developed a comprehensive model of ethical decision-making that is based on more than one

Table 10.1 Schumann's moral principles framework

Theory	*Questions*
Utilitarianism	What action will do the most good and the least harm for everyone who is affected? a. Who are the stakeholders? b. What are the alternative courses of action? c. For each alternative, what are the benefits and costs? d. Which alternative creates the most benefits and least costs?
Rights	What action do you have the moral right to take, that protects the rights of others, and that furthers the rights of others? a. Do you have a moral right to take the action in questions? Consider: • Are you willing to have the action in question done to you? • Are you willing to live in a world where everyone did the action in question? • Are you treating people with respect? b. What moral rights do other stakeholders have? c. Are there conflicts between your moral rights and others? Which take precedence?
Distributive justice	What action produces a fair distribution of benefits and costs for all of the stakeholders? a. Egalitarianism: What action produces a fair distribution of benefits and costs? b. Capitalism: What actions produce costs/benefits based on the contributions of actors? c. Socialism: What action distributes benefits based on need and costs based on abilities? d. Libertarianism: What action has been freely chosen by the stakeholders? e. Rawls' principles: What action provides all with equal liberties and opportunities while helping those in need to the greatest extent possible?
Ethic of care	What action cares for those people with whom you have special relationships? a. What action cares for your own needs? b. What action cares for the needs of the others who have a relationship with (e.g. family, friends, co-workers, competitors)?
Virtue	What action displays virtuous character traits? a. Does the action display virtues such as benevolence, civility, compassion, courage, fairness, conscientiousness, cooperativeness, generosity, honesty, industriousness, loyalty, moderation, self-control, self-reliance, or tolerance? b. Or does the action display vices such as cowardice, deceit, dishonesty, laziness, neglect, orselfishness? c. Take the action that displays virtues, not vices.
Resolve the conflict	Do all five moral principles reach the same conclusion, or do they reach conflicting conclusions? a. If in conflict, then, examine the nature of the conflict to see if it can be resolved by choosing a previously unconsidered course of action. b. If 'a' is not possible, then collectively decide which principle(s) should take precedence in light of your values.

Source: Schumann (2001).

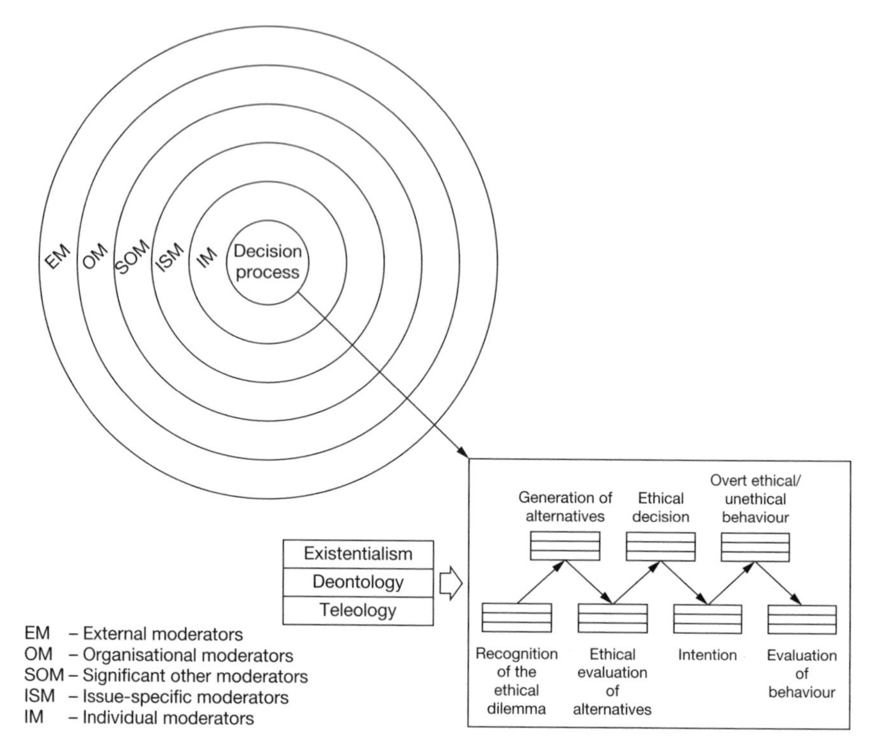

Figure 10.5 Ethical decision-making approach (comprehensive)
Source: Adapted from Malloy *et al.* (2000).

ethical theory (Figure 10.5). Their theory incorporates deontology (right behaviour), teleology (good behaviour) and existentialism (authentic behaviour) for effective ethical decision-making (see Chapter 3 for a description of these theories). The second aspect of the model includes the consideration of five different types of nested moderators ranging from the individual to the external environment. The inclusion of these reinforces the notion that the process of moral reasoning is a cognitive one. The last stage of the model, termed the 'process of ethical decision-making', includes a systematic procedure ranging from problem recognition to problem evaluation. The moderators and the processes that form the basis of their model are highlighted below.

The moderators

External. This includes a range of different influences, including technical, political, economic and societal which frame the context of a situation and influence the manner in which decisions are made.

Organisational. This moderator includes the ideology of the organisation, including the organisational culture and climate surrounding what constitutes ethical and unethical behaviour for those who participate, work or volunteer.

Significant other. This includes personal, interorganisational and extraorganisational moderators, such as family and friends, peers, group members and individuals from other communities and organisations.

Issue-specific. This includes consideration of the severity of the ethical issue under scrutiny. This is tempered by the normative consensus around the issue, physical and psychological 'distance' from the issue, the magnitude of the consequences of the issue, the probability and concentration of the effect, and the immediacy and strategic or tactical orientation.

Individual. This includes the ethical and values orientation of those involved, including demographic variables (age, education, gender) and moral development. (See the Kohlberg and Gilligan models, above, for an appreciation of the magnitude of internal moderators for ethical decision-making.)

The process

(1) Identify the problem from each perspective.
(2) Develop alternatives from each perspective.
(3) Evaluate each alternative from each perspective.
(4) Select the ideal solution.
(5) Determine intention to act upon the ideal solution.
(6) Make an actual decision.
(7) Evaluate the actual decision from each perspective.

The strategic process used in this model, along with the emphasis on an examination of the range of influences, allows for a more comprehensive template from which to consider the three ethical theories. The authors argue that a triangulated approach to ethical decision-making allows those who are charged with the responsibility to make decisions and to pose certain questions that are central to these theories. In respecting the teleological position, we must determine whether the issue prevents the ends from being realised for individuals and groups. From the deontological position, we are able to gauge whether rules, policies or standards have been broken and the implicit and explicit duties of individuals and groups. And, lastly, a respect for the existentialist perspective allows decision makers to decide whether the freedom of individuals and groups has been denied or controlled. These perspectives can be more closely examined in the scenario in Box 10.3.

Theerapappisit's model of Buddhist ethics

Researchers have documented many specific differences in Eastern and Western philosophies related to business (see Koehn, 1999). The Western approach is geared primarily towards the generation of profits, with a later emphasis on fitting business back into society *post profit* through efforts geared towards social responsibility. This differs from the Eastern conception, which is based on the treatment of business as just one of many institutions embedded within society. From the Eastern model the question: 'What is the good of

Box 10.3 The hockey game

You have been invited by your friend to attend an ice hockey game while on vacation in Canada. Since you have never seen a game you are a bit sketchy on the rules and actions of the players. A player on Team A is about to skate free on a breakaway with the likelihood of scoring a goal. A defender on Team B deliberately trips the opposing player to prevent the goal and your friend, who is a hockey fanatic, adds that it was a good foul. Phrased in this manner, your friend appears to be praising a violation, as you've learned, of the rules. If a rule-based orientation is the source of the ethical maxim, then clearly the defender deliberately violated a rule and, therefore, it is judged as wrong. However, if one takes a consequentialist approach, one can argue that the defender, by tripping the opponent, ensured a win for Team B. This, thereby, produces more good/happiness for the defending team than bad/sadness for the opposing team. Since the good/happiness for one team outweighs the bad/sadness of the other team, the deliberate rule violation was, in reality, a good foul. From an existential standpoint, the player committing the foul would need to examine his or her conscience to decide whether a deliberate violation of the rules is an instance of authentic behaviour.

The authors write that, when faced with a choice, which do you value more highly: obeying the rules under any circumstances? Insisting that what is good is that which creates more goodness than badness, no matter what behaviour was used to generate the outcome? Is being true to yourself always (authenticity) what matters only in deciding what is right or good, regardless of pre-established rules? Malloy *et al.* suggest that, in ranking these three options, the individual is engaging in an exercise of values clarification, which enables you to (1) learn more about yourself, and (2) gain insight regarding which of the three ethical theories (deontology, teleology, or existentialism) is more likely to influence your decision-making. Granted it is only one example, but it does serve to illustrate that, under a set of conditions or circumstances, one needs to be able to articulate a response based on the values that holds to be true and the ethical maxim that is most applicable.

Source: Adapted from Malloy *et al.*, 2000: 49–50.

the larger whole and how can I behave in such a way as to contribute to that whole?' (Koehn, 1999: 77), compels us to look at society as a whole in deciding whether products are able to contribute to the moral development of society. Koehn uses the example of North American cars in explaining that these, if sold to China, would not serve the country well at all in their moral development. This is because China already has severe pollution problems and there is not the infrastructure in place to accommodate the influx of automobiles.

Although a review of literature on tourism and ethics yielded little in the way of non-Western ethical decision-making systems, the work of Theerapappisit (2003) emerged as an important contribution. It stands as an example of how Buddhist ethics has been useful in addressing questions related to the balance of conflicts between stakeholders at various levels in tourism planning, and the difficulties in choosing between capital or social mobilisation. (Balance may be taken in the Aristotlean context of virtue ethics and the Golden Mean, according to Whitehill, 2000). This is evident in the development of tourism in the Makong region. While the traditional model of tourism development in the region was based on increasing tourism numbers and international economic cooperation, a renewed philosophy of 'sufficiency', as advocated by His Majesty the King, has taken precedence in the new economy of the region. Sufficiency is defined as:

> the middle path as the overriding principle for appropriate conduct by the populace at all levels. This applies to conduct at the level of the individual, families and communities, as well as to the choice of a balanced development strategy for the nation ... [it] means moderation in due consideration in all modes of conduct, as well as the need for sufficient protection from internal and external shocks. To achieve this, the application of knowledge with prudence, especially from various sources of local wisdom, is essential. (Theerapappisit, 2003: 49)

Through his efforts in interviewing key informants (government, industry and NGO stakeholders) about policies in tourism planning, Theerapappisit discovered that those who were seeking to marshal the social mobilisation platform (at the grassroots level) felt that ethical principles had to be introduced in broadening the awareness of policy makers regarding the impacts of tourism. The ethical principles the author has used are combined from many sources (including Buddhist scholars) and based on the philosophy of balance between problems and benefits. The six selected are articulated at three different levels, including the individual (self-development and control), community (deemed an influential aspect of the social system) and the external (considering the whole environmental system), with the understanding that balance must take place among all levels (Figure 10.6). These levels are elaborated upon below.

Self-development (individual learning)

This includes:

Morality. Balances **altruism** (giving/sharing, honesty, dignity and philanthropism towards accountability and transparency) and **greed** (individualism/egoism towards own benefits and corruption).

Wisdom. Balances **impartiality** (an understanding of the truth with impartial consequences leading to enlightenment) and **bias** (inequity, ignorance, carelessness, making someone unjustly suffer).

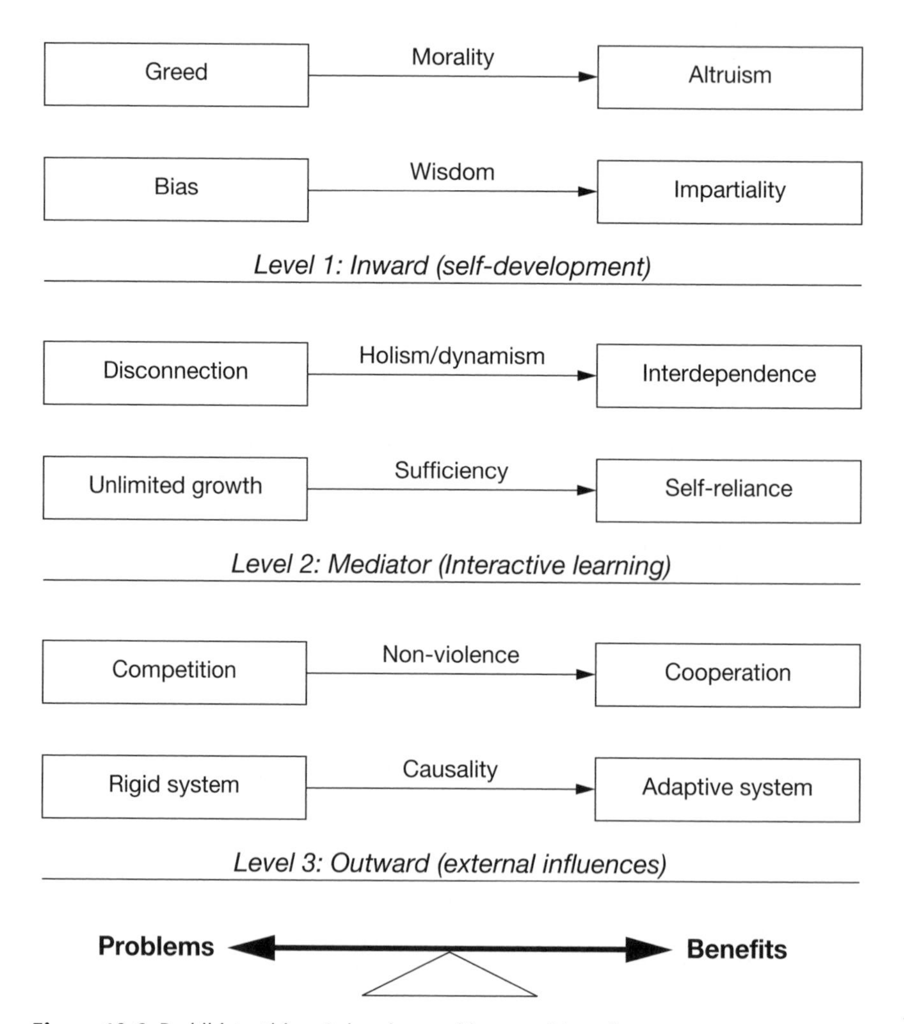

Figure 10.6 Buddhist ethics: Balancing problems and benefits
Source: Adapted from Theerapappisit (2003).

Interactive (social organisational learning)

This includes:

Holism and dynamism. Balances **interdependence** (understanding of multidimensional phenomena through life cycle, integrity values, uncertainty/interrelation of multi-faceted social networks) and **disconnection** (objective-oriented value, singular concern, attraction/interest).

Sufficiency. Balances **self-reliance** (sufficiency economy with a civil society concept – trust in local action/production) and **unlimited growth** (productive economic efficiency, belief in concept of profit-driven operations).

External influences (macro social environment learning)

This includes:

Non-violence. Balances **cooperation** (compromising with forbearance/concord, having democratic public consensus without oppression) and **competition** (command/control/fighting for victory).

Causality. Balances an **adaptive system** (causes and conditions towards alternative options/integrated solutions with systematic/critical attention to both the pros and cons arguments through the processes) and a **rigid system** (fixed output/timeframe, extreme ultimate/closed solution focusing on 'end-product' rather than 'process').

Theerapappisit says that the aforementioned six Buddhist ethical principles serve as a code of ethics in bringing groups together for policy making with less conflict. There will be a moral crisis in tourism development – the system falling out of balance – if decision makers who favour capital mobilisation fail to give enough credence to these non-economic values. Sustainable development will thus only emerge if the contradictions between social and capital mobilisation can be ameliorated.

Whistleblowing

An organisational issue that has generated a great deal of exposure in the new millennium is wrongdoing and the whistleblowing that often results from its identification. Defined, wrongdoing is 'behavior which is morally wrong as well as behavior which is illegal, unethical, wasteful, inefficient, neglectful, an abuse of power, or violates organisational rules or professional standards' (McLain & Keenan, 1999: 256). For example, not adhering to the explicit codes of ethics within the organisation, using one's station in the firm to further oneself or friends, or spending the firm's money on unnecessary resources can all be cited as wrongful in the context of the group.

The decision to respond to what may be viewed as wrongful, or to whistle-blow, comes as the result of the feeling that public interest is more important than the interests of the firm that one is tied to (Guy, 1990). Whistleblowing occurs when an employee recognises that an ethical transgression has taken place in the firm, and he or she acts on it. While often whistleblowing results in the reprioritisation of organisational rules and procedures, there is also a collegiality factor to be considered (unwritten norms and professional conduct), suggesting that individuals should consult with colleagues before approaching upper management or external sources. In a study of whistleblowing among lower-, middle-, and upper-management staff, researchers corroborated previous studies, which found varying degrees of support for the importance of whistleblowing at all levels of the management hierarchy. More specifically, upper management was found to be the most supportive if they believed they had greater degrees of discretionary power and less pressure to conform. Middle managers, in contrast, felt more at risk by speaking out and thus more compromised, while first-level managers were more likely to try to fit into the status quo (Keenan, 2002).

More specifically, researchers have found that whistleblowing is subject to a three-stage process: awareness, judgement and the decision to respond (McLain & Keenan, 1999). Foremost among these are issues related to the judgement of who is harmed by the action or inaction, the personal association with those who are wronged, the reward structure for reporting wrongdoing, the organisational culture of the company (i.e. the acceptability of wrongdoing or vice versa), the quality of the information related to the awareness of the wrongful act, the cognitive moral development (see Kohlberg, above) of those individuals involved in the scenario (suggesting that those who have a higher cognitive moral development are more willing to report the wrongful act), and the motivation to pursue an alternative action and the degree of support for reporting wrongdoing. The whistleblower who is witness to wrongdoing within the firm must act mindful of a series of pros and cons. As McLain and Keenan suggest, whistleblowing may risk damage to relationships with co-workers and may go against the organisational culture of the firm. In negotiating what appears to be an extremely touchy subject, the authors suggest the following five risk and response options to wrongdoing outlined in Table 10.2.

This framework – responses, gains and losses – may provide the needed level of information required for workers to consider the pros and cons of whistleblowing. Obviously the severity of the situation surrounding the wrongdoing may weigh heavily on the decision by the individual to bring this to the attention of his or her superiors. The authors note that there continues to be a

Table 10.2 Risk and response options for wrongdoing

Response	*Potential gains*	*Potential losses*
Join in	Rewards from wrongdoing; social acceptance among wrongdoers	Wrongdoing not stopped with harm to organisational effectiveness and to self; loss of dignity, esteem, trustworthiness among co-workers; legal punishment by company
Do nothing	Nothing; those who benefit from the wrongdoing will continue to benefit	Those harmed by the wrongdoing will continue to be harmed, including self
Direct attempt to stop the wrongdoing	Wrongdoing stopped, intrinsic reward; keep information about the wrongdoing and wrongdoers 'local'; often the fastest solution; control, personal responsibility for response and outcomes	Retaliation from wrongdoers; labelled untrustworthy; extrinsic rewards less likely than with procedural reporting
Report wrongdoing 'by the book' (procedural reporting)	Possible extrinsic rewards for stopping wrongdoing; wrongdoing stopped and deterred; anonymous reporting systems protect reporter	Negative reaction from organisation management; lack of recognition if anonymously reported
Non-procedural reporting	Wrongdoing stopped and deterred; intrinsic reward, sense of accomplishment; extrinsic reward (public recognition)	Retaliation from wrongdoers and organisation management; some people disdain a 'squealer'

Source: McLain and Keenan, 1999.

paucity of research on ethical dilemmas and decision-making in organisations, especially as they relate to the five options in Table 10.2. What organisations may wish to do is modify or extend not necessarily the number of response options, but the potential gains and losses in the context of their own organisational culture. It is more desirable to weigh more heavily on the potential gains of whistleblowing as long as these are seen to be constructive.

Methods of Moral Decision-making

One of the most comprehensive evaluations of scales for ethical decision-making in business and marketing was developed by Vitell and Ho (1997). These authors group scales into eight different categories: (1) personal characteristics, including studies of one's value systems or cognitive moral development (e.g. Hunt & Chonko, 1984); (2) organisational environment, including informal norms and codes of ethics that impact decision-making within the organisation (e.g. Hunt & Vitell, 1986); (3) deontological norms, which include scales designed to measure beliefs and norms (e.g. Mayo & Marks, 1990); (4) stakeholder scales, including studies that measure the importance of groups, such as organisations, clients and peers (e.g. Vitell & Singhapakdi, 1991); (5) deontological and teleological evaluations (e.g. see the bipolar scale of Reidenbach and Robin (1990), which is premised on the notion that agents use a variety of different ethical philosophies in making decisions); (6) ethical judgements (e.g. Hunt & Vasquez-Parraga, 1993); (7) intentions, which involves a method to measure the actual intention to make ethical decisions (e.g. see Mayo & Marks, 1990); and (8) ethical behaviour, which includes methods to measure the ethical behaviour of different groups (see Ferrell & Skinner, 1988; Fraedrich, 1993).

In view of the many studies listed above, the work of Rest (1979, 1986) and Victor and Cullen (1988) is perhaps most often cited or duplicated in related research. Rest developed the Defining Issue Test (DIT), which is a non-interview instrument designed to assess moral reasoning without a reliance on the verbal skills of respondents. The test is organised around six hypothetical dilemmas (e.g. saving a life, discontinuing a school newspaper because of its disturbing influence), which are used to assess levels of moral reasoning. In this way it has been implemented to alleviate some of the methodological concerns identified in Kohlberg's work. Although the test has been used almost exclusively in a business context, the scenarios are generic enough to be used for research in other areas. The reliability and validity of the DIT have been well established through a number of applications and range from 0.70 to 0.80. Conversely, the Ethical Climate Questionnaire developed by Victor and Cullen is a Likert-style survey instrument containing 26 items designed to assess a person's perception of the ethical environment of the organisation in which they work. The perception is based on the combination of philosophical decision criteria and the level at which the criteria are applied to ethical issues in the workplace.

These two instruments, in addition to the ones listed above, are thought to have excellent application to tourism in a variety of different capacities (e.g. see the work of Upchurch & Ruhland (1995) in Chapter 6, who used Victor and Cullen's work on ethical climate types in tourism).

The multidimensional ethics scale

One ethical scale that has been widely used and supported in business is the multidimensional ethics scale (MES), developed by Reidenbach and Robin (1988, 1990). This scale was developed primarily to improve the measurement of ethical evaluations of marketing activities. The MES is a semantic differential scale consisting of eight items representing three dimensions of ethical behaviour. These three dimensions are the result of a factor analysis of 33 items developed a priori from deontological, justice, teleological and relativistic theories of ethics. The analysis consistently yielded three dimensions, including broad-based moral equity, relativism and contractualism.

Moral equity includes theoretical perspectives such as deontology, justice and relativism. In this regard, 'decisions are evaluated in terms of their inherent fairness, justice, goodness and rightness. Moreover, this dimension incorporates the idea of family acceptance' (Reidenbach & Robin, 1990: 646), reflecting our early socialisation by family and religion to understand the right and the fair. The relativistic dimension allows us to better understand the sociocultural milieu of the system in which the moral agent is acting. Fundamental to this construct is the belief that 'tradition and culture shape our beliefs, values, and attitudes in all aspects of life and certainly influence our notions of what is right or wrong' (Reidenbach & Robin, 1990: 646). The third dimension, deontology, emphasises one's obligation to abide by rules, contracts and duties:

> business exchanges involve a quid pro quo wherein one party is obligated to provide a product, service, employment, or perform some action in return for something of value. … Violation of these implicit ideas would result in the condemnation of the exchange process or at least part of the process as unethical. (Reidenbach & Robin, 1990: 647)

Schwepker and Ingram (1996) used the multidimensional ethics scale to investigate the relationship between a salesperson's moral judgement and job performance (see also Barnett *et al.* (1996), Clark & Dawson (1996), Cohen *et al.* (1993) and Hansen (1992) who have used the scale in a number of business and marketing contexts). Their findings lend support to the belief that salespeople who make more ethical decisions with regard to selling practices are those who perform at higher levels. As a result, the authors suggest that ethical decision-making should be fostered within the sales force given the relationship to performance. Schwepker and Ingram view these results in the context of what Labich (1992) reported on in a survey of *Fortune* 1000 companies, which found that just over 40% of respondents were involved in conducting ethics seminars and workshops, while one-third of participants had established an ethics committee. Further consideration for operationalising ethics in the workplace may take the form of what Trevino has suggested with regard to ethical disequilibrium:

> Salespeople's cognitive moral development can be enhanced by engaging them in challenging moral decision making, particularly by exposing them to forms of reasoning one stage higher than the one to which the individual is accustomed. As a result, cognitive disequilibrium develops and the individual questions the adequacy of his/her own level, and considers the merits of the other. (Trevino, 1986, as cited in Schwepker & Ingram, 1996: 1157)

In a tourism context, Fennell and Malloy (1999) adapted the MES to gauge the ethical nature of selected tourism operator types. Ecotourism, adventure, fishing, golf and cruise line tourism operators were selected in order to represent certain activity-based segments of the tourism market. Ecotourism was selected due to its non-consumptive, educational focus; adventure tourism for its activity-risk orientation; fishing as an example of consumptive outdoor recreational use; while the cruise line/golf industries were included in the study (and combined as one group) based on: (1) their mass or mainstream tourism market focus; and (2) their larger organisational structures, in comparison to the more specialised niche markets of the other groups. In total, 167 surveys were returned, representing a response rate of 23.4% (operators: ecotourism, $n = 39$; adventure, $n = 49$; fishing, $n = 45$; golf/cruise line, $n = 34$).

For the purposes of this study, three new scenarios were developed by the researchers that were tourism specific, each of which was designed to provoke 'real-life' responses from the sample (see Box 10.4). Respondents were asked to respond to three tourism scenarios – economic, social and ecological – using the scale. In general, ecotourism operators were found to be more ethical than their counterparts on the basis of the responses to these scenarios. Upon further analysis, the authors discovered that ecotourism operators were found to be more ethically oriented than fishing operators, with the appearance of a trend between ecotourism and cruise line/golf operators (ecotourism operators still being more ethically oriented in their responses). In addition, ecotourism operators were found to: (1) use a corporate code of ethics more often than other tourism operators; (2) have higher levels of education (corroborating the theoretical and empirical work of Kohlberg); and (3) have smaller organisational sizes than other operators. All three of these measures have been found to relate to higher ethical standards in business practice.

The authors concluded by suggesting that, although the ethical practices in business contexts generally have received considerable attention from practitioners and researchers in recent years, the investigation of the ethical aspects of tourism is at a relatively early stage. The challenge for tourism researchers, therefore, is to use and interpret the growing volume of literature on ethics outside the tourism discipline in a way that is meaningful to both researchers and the tourism industry. For example, the measurement of tourism operator ethical viewpoints may be deemed important as a means by which to further examine differences between types of tourism. While tourist typologies abound in the literature, there is very little that exists on typologies of tourism. Ethics, in the way it could be addressed through the application of any of the aforementioned scales or techniques, holds tremendous potential. The knowledge that different types of tourism have different ethical viewpoints – social, economic and environmental – may be used to arm decision makers with the ability to foresee potential benefits and

Box 10.4 Ethical scenarios

Scenario 1

A young man, recently hired to develop strategies to attract visiting international tourists to a day-use tourism site, has been working very hard to favourably impress his boss with his selling ability. To lure in the tourists, he exaggerates the number and quality of attractions at the site and withholds relevant information concerning the tourism product/ experience he is trying to sell. No fraud or deceit is intended by his actions; he is simply over-eager.

Action: His boss, the manager of the site, is aware of the salesman's actions, but has done nothing to stop such practice. (Use the eight-item scale, p. 282.)

Scenario 2

An international resort development firm has recently built a large resort on an island (a small underdeveloped country), which has not experienced a high level of tourist development in the past. One of the policy conditions of the development was that the resort be fairly self-contained so as not to allow the expected high use to adversely affect the traditional lifestyles of aboriginal people living and working in the more peripheral regions of the island. Despite the initial attempts at control, there has been significant contact between tourists and local people all over the island, based primarily on the selling of trinkets and souvenirs at local markets.

Action: Although the resort's management realises the situation, nothing is done to curtail the emerging pattern, which is seen as having both positive and negative spinoffs. (Use the eight-item scale, p. 282.)

Scenario 3

A community organisation that operates a spectacular private mountain nature reserve, on land it owns, has the dual mandate of protecting the resource and allowing some visitation. In recent years it has been experiencing an increasing amount of tourism to the site. One of the main problems in reaching the reserve is a lengthy, rough unsurfaced road. As a result, the organisation is entertaining the notion of paving the road in an attempt to provide better access to the site. In doing so, they anticipate even more of an increase in tourism to this sensitive ecosystem in coming years. The economic benefit from this increased visitation will, they believe, provide more money in order to upgrade the quality of trails, facilities and overall conditions for both staff and tourists, despite the fact that there are some concerns that, at present, the reserve seems too crowded.

Action: The organisation goes ahead with the plans to develop. (Use the eight-item scale, below.)

Your response to this action is that it is...

Fair	1	2	3	4	5	6	7	*Unfair*
Just	1	2	3	4	5	6	7	*Unjust*
Morally right	1	2	3	4	5	6	7	*Not morally right*
Acceptable to my family	1	2	3	4	5	6	7	*Unacceptable to my family*
Traditionally acceptable	1	2	3	4	5	6	7	*Traditionally unacceptable*
Culturally acceptable	1	2	3	4	5	6	7	*Culturally unacceptable*
Violates an unspoken promise	1	2	3	4	5	6	7	*Does not violate unspoken promise*
Violates an unwritten contract	1	2	3	4	5	6	7	*Does not violate unwritten contract*

Source: Adapted from Fennell and Malloy (1999).

problems in the development of tourism in local and regional contexts, and to act proactively instead of reactively.

Conclusion

The many methods and models included in this chapter serve to illustrate that there are a number of ways in which to focus on ethical decision-making. It should also be noted that the move towards comprehensiveness in ethical decision-making has opened the door for an ethical pragmatism among scholars and practitioners. This means that philosophers are sometimes not the best or only people who should be involved in problem-solving. While philosophers employ a certain skill set in the resolution of particular issues, other stakeholders, including tourism operators, park staff and government officials may aid in the process of problem-solving based on their 'closeness' to an issue (Fennell, 2000b). Wrongdoing and whistleblowing were included in this chapter, along with a model of risks and responses to wrongdoing, for the purpose of linking this contemporary concern with the other forms of moral reasoning. The breadth of options for ethical decision-making opens the door for a style or process that may be adapted to many different circumstances or settings in the tourism field. A selection of the models discussed in this chapter are used to address a series of different ethical dilemmas in the tourism field in the next chapter.

11 Case Study Analyses

The suggestion that in doing philosophy one should not try to banish or tidy up a ludicrously crude but troubling thought, but rather give it its day, its week, its month, in court, seems to me very helpful
Foot (2003: 1)

Introduction

In this chapter an effort is made to answer the question: 'How?'. This takes place through the implementation of a three-stage method for addressing different tourism industry issues, according to: (1) the basic dilemma; (2) the specific ethical issue; and (3) the application of a moral decision-making framework to cast a different light on how we might address the situation through ethics. A different decision-making framework from the preceding chapter will be applied to each in an effort to accomplish this end. Six main issues will be dealt with, including sex tourism, ecotourism, all-inclusives, the Holy Land, volunteerism and cruise ship tourism.

Sex Tourism

Basic dilemma

Canada's *National Post* newspaper reported that the Outdoor Life Network, also in Canada, has recently come under fire for airing a new TV show in 2003 called *Red Light Districts*. The show highlights some of the many sex tourism destinations around the world by exposing viewers to 'adventurous activities that stimulate the human spirit', as well as depicting 'intriguing people, places, and subcultures', as described by Blackwell (2003: A5). One of the initial shows was about the sex tourism industry in Bangkok and Pattaya, Thailand. While offering no criticism of the practice of sex tourism in these areas, the show unabashedly illustrated young girls in school uniforms who openly negotiate sexual services for money. More explicitly, the show recorded, in an electronically distorted set of images, a Scottish man having sex with two women. The host of the show explains that tourists need only negotiate, be friendly and have a good time. Spokespersons for the network claim that the show is nothing more than a documentary, which reflects reality in the region, but does not promote prostitution in any way. Blackwell quotes a spokesperson for Beyond Borders, an organisation that fights child prostitution and pornography, who was outraged at the show. She suggests that 'Presenting young women and young men ... as sexual objects, as people you should come and be sexually involved with it because they're good at it, they're cheap and they're Asian, is positively, not only offensive, but un-Canadian' (Blackwell, 2003: A5). A major concern to Canadians was that the show received funding from the Canadian Film/Video Production Tax Credit Programme, and features the Canadian flag logo at the end of the show.

At the heart of the sex tourism issue is the power and control that dominant parties have over children, women and families, compelling these individuals or dependents to be sold to the sex trade in the most abusive and degrading ways imaginable. It is an industry fuelled by corruption, increasing demand and tradition. Sex tourism has grown astronomically over the course of the last two decades. Participants (tourists) are not simply some of the worst offenders in society, but also businessmen, husbands, brothers, fathers, next-door neighbours and community leaders (Beddoe, 2003; Garely, 2004). They travel not only to traditional sex tourism places, such as Thailand, where the practice is firmly entrenched within religion and society (Leheny, 1994), but also to a widening array of destinations that have been nurturing the sex tourism trade. In fact, there are over 1 million children involved in the sex trade in China, the Dominican Republic, India, Pakistan, the Philippines, Sri Lanka, Taiwan, Thailand and Venezuela alone (Garely, 2004). The commercial sexual exploitation of children, defined as the 'use of children for sexual gratification by adults for remuneration in cash or kind to the child, or a third person(s)' (Leheny, 1994: 1), has become a multibillion-dollar industry facilitated by the tourism industry to its own advantage. This has taken place through the efforts of tour operators who arrange hotels and ground transportation, tour agents who sell packages to tourists and hotels that allow adults to solicit women, men and children at their bars. The smaller hotels frequently use sex tourism, and the money it generates, as a means to compete with the larger hotels in these destinations (Ryan & Kinder, 1996). What has further spawned this industry is a legal system that does not punish those who abuse children and where bail is often a euphemism for bribery (Barr, 1995). Interestingly, an International Labour Organization study says that eliminating child labour in the developing world, and placing kids in schools, would produce economic benefits (e.g. improved productivity, decreased health costs) of approximately $5 trillion by the year 2020, while the costs of removing children from work would be about $760 billion (Williams, 2004).

Sex tourism has been explored in detail by McKercher and Bauer (2003). These authors contend that child sex tourism and commercial sex tours are the most negative and exploitative of a range of different forms of sex tourism. The main motivation for such travel is to engage in sex, which is, they characterise, negative and exploitative rather than positive and mutually beneficial, as in the case of a family holiday or a honeymoon.

Specific ethical issue

When we place a monetary value on people, even our friends, children or our country, we reduce humanity to the lowest level imaginable. This is what happened in slavery, where perpetrators essentially removed one's moral identity (Tetlock, 1999). By treating women and children, the main protagonists in the sex trade, as means to an end, the industry diminishes their moral identity by reducing these individuals to objects over which to exercise violence, control and exploitation.

Moral decision-making framework

Schumann's (2001) moral principles framework is used to examine sex tourism from the vantage point of five unique normative theories. The example of sex tourism is close to the example of why sexual harassment is unethical in the work environment, as described by Schumann. Each of the five theories is described below and used to explain why sex tourism is morally wrong.

Utilitarianism

As examined in Chapter 3, utilitarianism focuses on the ends or consequences of action or inaction, so we must decide what the consequences are from sex tourism. The utilitarianism principle is summarised by Schumann as follows: 'The morally correct action is the one that maximizes net social benefits, where net social benefits equals social benefits minus social costs' (2001: 97). The focus here is both on sexual gratification and financial benefit, both as ends, which can come at the cost of human dignity and the dislocation of family, as outlined earlier. Considered from another angle, human relationships based on sex alone and not on any other personal contexts are diminished. For these reasons, net social benefits (it is argued that there are more social costs than benefits from this form of tourism) are reduced.

Rights

While utilitarianism focuses on the ends or consequences of action, rights, as a deontological theory, focuses on the means or methods, by ascertaining whether the person committing the act has the right to do so or not. The concept of rights can be examined as follows:

> The morally correct action is the one that the person has a moral right to do, that does not infringe on the rights of others, and that furthers the moral rights of others. (Schumann, 2001: 99)

From this perspective we can examine three concepts: reversibility, or the notion that the person performing the act would not want the act performed on him or her; universalisation, which considers if everyone in the world should perform the action all the time; and respect and free consent, which considers if we are treating people as means to an end. Each of these three has relevance to Kant's categorical imperative, as discussed in Chapter 3, and perhaps more directly to the practical imperative, i.e. to treat as an end and not simply or merely as a means to an end. People, Kant argued, contain not only theoretical reason, i.e. that which allows for our ability to think logically in regard to, say, mathematics, but also practical reason. This latter power is said to serve our goodwill, which is the motive that compels us to be morally good (Honderich, 1995). The rights principle fails in this case because the perpetrator would not wish to be the victim of sex tourism (reversibility); because sexual control and manipulation should never be universalised; and because sexual manipulation and control do not treat people with the respect they deserve, since it is often the case that the victim has not freely consented to his or her involvement.

Distributive justice ethics

As justice is the theory behind the assessment of what is fair or not in society, this third principle adopted by Schumann is summarised as follows: 'The morally correct action is the one that produces a fair distribution of benefits and costs for everyone who is affected by the action' (2001: 101). Here, Schumann speaks of a number of different aspects of distributive justice, including egalitarianism, or the notion that humans should be similar in all relevant aspects; capitalism, which is the belief that the benefits accrued to any individual are a function of the contributions that he or she makes; socialism, which supports the belief that those in need deserve more benefits; and libertarianism, which, as noted earlier, means that what is relevant comes about as a result of the free choices of individuals. Sex tourism fails because it cannot produce the same level of good vs. harm (egalitarianism); because the victim has often done nothing to earn the role as prostitute (capitalism); because the victim cannot shoulder the harm (socialism); because the victim has not chosen to be harmed (libertarianism); and because it does not bring about equal liberty or opportunity.

Virtue ethics

Virtue ethics differs from the preceding theories by focusing on the character traits of the individuals in question. Virtues include compassion, fairness, honesty, benevolence and so on. By contrast, vices include characteristics such as greed, selfishness and thoughtlessness. To Schumann this means that the 'morally correct action is the one that displays good moral virtues, and does not display bad moral vices' (2001: 103). Sex tourism exists as perhaps one of the most explicit examples of bad vices. In this regard, sex tourism fails the virtue principle because it is a vice and not a virtue.

Ethic of care

This principle emphasises the importance of special relationships that may emerge among individuals through an examination of the level of care demonstrated between interactants. This is summarised by Schumann as follows: 'The morally correct action is the one that expresses care in protecting the special relationships that individuals have with each other' (2001: 104). Care ethics suggests that although we have different biological leanings, we have the capacity to choose how to act; it provides an alternative perspective on how to reason through problems or situations. And the historical experiences of women's private sphere which may include compassion, trust, cooperation, and so on, may add complexity to other normative theoretical perspectives in arriving at a moral praiseworthy decision.

As social beings we live in a web of relationships that should be nurtured through love and other intrinsic values. Sex tourism can be viewed as the antithesis of this perspective, where short-term relationships are forged with little attempt by the perpetrator to nurture a correct and respectful relationship with the victim. It is the trade aspect of the sex industry that is most disturbing – turning a natural act into an industry, with the exploitation of individuals for the benefit of others. The practice often occurs even when tourists know that what they are doing is wrong. Fennell (2015a) used the concept of akrasia to explain why tourists are prone to transgressive acts like sex tourism. Akrasia is defined as weakness of will or why individuals behave irrationally, i.e. why we act against our better judgments. In cases like sex tourism and the purchase of endangered species souvenirs, the search or need for pleasure overwhelms good judgement.

Ecotourism

Basic issue

Ecotourism has been both heralded and denounced for its role as one of the greenest and most ethical forms of travel. Pundits argue that because of its: (1) focus on sustainability; (2) small-scale nature and ability to place money in the hands of locals; (3) conservation mandate; and (4) experiential and educational capacities, ecotourism is both a device to stimulate social and environmental learning, and a sustainable model of tourism development (Fennell, 2015b). On the other hand, critics argue that ecotourism has failed because it has not accomplished these ends. The argument is that, although there may be successful ventures that support these lofty goals, there is too much variability in programme content that exists across space and time, which naturally undermines the ability to instil positive, harmonious relationships. This has created structural inequalities, which plague ecotourism in the same way they plague other forms of tourism, as discussed earlier. Antarctica is included here as an example of an ecotourism destination that has wrestled with some very significant ethical dilemmas (scenario adapted from Fennell, 2003a).

As the largest remaining wilderness on earth, Antarctica is visited by some 5000 scientists each summer at one of 37 research stations operated by 18 different countries. Because of the unique location of Antarctica and its international significance, no one nation has sovereignty over the region. It is governed by the Antarctic Treaty System of 1959, which came into effect in 1961 to provide for the conservation, research, management and use of Antarctic resources. As regards tourism, the Treaty in Recommendation VIII-9/1975 acknowledges that 'tourism is a natural development in this Area and that it requires regulation' (Heap, 1990, cited in Bauer & Dowling, 2003).

The Argentinian ship, *Les Eclaireurs*, was the first to bring tourists to the Antarctic peninsula (a 1600 km spine jutting from the Antarctic continent towards south America) in 1957–1958 (Grenier, 1998). In total, four cruises from Argentina and Chile brought about 500 tourists in that season. Lars-Eric Lindblad is generally acknowledged to be the first to consistently run international ecotours to Antarctica from New York (he chartered an Argentinian naval ship), starting in 1966. By the end of the 1990s, 84,173 ship-based passengers had visited Antarctica, with 14,623 alone visiting the region in 1999–2000 (Bauer & Dowling, 2003). The importance of ship-based tourism remains today, accounting for up to 90% of all tourists to the area.

Most trips to the Antarctic peninsula depart from Ushuaia, Argentina, Punta Arenas, Chile or Port Stanley in the Falkland Islands, with the trip from the South American continent to Antarctica taking approximately two days. Ships accommodate anywhere from 35 to 400 passengers, and itineraries to this region of Antarctica vary according to the type of vessel used. Trips beyond the Antarctic peninsula require an ice-breaker. Those travelling to the Ross Sea section of Antarctica depart from Hobart, Tasmania, Invercargill, New Zealand or Christchurch, New Zealand, from late December to late February, with mid-January to mid-February reported to be the best time to view wildlife. Ecotour activities on such trips normally include activities like viewing penguins (some ships have helicopters with which to visit certain penguin colonies), whales, seals and sea birds, visiting historic sites, scientific and whaling stations, and appreciating the grand scenery of the region. Tourists typically spend about two hours on shore at a time walking around and photographing wildlife.

Due to the absence of legislation regarding tourism in the Antarctic treaty, tourism operators, through the efforts of Lars-Eric Lindblad, took it upon themselves to develop codes of ethics for operators as well as for tourists. Lindblad also inspired a coalition of cruise operators, called the

International Association of Antarctica Tour Operators (IAATO), to help administer these guidelines (Stonehouse, 2001). The Rules for Antarctic Visitors, as distributed by Marine Expeditions, are as follows:

- Do not leave footprints in fragile mosses, lichens or grasses.
- Do not dump plastic or other non-biodegradable garbage overboard or on the continent.
- Do not violate the seals', penguins' or seabirds' personal space.
- Start with a baseline distance of five metres from penguins, seabirds and true seals, and 18 metres from fur seals.
- Give animals the right-of-way.
- Stay at the edge of, and do not walk through, animal groups.
- Back off if necessary.
- Never touch the animals.
- Do not interfere with protected areas or scientific research.
- Do not take souvenirs.

Despite these guidelines, a number of ethical transgressions have taken place over the course of many years, as noted by Grenier (1998: 186). Some of these include:

- A musician scaring away a group of penguins by giving his 'first Antarctic flute concert'.
- Passengers collecting stones, feathers and bones as souvenirs.
- A passenger tossing stones at the foot of a penguin to improve a photographic opportunity.
- Visitors touching penguins.
- A tourist walking straight into a penguin rookery, despite warning, stating that he had 'paid for the right to go as he pleased'. The penguins were scared off, exposing chicks to predators and freezing temperatures.
- A captain positioning himself next to a seal for a better photo opportunity.
- A zodiac driver consuming alcohol before an excursion conducted after sunset.
- A crew member harassing juvenile penguins dying after having been abandoned by their parents.

Researchers note that the management of people in this type of wilderness is difficult because of: (1) the lack of sovereignty by any one country; and (2) because of the differing cultural views on wilderness management philosophies (Davis, 1998). For example, tourists have consistently favoured the development of more facilities on the continent, such as toilets, accommodations, a post office and gift shop, and of activities such as diving, skiing and camping. Davis' preference is to stipulate which activities are acceptable as opposed to regulating all activities equally. And because only

2% of the land is ice-free, there is a tremendous amount of competition for space and resources, not only between tourists and researchers, but also between humans and wildlife (Hall & Wouters, 1995). This is compounded by the fact that the tourism season coincides with the peak breeding season. The constant human intervention is thought to cause behavioural changes in animals, denudation of habitat and the spread of new animal and plant diseases. In 1997, for example, a poultry virus was found to have infected Antarctic penguins, brought to the continent either by scientists or tourists. The concern is not only what types of impacts are occurring on Antarctica, but who causes them. While scientists worry about the impact of tourism, conservationists say it is science that threatens the fragile Antarctic environment (Tangley, 1988). This issue has touched off a debate on the responsibilities and effects of both parties, with scientists most vociferous about the dangers of an ever-increasing presence of tourists on the continent.

Like the Galapagos Islands in Ecuador, Antarctica is truly one of the world's prime ecotourism hotspots. How it comes to be managed is absolutely critical as a case study that might provide meaningful leadership in the management of other tourism regions around the world. In considering this region, the question of humankind's obsession with visiting every corner of the planet comes to mind. While some argue that tourists have no reason or purpose to see Antarctica, others suggest that it is our right as humans to go where there is the opportunity to see and experience new lands. The question might become more important over time as we see tourism continually expand into the world's most pristine areas. The argument for tourism in these areas is the notion that it helps to support economically depressed communities. Antarctica is perhaps the only place in the world where this position holds no credence.

Specific ethical issue

This ethical issue involves two Rules for Antarctic Visitors, as defined by Marine Expeditions (above) and also two ethical transgressions that go against these rules (also above). These are as follows:

Rules. (1) 'Do not violate the seals', penguins' or seabirds' personal space'; and (2) 'Never touch the animals'.

Transgressions. (1) A passenger tossing stones at the foot of a penguin to improve a photographic opportunity; and (2) visitors touching penguins.

Moral decision-making framework

Malloy *et al.*'s (2000) comprehensive ethical decision-making model is selected for this dilemma. This framework involves three stages, including sources of ethical decision-making, moderators influencing ethical decision-making and the process of ethical decision-making.

Sources of ethical decision-making

In applying the Malloy *et al.* framework, the decision maker is urged to consider the analysis from three separate ethical perspectives, including teleology (good), deontology (right) and existentialism (authentic). As these have been discussed at length in previous sections of the book, they will not be elaborated upon here, except in cases that demand further interpretation.

Moderators influencing ethical decision-making

As an *external* moderator, the political situation tempers the extent to which ecotourism may develop in Antarctica. Those who sanction ecotourism in Antarctica, including the 'society' of tour operators, as one of the world's flagship ecotourism destinations, would view such behaviour as unacceptable in line with the management objectives of the regulatory (political) system in place. *Organisational* moderators include those attitudes, values and beliefs of the ecotourism organisation itself (i.e. the service provider). In general, most ecotourism operators in this region subscribe to a more robust ethical orientation, which will be reinforced not only within individual organisations, but also between service providers. This orientation, however, is only as valid as the individuals charged with the responsibility of upholding these values. Those who are only interested in the satisfaction of clients, at any cost, would turn a blind eye to such incidents, if detected at all. In general, however, it can only be assumed that the organisational climate of the service provider would be inconsistent with the actions of the transgressing ecotourist. *Significant other* moderators would include the actions, reactions and pressures that might result from other tourists or other participants in the area, including the scientific community who also occupy this space. It may be the case that other ecotourists observing the unethical actions of others may act to prevent such actions from occurring in the future by either alerting the tour guide or confronting the perpetrator him- or herself. Left uncontested, these actions may be seen to lie outside the value set of other tour participants, leading these others to ostracise the individual responsible for the transgression. Important in the decision to act or not against the transgressions are the *specific issues* related to the incident. If the incident is seen only to be a minor transgression with few consequences and a limited scope of impact, then observers may decide to do nothing. If, however, the act is seen to have a wider scope of impact, then the opposite may be true. Important here are the normative expectations of what is right or wrong for a particular situation in a given setting, and among certain groups. Again, given the mandate of ecotourism and ecotourists, actions that are contrary to the ideals of ecotourism may be viewed as punishable. Consistent with the previous discussion on moral development, there are many factors that play heavily on the ethical orientation of individuals. Finally, *individual moderators* of ethical decision-making are critical in assessing what is deemed right or wrong to the person. Individuals who maintain an existential view of life may focus on what expresses their freedom and creative will. The teleologist's position might be to get the picture (the best end) at any cost, whereas the deontologist might focus on obtaining photographs only by observing the rules of proper conduct. Tied to individual ethical conduct is the stage of moral development that the ecotourist has attained (see Kohlberg's and Gilligan's work, discussed in Chapter 10). It may be that those well-educated older females are those who are most sensitive to the natural world, while the less-educated younger males are most likely to commit an offence.

The process of ethical decision-making

As outlined in Figure 10.5 (p. 262), there are seven stages that comprise this component of the model. For this stage of the model it is significant to emphasise the importance of placing ourselves in the position of decision maker (e.g. guide), who has witnessed these transgressions or become aware of them through other means.

(1) *Recognition of the ethical dilemma*. Those who recognise an ethical dilemma from the teleological position will focus on the best ends for the group (utilitarianism) or for the individual (hedonism). The decision maker must determine whether the dilemma acts as a constraint to the goals of the group, or of conservation in general, given the example of ecotourism. Deontological recognition includes a focus on whether rules have or have not been followed, or duties assumed (or not). The authors (Malloy *et al.*, 2000) illustrate that, from the deontological position, certain questions need to be asked, including: (1) Has a rule, policy or code been broken? (2) Has a law been broken? (3) What was the implicit/explicit duty of the individual in this case? (4) Was a sense of duty followed? Existential recognition will entail a situation where some aspect of the individual's authenticity has been denied or restricted. This will prompt the decision maker to ask: (1) Is there a restriction on the person's freedom to choose? (2) Is there a restriction on the person's freedom to take responsibility for action? (3) Is there an attempt to control behaviour? (4) Is there an attempt to deny free will? These perspectives, and the questions they generate, provide the basis for a wide range of interpretation. After consideration of these perspectives, the decision maker must decide what is the actual problem and corresponding subproblems. The authors (Malloy *et al.*, 2000) further advise that it is important to consider all options, not just the first apparent problem, in order to organise issues and symptoms into their logical order.

In our situation, we can suggest that the decision to both toss stones at penguins and attempt to touch them is decidedly hedonistic, as it attempts to further the ends of the individual over the safety of the penguins. It also means that the group (ecotourists on the whole) are chastised by the actions of one, leading to costs rather than benefits for the entire group. Clearly, as illustrated by the code of ethics for Antarctica, there is a violation of pre-established codes of ethics that ask tourists not to violate the personal space of penguins and not to touch animals in general. So rules have been violated. From the existential perspective, the free will (if we can make this leap, freedom or instinct if we cannot) of the penguins has been violated by the actions of the ecotourists through the freely chosen actions of ecotourists, and perhaps by their lack of responsibility for any impacts that arise from this behaviour.

(2) *Generation of alternative solutions.* During this stage the decision maker develops plausible alternatives to the dilemma based on teleological, deontological and existential perspectives. This is important to avoid being locked into one theoretical perspective over others. Teleological alternatives will focus on the best ends for individuals, groups or the environment; deontological alternatives will focus on policies, codes of ethics or laws; and existential ones will focus on free will and authenticity. The generation of alternatives is said to entail a great deal of effort and creativity by Malloy and his colleagues, which may ultimately lead to an acceptable resolution.

The guide may decide that the only ends worth considering are those that benefit penguins and ecotourism as a responsible form of tourism. He or she may also suggest that pre-established codes of ethics must be adhered to in order to limit impacts and sustain ecotourism in the region for an extended period of time. A further choice would be to acknowledge that although it is important for people to enjoy themselves (free will and authenticity), there are only a few basic rules which need to be respected in light of the sensitivity of the environment in which they are situated.

(3) *Evaluation of alternatives.* Alternatives are assessed on the basis of the principles inherent in each of the three ethical theories, as outlined in stage 2. The alternative that best satisfies these criteria – in a comprehensive fashion regarding good, right and authentic – is said to be the one that is most appropriate for the ethical dilemma.

The decision may be to reinforce the tenets of ecotourism and the fact that the resource base must be left in an untrammelled state (respecting the greater good that may come to wildlife through responsible behaviour); that all guidelines must be adhered to no matter what the circumstance; and that, although it is important to obtain photographs, tourists must do so freely and in a responsible manner.

(4) *Selection of the ideal solution*. The identification of the ideal or comprehensive alternative does not always result in the use of this solution. The authors (Malloy *et al.*, 2000) are quick to suggest that organisations, human interactions and different settings are inherently chaotic. Such moderators, as identified above, are said to disrupt the ideal or theoretical process defined here in any number of ways.

(5) *Intention*. As noted in previous sections of this book, intention has a critical role to play in inducing action. Those who intend on taking certain actions are more likely to do so. However, the authors suggest that moderators such as peer pressure, organisational climate and moral intensity often have an effect on the ability to implement the ideal solution. The ideal solution thus gets modified into an actual decision – although it might not be the optimal one. Decisions then are made on the basis of extrinsic factors and not necessarily on the basis of what is right intrinsically. Using the example of the tour guide above, it may be that, because he wishes to project himself as one of the crowd instead of a guardian of wildlife, certain rules that should be upheld are not in order to allow people to gain closer access to these resources. This appears to be the case in Africa, where guides regularly drive closer to wildlife, especially when induced to do so through monetary rewards.

Returning to Antarctica, it may be that on the day of the tourists' departure the penguins, on the only day available to see this species, are relatively inaccessible (i.e. much further than five metres away as stated in the code of ethics). Acknowledging that ecotourists have paid many thousands of dollars to see and photograph these penguins, there is pressure on the guide to move closer with an attempt to minimise impacts. In this case the ideal situation is not feasible or practical.

(6) *The actual decision*. This becomes the product of arriving at a comprehensive decision (good, right and authentic), the discovery of an ideal decision, moderating variables that will allow for its implementation, and individual intent to carry it through. Although it may not satisfy all aspects of the dilemma, the decision maker can be assured that many different perspectives have been adopted in arriving at a resolution. In such a case, and considering the welfare of the penguins and the tenets of ecotourism, the rules of appropriate conduct, free will and responsibility as ecotourists, and the commitment and amount of money ecotourists have expended on this trip, the guide decides to alter the ideal situation. It is decided that ecotourists would be allowed to approach the penguins on a section of ground that will not be adversely impacted, if numbers are kept small. Ecotourists are asked to abide by the other rules as stated in the code of ethics.

(7) *Evaluation of the actual decision*. The decision maker must ask whether the actual decision has contributed to accomplishing desired ends, whether

it follows existing rules and whether it allows for free will and responsibility.

If the welfare of the penguins has been preserved, all ecotourists have been able to photograph and observe the penguins, the other rules have been observed (including staying at least five metres from the penguins), and the participants have been able to exercise their free will with responsibility, then the guide can claim that the actual decision was an ethical one. In observing the actions of ecotourists and reactions of penguins, the guide has the chance to evaluate formatively (i.e. during the course of the event) or summatively, (i.e. at the end of the event). If, for example, the penguins are being harassed, his formative evaluation would most likely lead him to rethink the decision to allow the tourists to visit the penguins before too much of a disruption takes place. If little or no damage has resulted from the experience for all involved, this may allow him to repeat this decision in the future if the same situation presents itself.

The ethical triangulation used above (i.e. the combination of teleology, deontology and existentialism in a comprehensive format for ethical decision-making) has been used in the context of ecotourism in the past (Fennell & Malloy, 1995). In this research, the authors argued that all stakeholders in ecotourism, including tourists, service providers and other decision makers, should moralise in a comprehensive fashion (see Figure 11.1). Although each advocates radically different perspectives, stakeholders in search of comprehensive ethical decisions may employ each of the foregoing positions to arrive at ethically good, right and authentic solutions in ecotourism. It was also suggested that it is improbable that people would rely exclusively on one pure form of ethics as a means by which to make decisions. As such, a triangulated approach would correspond to the many demands, both organisational and moral, that exist within the ecotourism industry.

All-inclusives

> All mass movements slip with the greatest ease down an inclined plane made up of large numbers. Where the many are, there is security; what the many believe must of course be true. (Storr, 1983: 377)

This quote emphasises that where there are many, there is the truth; and there is no greater truth than that which comes from money. This tack is defensible in discussing mass tourism, particularly from the perspective of the all-inclusive, which is seen by critics to have a great many social and ecological impacts on local and regional systems. I think of places such as Cancun, Mexico, which today is nothing like it was before tourism came to town. The residents of these towns are of two types: those who can see how

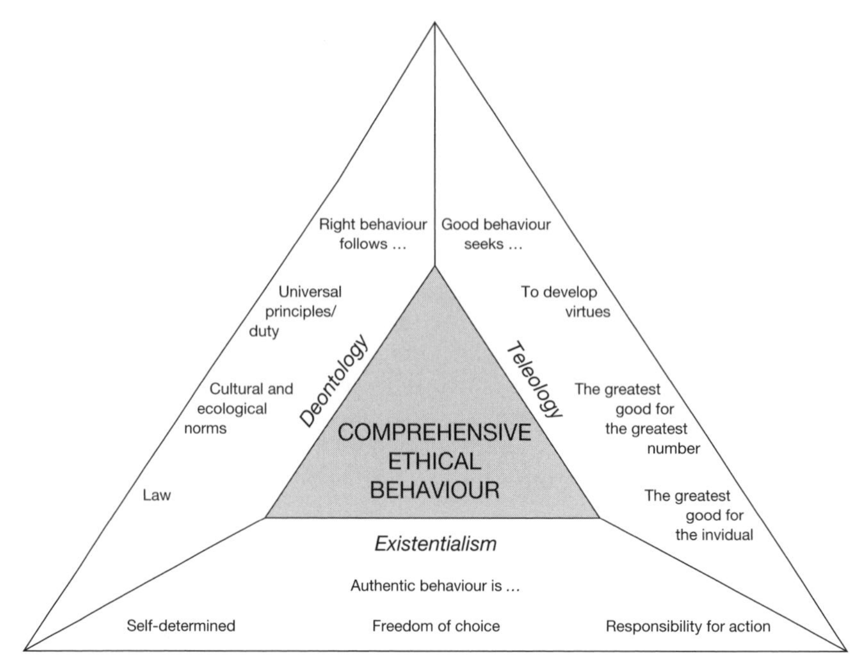

Figure 11.1 A model of ethical triangulation
Source: Adapted from Fennell and Malloy (1995).

things were before tourism arrived and those who have no recollection of these times but only what they have been born or migrated into. For the former, being on the outside, i.e. being marginalised to the outer fringes both geographically and socially, is to lose a continuity with the past and to know that it will never come back. This must be one of the worst forms of alienation. What is perhaps just as troubling is that the displaced are further marginalised because they stubbornly refuse to give up their cultural identities – refusing to step on the moving walkway of a political and economic ideology. The magnitude of the mass movement as something that is thought to be more superior forces the denial of the natural soul (Broom, 1987), and too frequently necessitates its replacement with an artificial one.

The all-inclusive resort concept first evolved in Jamaica in the 1970s, from: (1) the need for wealthy tourists to vacation in exotic destinations without the displeasure of witnessing the poverty that surrounds them; and (2) the belief on the part of hoteliers that there was the opportunity to make more money by selling guests extras (Josephides, 1997/8). The concept was so successful that it spread to most of the other islands in the Caribbean, contributing to the same pattern of disenfranchisement. This pattern includes: (1) the marginalisation of people being drawn to the larger tourism centres to eke out a living selling crafts, food, clothes and so on (Holland,

1997/8); (2) restaurateurs, shop owners and other local businesses going out of business because they are not associated with the tour operator or hotel (Jeng, 1997/8; Josephides, 1997/8; Pattullo, 1997/8); (3) hotel complexes marketed rather than the destination, with the local community losing tax revenues when locally owned hotels get pushed out of business (Farrington, 1999); and (4) guests feeling overly harassed by local people, many of whom work in the hotel (white-skinned guests are protected from the dark-skinned locals by armed guards), increasing the hostility within the local community. This latter issue is most effectively articulated in the following lengthy excerpt from the work of Pattullo (1997/8), which serves to illustrate a growing sense of uneasiness with the all-inclusive resort concept in the Caribbean:

> The popularity of *Alien* [a song about the negative aspects of tourism in St. Lucia] had also been fueled by a dispute between a well-respected St. Lucian musician, Ronald 'Boo' Hinkson, and the all-inclusive resort Sandals La Toye. Hinkson had gone to Sandals one evening to drop off two friends at the resort when a security guard demanded to search his car, saying that the procedure was company policy. Hinkson refused to give his permission, was held against his will and, as a result, sued the company for wrongful imprisonment. Hinkson expressed his outrage at what had happened at Sandals in an open letter to the Minister of Tourism, Romanus Lansiquot: 'No matter how many jobs, airline flights or US dollars Sandals brings into the country, they cannot buy the authority to trample on the rights and human dignity of St. Lucian people.' (Pattullo, 1997/8: 7)

A similar episode from another island touched the same nerve. The Prime Minister of Antigua, Lester Bird, was apparently barred from the all-inclusive Club Antigua in 1994 when a zealous security guard, not recognising him, refused him entry because he had no pass. The opposition newspaper *Outlet* (usually no friend of Lester Bird) commented:

> Antiguans and Barbudans who literally give millions of dollars to these very all-inclusive hotels in customs duties and taxes, do not, cannot, will not, and must not be asked to accept by those to whom we give, that we are not allowed in these halls, except accompanied by a trailing security guard. There was a time, in the long and far off times, when people of this country had to give a pass to be out of doors after the ringing of a church bell at night. We overcame that. We did not overcome that in those long and far off times to have it reimposed in another form in these present times. (Pattullo, 1997/8: 7)

The issue here is one of race relations, which places disadvantaged groups in a position of subservience. What are the grounds for such a separation? It used to be assumed that people of different regions and origins, and who looked different either by type of hair, skin and social habits, were in fact very different biologically. In fact, phenotypic differences have been used over millennia, and especially through the last 300 years, to classify humans as superior or inferior. However, Lewontin (1972) has put this to rest in his work on the apportionment of human diversity. What he found was that there is no scientific basis from which to make claims of racism, i.e. that the races were different and that some are more superior than others. He found that only 6.3% of human diversity is assignable to race, and concluded the following:

> It is clear that our perception of relatively large differences between human races and subgroups, as compared to the variation within these groups, is indeed a biased perception and that, based on randomly chosen genetic differences, human races and populations are remarkably similar to each other, with the largest part by far of human variation being accounted for by the differences between individuals.
>
> Human racial classification is of no social value and is positively destructive of social and human relations. Since such racial classification is now seen to be of virtually no genetic or taxonomic significance either, no justification can be offered for its continuance. (Lewontin, 1972: 397)

More recently, other theorists have shown that race can be eliminated as a factor in categorising people, when exposed to a different social world, in less than five minutes (Kurzban *et al.*, 2001). This work shows that even though subjects had experienced ethnicity as a prime predictor of social alliances and coalitions, this could be easily overridden by new circumstances occurring within one's social world. This is important because it illustrates that morality is an individualistic phenomenon as much as it is a social one. But even though we have moved beyond slavery, industries like tourism tend to perpetuate and exacerbate race inequalities. It then becomes a matter of our attitudes towards ourselves as members of one social group in comparison to others. In order to think outside this cultural membership, and of ourselves, we need first be able to ideate or imagine others.

Another prevailing theme perpetuated by all-inclusives as a form of mass tourism is child labour. We see this at airports, where the toting of tourist luggage by children may be viewed as a chronic problem. Tourists freely give their bags, and money, almost as a matter of ritual and without question or thought – because it is cute and because we are contributing to the welfare of the unfortunate fellow and his family. Logically, we can deduce that their parents either do not work or, if they do, earn very meagre wages; after all, why else would you enlist your child in such acts? But is it just a matter of parents wanting to earn extra wages, or does it cut deeper? Not unlike the sex tourism trade examined previously, there is often a third-party undercurrent to this phenomenon to which most of the money collected from this enterprise flows. In participating in this micro industry, tourists should be wise to the words of Oscar Wilde (1895), who said in his *The Soul of Man Under Socialism* that 'charity creates a multitude of sins'. The industry that has been created through children as 'frontmen' is most definitely a sin. So, as Mary Wollstonecraft (1792) once quipped in regards to the rights of women, 'It is justice, not charity, that is wanting in the world.' Giving money to children at airports is not justice and it is not equity.

Ethical issue

Although there are many social and environmental issues drawn from the literature on all-inclusives, I shall focus on bag-toting children as a means by which to exercise Haidt's (2001) social intuitionist model. The reader is asked to place him- or herself into the position of being confronted by a child at an airport in an LDC who wants to carry your bags to a waiting car.

Ethical decision-making framework

What complicates this ethical dilemma, especially for tourists who have not visited the hypothetical country before, is the influence of cultural relativism. That is, unless the tourist has done some extensive reading on the customs and practices of the people of the destination, it may not be known what the expectations are around the carrying of bags. It may also not be known what the repercussions are around refusing to have one's bags carried. But not unlike the passionate situation involving the brother and sister in Chapter 10, we may be confronted with the sense or feeling that having an eight-year-old carry our heavy bags is intuitively wrong. We may not know why it is wrong; we just feel it is inappropriate. In this case, if we cannot find any a priori accepted norms of society to help us make a judgement, we have to rely on post hoc reasoning, which is said to take place only after a judgement has been made on the basis of moral intuitions. Our moral intuition allows us to make a moral judgement on what is right or wrong, which we later reason through. As stated in the previous chapter, intuitionism has become more helpful in moral decision-making because it has been found that many of our cognitions take place outside our consciousness. People often cannot tell us why they think something is right or wrong. Emotions are not as irrational as once thought, and reasoning is not as reliable as once thought. The framework is based on the following six links, which are adapted to our ethical situation:

(1) *The intuitive judgement link.* What we intuitively feel is right or wrong, our moral intuition, allows us to make an automatic conscious judgement (i.e. we are provided with the ability to think about what is right or

wrong). This process is inaccessible, allowing only results to enter our consciousness. In our case, our intuition may compel us to decide that it is fine for the child to carry our bags.

(2) *The post hoc reasoning link*. In this link we are compelled to place effort into the search for reasons why we have taken this particular stance. It may be that we feel sorry for the child and want to help him and his family financially. It may also be that this provides us with our first authentic interaction with the people of the destination.

(3) *The reasoned persuasion link*. In this stage we verbally communicate to others who are with us or involved in the situation why we have made the judgement and why it is acceptable. This means that the tourist has explained to her travel partner why she thinks that allowing the boy to carry the bags is acceptable (i.e. that it will help him and his family, or other stated reasons).

(4) *The social persuasion link*. The decision by the tourist to allow the boy to carry her bag has a direct influence on her travel companion, who in turn accepts the argument and therefore allows the boy to carry his bag too.

(5) *The reasoned judgement link*. Sometimes people will use logic to arrive at a judgement, but the model supports the premise that this only occurs when the moral intuition is weak. In this case, and stage, there is no reason to suggest that logic has been able to overcome the moral intuition that says that it may be acceptable to have the child carry the tourist's bags.

(6) *The private reflection link*. After consideration of the situation, the agent may later spontaneously activate a new intuition that is in opposition to the original judgement. This can take place through role-playing or placing oneself in the shoes of others. In our case, we may arrive at a moral intuition that contradicts our original position. This may come on the back of our worries that the child, on a Tuesday morning, should be in school or, if not in school, doing what children of that age do naturally with their friends. This may compel us to consider our behaviour as wrong, and compel us to avoid this situation altogether in the future. We would not want to be in the shoes of the child, nor would we want, as parents, to send our child out in an effort to make more money for the family, because it seems to go against our beliefs about child-rearing and child happiness.

As a departure from the normative a priori approach to reasoning, this model has potential because of its ability to tap into the emotive side of human agency on the basis of moral intuition. Although it may be criticised because of its divergence away from societal norms, it may perhaps be useful because it allows us to tap into the thoughts or cognitions that take place outside our consciousness, and because emotion is not as irrational as once thought. Balance between reason and emotion may be defendable as a means by which to determine ethical action.

The Holy Land

Basic dilemma

Vicar Stephen Sizer (1999) of Virginia Water, Surrey, has identified a series of ethical challenges that have emerged in managing pilgrimages to the Holy Land. As a leader of such journeys from the UK to the Holy Land for 12 years, he suggests that the Israeli government and many tour operators have perpetuated political tensions that divide Jews and Palestinians because of policies that restrict the ability of Western Christians to interact with the indigenous Christian community. The implications of this are said to be considerable. The sensitivities of tour operators towards Palestinian Christians are reflected in the four categories of service providers as established by Sizer (Table 11.1).

Sizer notes that, while the first of these 'are benign, and the second blind, the third appear bigoted, and only the fourth offers any genuine dialogue between pilgrims and Palestinian Christians' (1999: 86). This led Sizer to contend that most tour operators fail to recognise how they are being manipulated by the Israeli government and the effects on Palestinians (one of which is the inability to make money through tourism). Other deleterious effects noted by the author include dispossession of their land, racial discrimination through the denial of basic human rights, being hidden from the Western touristic gaze, misrepresentation through Israeli media and propaganda, and depletion of the Palestinian community in Israel and the West Bank. This situation also led Sizer to postulate that the Holy Land is going the route of a religious theme park, hastening the day when Palestinian Christians disappear from the region entirely, and their churches are turned into museums. It is suggested that, in order to avoid this situation, tour operators must make some ethical choices about how

Table 11.1 Categories of Holy Land tour operators

Type of tour operator	*Nature of tour offered*	*Effects on indigenous Christians*
Secular	Specialist package	Irrelevant
Christian	Biblical archaeology sites	Ignored
Israeli or Zionist	Bible from Jewish dimension	Antagonistic
Living stones	Encountering the people	Encouragement

Source: Sizer, 1999.

to programme their services in a responsible and inclusive manner. These include:

- Preparation before visits regarding the history of the Arab–Israeli conflict.
- Meetings with indigenous Christians during pilgrimages.
- Visits to Christian projects, including schools and hospitals.
- Worship on Sundays with local Christians.
- Employment of indigenous guides and Palestinian services.
- Staying at Christian hospices and hostels or Palestinian hotels.
- Meeting with Jewish and Moslem representatives working for peace through justice.
- Seeking to develop long-term relationships with Christian communities.
- Becoming advocates for the indigenous Palestinian Church on return from pilgrimage.

Specific ethical issue

The division of Jews and Palestinians is being perpetuated by both governments and tour operators. In the former case, policies have been constructed by the Israeli government, which manipulates certain tour operators into restricting the ability of Western Christians to interact with the indigenous population. In the latter case, Israeli or Zionist tour operators are offering tours that interpret the Bible from the Jewish dimension only (antagonistic). Palestinians have thus been unable to make money through tourism, they are being dispossessed of their land, and there is racial discrimination, as well as misrepresentation of the Palestinian population through media propaganda.

Ethical decision-making framework

Theerapappisit's (2003) model of Buddhist ethics holds value in addressing the issues described above in the Holy Land. The Eastern conception of ethics is based on the notion that business is just one of many institutions that are firmly embedded within society. As a significant aspect of the economy, tourism should be developed and managed according to how it can contribute to the good of the larger whole. That is, tourism is considered good if it can contribute to the moral development of Eastern society, beyond simply making profits for a few at the expense of many. Central to the foundation of Buddhist ethics are the aspects of balance and scale, and these may be achieved through the concept of sufficiency, which advocates taking the middle road or moderation between extremes. The implications for tourism planning and tourism service provision are such that balance must be achieved in full view of the costs and benefits of tourism, at individual, social and external scales. The three scales and six ethical principles related to the issue are as follows.

Self-development (individual learning)

At this scale, balance is achieved through morality and wisdom. For morality, this involves mediation between altruism and greed, as two polar extremes. Too much altruism would perhaps mean that no one would benefit from tourism directly, including Israelis and Palestinians. Conversely, individuals who are responsible for the development of policies and those who conduct tours must strive to avoid their own selfish ends through tourism (e.g. marginalising the Palestinian population and structuring the tourism industry so that it benefits some and not others). The employment of indigenous guides and the use of Palestinian services, as noted by Sizer, may be a way of balancing greed with altruism in generating a moral tourism industry.

In further recognising the importance of tourism as an industry, too much impartiality would perhaps diminish the ability to protect the interests of the tourism industry in the spirit of healthy competition. However, weighing too heavily in the direction of bias would create massive inequities regarding what is fair and just across the full spectrum of human interest groups, including many interpretations of Christianity. Providing the opportunity to have tour group members attend any service of their liking may be a way in which to balance impartiality and bias in achieving wisdom in tourism.

Interactive (social organisational learning)

At this second scale, the focus is on the ethics of holism/dynamism and sufficiency. Regarding the former, interdependence is contrasted with disconnection in attempts to find balance. Too much interdependence (sharing the same values and networks) would perhaps take away from the special characteristics that make different religious sects unique. On the other hand, too much of a disconnection leads to singular concerns and interests that benefit only one group over others, when we should be celebrating our diversity instead of extinguishing it. Meeting with Jewish and Moslem representatives working for peace through justice, as noted by Sizer, may be one way in which to balance interdependence and disconnection.

Central to this social organisational scale is the aspect of sufficiency, which balances self-reliance against unlimited growth. Too much self-reliance, especially from the economic and cultural standpoints, forces us to rely only on the resources and values that we have within a narrowly depicted social network. Too much of a reliance on unlimited growth, however, will sacrifice the unique characteristics of the region, making it easier to favour certain individuals or religious groups over others. In this case, 'balance' may take the form of a heavier weighting on self-reliance, with an emphasis on the social mobilisation model instead of unlimited growth. This differential weighting may hold true for each of the other ethical perspectives as well. Seeking to develop long-term relations with all Christian communities, through tourism, may be a way to discover what is sufficient for all groups.

External influences (macro social environment learning)

As noted earlier in this book, cooperation and competition are central aspects of human nature. In achieving the ethic of non-violence there must be balance between these two dichotomous forces. Unhappily, this ethic is heavily weighted towards competition, leading to violence, which has at its root fundamental differences on the basis of religious interpretation. Buddhist ethics would thus suggest that this has created imbalance, which may only be restored through compromise, forbearance and democratic public consensus without oppression. Tourism may be a catalyst for this reassessment of values by encouraging tour operators to promote interaction with indigenous Christians, which may thus generate more interest in travel to many parts of the region and also generate more evenly spread spending patterns. This, in turn, may motivate policy makers, who recognise the value of tourism as a foreign exchange earner, to support cooperation over competition.

We also discovered earlier that religion, as a deontological form of ethics, is not always ethical. What is perpetuated by individual religions is often a fixed system that allows for little adaptability. This is represented in the sixth and final Buddhist ethic of causality. This ethic attempts to balance an adaptive system (a system that learns through integrative and systematic processes) with a rigid system, which is defined as one that is fixed and focuses on ends rather than processes. Tourism may thus be seen as an end that reinforces the differences in religious interpretations through policies and the actions of tour operators, who may either know or be ignorant of the manipulations of government. Contemporary research supports the value of adapted systems, which continue to learn through feedback mechanisms that instil resilience and equilibrium (Berkes & Folke, 2000). In the manner discussed above, and weighing more heavily in favour of adaptive systems in generating balance, positive change may be achieved through a form of tourism that allows for meetings with indigenous Christians, encourages visits to Christian schools and hospitals, and strives to encourage tour operators to include both Israeli and Palestinian accommodations in the itinerary.

Volunteerism

Basic dilemma

Tourists often volunteer, which means spending their holiday time and money involved in one or more altruistic endeavours, including assistance with any number of social or ecological issues within communities (Singh, 2002). In one of the first and most comprehensive treatments of volunteer tourism, Wearing (2001: x) defined the volunteer experience as 'a direct interactive experience that causes value change and changed consciousness in the individual which will subsequently influence lifestyle, while providing forms of community development that are required by local communities'. This form of tourism has direct benefits to both the community as well as the volunteer. After all, volunteerism is a selfless act (time, money and effort), where the benefits to one or many outweigh the benefits to the donor (altruism). At the same time, it is encouraging to read that volunteers do get something in return for their actions. This comes, as noted by Wearing, in the form of personal awareness and learning, interpersonal awareness and learning (e.g. appreciation of other people, conflicts), confidence and self-contentment.

Although it might be viewed as sacrilege, it can be argued that there are perhaps just as many selfish reasons for participating as a volunteer as altruistic ones. That is, could the act of volunteering be considered less virtuous if people did so for non-intrinsic reasons, such as social status or to place oneself in the position of securing a job in the future? Here we need to revisit the words of Oscar Wilde, who felt that charity creates many sins, and Mary Wollstonecraft, who argued for justice, not charity, in making the world a better place. The theme, which underscores the notion that acts of charity give power to those who execute them, has therefore long been recognised and reinforced in fiction and non-fiction. This has been emphasised by Myers (2003), for example, who writes that he has seldom met a volunteer in the conservation field who has not sought some personal reward from their conservation efforts. These rewards include feelings of superiority, opportunities to fish and hunt with people in power, accessing expertise or goods from people who work in the field, and taking advantage of business opportunities if the volunteer happens to be a service provider and in need of more business. Even when subjected to an educational regime, volunteers like famine workers (arguably the most devoted of volunteers), say that the intervention was not enough to induce them to sacrifice themselves to a point where their marginal loss equalled the victim's marginal gain. There is no doubt that such workers go to great lengths to alleviate the problems in question, but not so far as to disrupt the disequilibrium of gains. The argument put forward by those in helping positions is that they must keep healthy in order to maximise the gains for the less fortunate (Griffin, 1997). Keeping healthy, getting a job and increasing one's social status are all important in this calculus.

The tendency towards working for oneself under the guise of working for others is reinforced in biology, where all altruistic behaviour is said to result from either manipulation or self-interest. Regarding the former, and in reference to our discussion in Chapter 2, there are numerous cases where a member of one species exploits another for its own benefit, as observed by Williams:

> As a general rule today a biologist seeing one animal doing something to benefit another assumes either that it is manipulated by the other individual or that it is being subtly selfish. Its selfishness would always be defined in relation to its own genes. Nothing resembling the Golden Rule or other widely preached ethical principle is operating in living nature. It could scarcely be otherwise. Evolution is guided by a force that maximizes genetic selfishness. (Williams, 1988: 391)

In humans, although manipulation is selected in the interests of selfishness, it may also be used as a benefit to society. Williams observes that anyone who gives blood or donates money to any of the numerous public campaigns in existence is a victim of manipulation. Truly altruistic acts are not favoured by selection, but the ability to manipulate others to act altruistically can be. But Williams has also shown that manipulation can benefit both the sender and the receiver in an adaptive fashion. Vocal signals may be used by a hen that serve her own genetic interests but that also prompt obedience, which will serve the interests of her chicks (in the same way verbal exchanges in human families may serve the interests of children and parents, for example, in how to cross busy streets). It has also been argued that the label 'altruistic' should be given to those who raise money for a needy charity even though their status is raised within the community. The argument lies in the belief that it is the action or behaviour itself that is most important to society and not necessarily the motives that run behind it (Ridley, 1998).

This perspective flies in the face of hundreds of years of research on the importance of positive sentiments in acts of kindness, as well as a great deal of contemporary social science research on intrinsic motivation. If volunteerism equals better jobs and higher social status, or if there is an external expectation that 'as part of the "Smith" family, all children must volunteer because it builds character', the act may not be as genuine. This holds true in leisure research, where one is said to have attained a higher state of leisure the more one is able to participate for intrinsic motivations (as opposed to extrinsic reasons, such as participating because it increases your social status), and if there is the perception of freedom (as opposed to being controlled by others) (Neulinger, 1974). The implications for volunteerism as an altruistic act are apparent. This is not to say that people cannot change their motivations or seek deeper significance through volunteer experiences. We cannot paint all volunteers with the same broad brush. When Frankl (1985) asked people who were heavily depressed over major events in their lives to volunteer their time to some worthwhile organisation, their depression disappeared because they were able to find deeper meaning. This led Frankl to observe that those who were unwilling to help others were trapped in an existential box, which holds the individual hostage, leading to depression and angst. These negative feelings could not be alleviated by the search for more personal wealth or prestige, for, although these individuals had the means, they still lacked meaning.

Specific ethical issue

Three individuals, all of whom are friends and who have just graduated from the same high school (two girls and one boy), have decided to volunteer for an international agency in an LDC. The idea to volunteer for this agency came about from girl 'A', who had read about the plight of this particular group and, by herself, decided to travel at her own expense to make a difference. Hearing of this decision, the father of girl 'B', seeing that his daughter had no summer commitments, decided to pay for her ticket to go – arguing that such an experience would be good for her. The boy, who romantically likes A and is a cousin of B, decides to go as well with the belief that this may advantage him in his pursuit of A's affections.

Ethical decision-making framework

In helping us to more fully examine the motivations behind these three acts of volunteerism, we can use the two seminal theories of kin selection (inclusive fitness), and reciprocal altruism, identified in Chapter 2. As explained by Hamilton (1964), the theory of inclusive fitness posits that altruism takes place among those who are related genetically. That is, 'altruism would be more likely to spread if it were directed at close relatives, because relatives share genes' (Hamilton, 1964: 38). In contrast to this theory, Trivers (1971) illustrated that, although animals and people are often self-interested, they periodically demonstrate a willingness to cooperate, and this cooperation occurs if there is a chance that the beneficiary might reciprocate in the future. However, far from being completely altruistic, it was argued that both individuals would be doing each other a favour, as long as the benefit of receiving the favour was not outweighed by the cost of returning it.

The work of Alexander (1987) is helpful in this regard, through his comprehensive matrix of interactions of motivations and outcomes in determining the morality and immorality of social acts (see Table 4.1). The outcomes of social acts may be seen to include the aforementioned theories, moving from inclusive fitness to reciprocal altruism, and later to pure altruism as one moves down the options. His work is based on the notion that people are not always 100% selfish or 100% altruistic. In this regard, the belief is that in reality it is possible to combine selfishness and altruism in a way that illustrates many different positions depending on circumstance. This means that there is not always universal consistency in the moral behaviour of people all the time. For our purposes, it quite effectively shows contrasting positions of morality, which can be ascribed to the three individuals identified in our ethical dilemma. For example, we can say that the motivations for participation as volunteers are all different. For A the motivation is truly intrinsic and altruistic, with the intent of helping others and no expectation of a return. The motivation of B is to obtain a greater degree of social status within her family by winning the approval of her dad, without any further motivation to make a difference in the LDC community. She is aware of the motivations of her cousin, C, and will help him and, at the same time, not disclose his intentions to A. The motivation of C is to win the affection of girl A by using his cousin, who happens to be a good friend of A, with the intent of working as hard as possible in the community to achieve his ends.

The motivations and outcomes for A

It is clear that the decision to volunteer is a deliberate one, which is representative of the right side of Table 4.1. This individual would likely believe that her acts are altruistic, otherwise she might not do them, and so may expect to be successful in her venture (to win) because of altruism (column E). Regarding the outcomes of her social acts, it is believed that she would help anyone who needs it even if the immediate cost is great (second last option from the bottom). If we cross the column with the row, we see that her act is truly moral (cell 50). This means that her act is neither an act of inclusive fitness nor an act of reciprocal altruism, but rather an act that approaches true altruism.

The motivations and outcomes for B

In the case of girl B, it is clear that she is both helping herself in the eyes of her father and family, but also her cousin. These are the two outcomes of social acts that are found as the top two options (rows) on the left side of the table. In helping herself, and if she believes that she is selfish and expects to win because of her selfishness (gaining approval from her father, and seeing this way of life as a burden (column D)), then it is unlikely that her act is moral (cell 4). Conversely, she may view herself as being altruistic by helping her cousin win the affections of her friend. In this she wins because of this altruism, in the eyes of her cousin and perhaps friend (column E). But in helping herself and the relative in this way, the table illustrates that there is the probability that her behaviour is seen as being immoral because of the outcomes and despite motivations (cell 14).

The motivations and outcomes for C

The boy may also believe he is being selfish (soliciting the help of his cousin and going on the trip only to win the affections of another, but yet willing to work hard in view of this) and expect to win because of this selfishness. He may also see his way of life as satisfying, acting this way because he enjoys it (column C), or conversely, he may view this as a burden (column D). It is also clear that the outcomes of the social act relate to the fact that he is helping himself and others (the people of the community) in the presence of reciprocators (his cousin) (fourth option of 'Outcomes of social acts'). In the event that he is satisfied with his way of life, his choice is considered to be immoral (cell 30). If he considers his life to be a burden (column D) then it is also unlikely that his act is a moral one.

This moral decision-making framework provides scope on what is considered to be moral and immoral in light of the biological theories of inclusive fitness and reciprocal altruism. Structuring motivations and outcomes in this way helps us to shed some light on difficult dilemmas such as volunteerism and the range of options that might follow from the decision to volunteer, based on these motivations. In general, Table 4.1 illustrates that those acts that emerge from the inclusive fitness platform are found to be the ones that are most immoral. Acts that emerge from the reciprocal altruism standpoint are found to be slightly less immoral, while those acts that are altruistic in nature are the most moral of all.

Cruise Line Tourism

Basic dilemma

One of the biggest players in tourism is the cruise line industry. As of 1998, there were 223 cruise ships operating around the globe, and by 1999, 9.5 million people had taken a cruise (Dobson *et al.*, 2001). The magnitude of the industry is explained by Klein (2002), who notes that the number of people taking a cruise since 1970 has increased by more than 1000%. In North America alone the cruise market went from 1.4 million to 7 million between the years 1980 and 2000. Cruise ships are also getting bigger. In 1970, a typical ship accommodated about 650 passengers. At present, ships like *Voyager of the Sea* carry as many as 3840 passengers and 1180 crew (Klein, 2002). In recent years there has been an expanding base of literature on the cruise line industry which discusses economic, social and ecological impacts, as well as regulation, employment and racism.

Pundits of the cruise line industry argue that it contributes significantly to the welfare of both communities and tourists. Such contributions include: foreign exchange earnings for the economies of hundreds of destinations; a furthering of tourists' understanding of the need for conservation; an increase in environmental awareness, environmental responsibility, ecology and knowledge of indigenous culture; time, money and materials for programmes such as the Alaska Raptor Rehabilitation Centre; and portions of shipboard sales on certain items to benefit some threatened or endangered species (Davies & Cahill, 2000). However, the majority of the academic literature leans towards recognition of the many negative economic, social and environmental impacts of the industry.

Some of the most significant ecological disturbances from the cruise line industry include: solid waste (wet waste [kitchen wastes], recyclables, sewage, dry goods and paper); air pollution; oil and chemical effluent; introduced species; tourist activities that damage coral reefs; and fragile ecosystems through, for example, trampling of vegetation and habitat (Davies & Cahill, 2000). A more comprehensive list of impacts has been developed by the Bermuda National Trust (1999), including: a strain on transportation resulting in congestion; a strain on sightseeing facilities and hospitality services; a strain on sewage treatment plant (although ships are becoming more efficient at treating their own wastes); the accumulation of sediment in coral reefs, cutting down available light necessary for photosynthesis; the build-up of toxins that leach out of the painted hulls of ships, accumulating in sediment (not only cruise ships but also ocean-going vessels); the practice of dockside cleaning of ships, which has resulted in an 'abundance' of paint chips being found in the harbour settlement; an increase in oil spills (Royal Viking Sun taken to court in Bermuda in 1999); smoke stack emissions; and noise pollution due to increased traffic when ships are in port. According to Shaw (2000), large ocean vessels are the dirtiest source of transportation, leaving huge trails of smog across the major shipping routes, and are leading emitters of nitrogen (273,000 tons of nitrogen oxides in the US per year alone), sulphur oxide and diesel particulate. Tests conducted on cruise ships plying the Alaskan waters in the summer of 2000 indicate high levels of faecal coliform in treated sewage (black water) and untreated dishwasher and laundry water (grey water) (Bermuda National Trust, 2001). Added to this is the disturbance to residents, who in some cases must endure noisy helicopter tours, as well as an increase in prices of waterfront land and a decline in local fishing activity (Dwyer & Forsyth, 1997).

In some cases, the cruise line industry has been held accountable for their actions, including $10 million in fines over the last three years for Holland American Line and Royal Caribbean for polluting Alaskan waters (Bermuda National Trust, 2001). In 1998, Royal Caribbean Cruises Ltd pleaded guilty (fined $9 million) to dumping waste in the ocean in an effort to save millions of dollars, by admitting to rigging pipes to bypass anti-pollution equipment (Frantz, 1999). A federal indictment accused Royal Caribbean of having a fleet-wide conspiracy to illegally dump oily waste at sea. Government prosecutors stated that the dumping was so extensive and so long-standing that the practice had become 'routine'. In response to many of the transgressions, the cruise line watchdog Bluewater is pressuring the Environmental Protection Agency (EPA) to set 'feasible and enforceable standards on all future ships built in US' (Shaw, 2000). Furthermore, Alaska has passed a bill that will institute regulatory powers over cruise ships, setting strict standards and giving inspectors powers to board ships to take samples. In response, the cruise industry has hired four of Alaska's highest-paid lobby groups in an attempt to kill the state's proposed regulatory bill (Bermuda National Trust, 2001). These actions will be difficult, some contend, because of the strength of the cruise line lobby in Washington, which contributed $1.2 million between January 1993 and October 1999 to candidates in federal office in Washington. Most of this money went to Alaskan and Floridian delegations, where the majority of US-based cruising takes place (Fountain, 2017).

Although it has been suggested that the cruise industry contributes economically to destinations, the contribution is far less than land-based tourism (Bermuda National Trust, 1999). Cruise ships actually cost Caribbean islands more money than they bring in, through costs related to garbage, crowding, traffic and pollution, all of which strain local government, police and infrastructure. In addition, hotels are taxed excessively, while the cruise industry is essentially 'tax exempt'. Only about 1% of total taxes paid by the cruise lines go to the Caribbean islands (Wise, 1999). In 1993, 13 Caribbean islands attempted to form a coalition that would have imposed a $15 head tax on cruise passengers. In response, the cruise industry confronted the governments of each island and threatened to avoid their islands altogether. The sad reality of this control and manipulation, Wise observes, is that, although there are some benefits accrued to local people, cruise tourism is becoming a substitute for land-based tourism, resulting in a loss of revenue for islands.

Cruise lines also use flags of convenience to access low-wage labour (Mentzer, 1989, as cited in Bull, 1996). Crew often work 100 hours per week, usually with no health insurance or benefits, averaging in some cases only about $1 an hour (Nevins, 1989, as cited in Hobson, 1993). Corroborating this, a *Wall Street Journal* article reported in 1997 that a Haitian brass buffer earns only about $1.55 hourly, while a Croatian busboy earned only $1.95. In response, a cruise industry spokesperson emphasised that these are considered 'princely wages' when compared to what these individuals earn at home (Prager, 1997, as cited in Wood, 1999). Furthermore, most employees work for seven days a week, for six-month periods, with very few hours off at a time (Wood, 1999). Terry (2009) argues that wages have not greatly improved. Low-level workers are earning between US$ 400–500 per month under similar contracts and conditions. Although these wages are meagre, Carnival Cruise Lines points out that their catering employees often stay with the company for as long as 12 years (Rice, 1989, as cited in Hobson, 1993). As catering is an essential front-line aspect of cruising, it makes sense that a good portion of wages, relatively speaking, be allocated to this group. The reality is that most cruise ships have clear horizontal and vertical lines of ethnic stratification. This has created heightened levels of on-board racism, where white European and American crew receive better accommodations, wages and working conditions than the non-Western crew. An instructor with Royal Caribbean has stated that there is minimal contact between some staff and passengers because contact is disturbing to the passengers (Klein, 2002).

Specific ethical issue

There are several ethical issues that emerge from the discussion above. Perhaps one of the most disturbing is the number of companies that frequently discharge materials such as plastic and oil into the environment, as noted by Klein (2002). The basis of the ethical issue is, therefore, the practice of illegally dumping materials at sea for the purpose of saving money.

Ethical decision-making framework

This section references the work of Malloy and Fennell (1998b), who examined moral development in ecotourism cultures through the work of Schein (1985) on organisational culture, and through Kohlberg's (1981) and Gilligan's (1982) work on moral development. Through a combination of these models, it becomes easier to understand the culture within the organisation, as well as the stages of moral development of those working for a business enterprise or the industry.

Schein's 'peeled onion' comprehensive model of organisational culture is described in terms of four constructs, including artefacts, behaviours, values and basic assumptions or meta-values. Applied to tourism, artefacts encompass the physical characteristics of the organisation, such as its reward and punishment structure, the advertisements and brochures it produces, the tours it promotes, and the codes of ethics it stipulates to employees and to clientele. The second layer, behaviours, consists of the actual behaviours that are observable to employees, to clientele and to the public at large. For example, do individuals in a tourism organisation behave in a manner that is consistent with its written mandate as outlined in brochures and company literature? The third layer, values, can be either internalised by employees in a strong or clan-like culture, or they may be less strongly held or internalised in a transactional or exchange-oriented culture (Tichy & Devanna, 1986; Wilkins & Ouichi, 1983). If a value is a concept of the desirable with a motivating force (Hodgkinson, 1991), then this layer describes those concepts that activate or mobilise the kind of conduct that is manifested in the second layer of the 'peeled onion'. The core of the model, basic assumptions, consists of the basic ontological realities of the organisation. This core describes the meta-values (i.e. unquestioned tacit beliefs) that have come about as a result of the ability of these concepts to traditionally assist in the resolution of organisational dilemmas. This layer differs from the previous one by being much more deeply and latently held by employees. While the external layers of the model may be more readily subject to change and easily observed empirically, the more central layers and core are more difficult to change and require deeper subjective/qualitative methods to reveal their ontological nature. This model allows for observers to dissect and categorise the multidimensional nature of an organisation in order to explain why individuals do what they do – morally or immorally. Culture in an organisation, therefore, provides a template to foster organisational socialisation. It sets the expected values and norms for the behaviour of all members and provides the guidelines, written and unwritten, to distinguish the ethical from the unethical and the acceptable from the unacceptable, as well as the rewards and punishments for observing or neglecting these practices (Sinclair, 1993; Sridhar & Camburn, 1993; Trevino, 1990).

In the previous chapter, we examined Kohlberg's model of moral development, which described morality according to three levels and six stages. The first level, termed pre-conventional, reflects moral reasoning based upon highly egocentric rationale (avoiding punishment and seeking pleasure); the second or conventional level describes reasoning based upon the desire of the decision maker to receive approval from significant others or from society in general. It is important to note that, for both the pre-conventional and conventional levels, the locus of control for the decision maker is external. This, however, is not the case for the third level of Kolhberg's model, the post-conventional level, which reflects a mode of reasoning that has progressed beyond the influence of external sources and is internally driven and self-regulating.

Individuals base their reasoning on notions of community-based justice (i.e. social contract). What is right, therefore, is what the community has deemed to be right as opposed to what has been externally imposed. The final stage moves beyond the locus of the community to a universal perspective. What is just is that which is just for all of humanity and, more broadly, for the biosphere.

Malloy and Fennell juxtaposed Schein's and Kohlberg's models, substituting Kohlberg's pre-conventional, conventional and post-conventional levels of moral development with market, sociobureaucratic and principled organisational cultures, respectively, to more accurately reflect the tourism context (Figure 11.2). They argued that ethical decision-making behaviour in these organisations can be understood by identifying basic assumptions, norms, values, behaviours and artefacts (Carlson & Perrewe, 1995; Schein, 1985; Sinclair, 1993; Sridhar & Camburn, 1993), in association with an understanding of the cognitive environment in which advanced stages of moral development can be actualised.

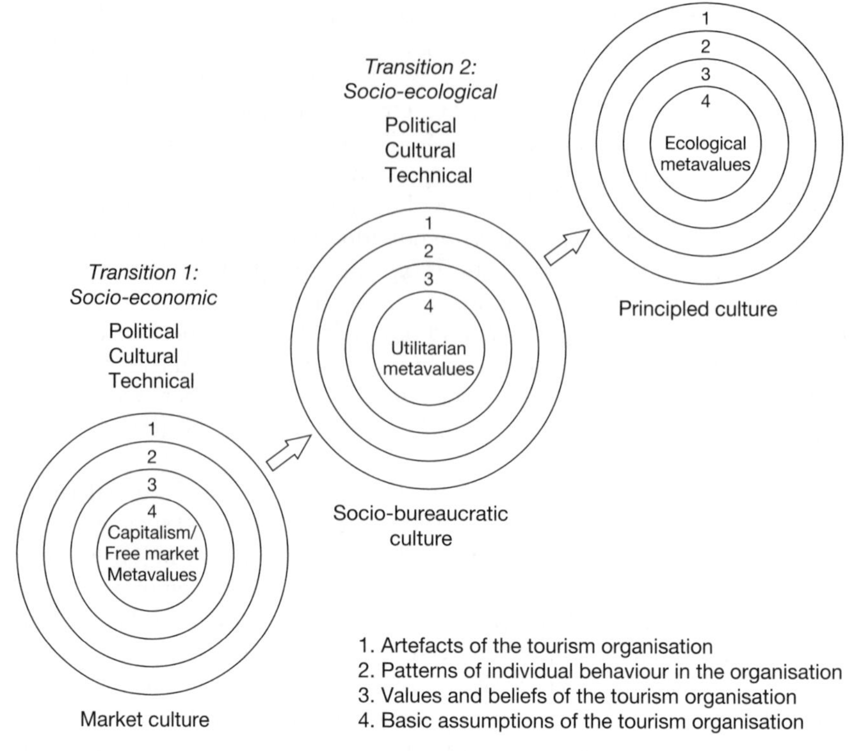

Figure 11.2 Moral development in tourism organisational cultures
Source: Adapted from Malloy and Fennell, 1998a.

The market culture

The market archetype (i.e. the pre-conventional orientation) may have as its basic assumption or business ontology the free market philosophy of capitalism. The view of ecology or nature may be Hobbesian, whereby sustainability may be obfuscated by the desire to exploit for profit. Organisational culture is defined by survival/profit of the individual employee, shareholder or member. Behaviour is geared towards achieving individual or organisational goals at the cost of the clientele, society or ecosystem. Ethical conduct may exist in this culture; however, it does so only as a means to avoid external punishment (i.e. laws – domestic or foreign or bad business reputations) or to seek external reward (i.e. good ethics is equated with positive business reputation and performance).

The sociobureaucratic culture

This culture (i.e. the conventional orientation) is equally concerned with profit, however the manner in which goals are sought may differ significantly from the pre-conventional counterpart (i.e. the market culture). Here there is respect for organisational policy as well as societal norms and expectations. Local customs of host countries or regions may be foremost in the minds of industry representatives as they seek to achieve their ends through organisational and societally sanctioned means (Van Buren, 1996). We can say that most responsible tourism organisations are likely to fall within the sociobureaucratic culture in terms of their moral behaviour. This does not include the cruise industry.

The principled tourism culture

The principled or post-conventional culture represents the most advanced moral archetype in Malloy and Fennell's organisational framework. The basic assumption here may include the notion of an ecological holism, where the organisation operates not only with region-specific laws and codes in mind, but also with the global ecology as a primary guide. In this regard, tourism service providers would *not* exploit biodiversity to gain profit for themselves or for the local inhabitants. Behaviour may result in profitable returns, but the means used by individuals operating in this culture would be based upon an ecologically sound and universally just rationale. If stakeholders institute values that go against the principled operator, he or she would cease to rely on these as they would fundamentally go against the values of the organisation.

In reference to the preceding three cultures, and the literature review on the cruise industry, we can be safe in suggesting that the culture that runs this industry is very much a market culture, where organisational goals are achieved at the expense of the environment. If there is an attempt to operate on an ethical level, it is based on the need to avoid external punishment. However, the willingness to even operate on this level is arguable by virtue of the number of pollution-related fines against these organisations, and for some of the unethical practices that have attempted to bypass regulations. So it is not only that they are not operating within the law, but also that they are deliberately going against these laws in order to be more profitable at the expense of the environment. If the cruise industry is to reach a heightened level of moral development within its organisational milieu, it will need to challenge the existing cognitive scheme or mindset according to political, cultural and technical processes (Cohen, 1995). Two transitional phases for cultural change were presented by Malloy and Fennell (1998b). These phases, not conceptually unlike those proposed by Gilligan (1982), represent the movement or evolution of collective cognition from market to sociobureaucratic culture (Transition I) and from sociobureaucratic to the principled culture (Transition II).

Transition I: Socioeconomic change

The transition from market to sociobureaucratic cultures involves an enhanced awareness and involvement of internal and external stakeholders. This transition comes about as the organisational leadership and/or membership begins to appreciate the limitation of the market perspective (i.e. basic assumptions of the market culture no longer sufficiently represent the ontological position of the leadership or membership) and begin to recognise the reciprocal relationship that exists and the subsequent influence that occurs between the tourism organisation and the social milieu in which it operates (Payne & Dimanche, 1996). Social consciousness is expanded and new demands are placed upon the behaviours, norms and values of the organisation. The following processes for change are suggested:

Political:

(1) Encourage a greater range of stakeholder participation in strategic decision-making (e.g. internally through employees, management, union, volunteers; externally through allied organisations, lobby groups, suppliers, clientele).
(2) When making decisions, include not only economic but also societal variables in the process (triple bottom line accounting, for example).
(3) Where possible and appropriate, attempt to decentralise or distribute meaningful authority and responsibility to staff, who may be able to react according to site-specific circumstances in a more ethical way.

Cultural:

(1) Through consultation with relevant stakeholders, develop an organisational vision that reflects a socially responsible and economically viable perspective.
(2) Communicate this vision through the development, promotion and implementation of organisational mission and value statements as well as enforced codes of ethics.
(3) Foster a sense that economic and socially responsible behaviour is a basic assumption of all members of the organisation.
(4) Develop a system of promotion that demonstrates that efficient and effective performance and socially responsible conduct are fundamental to employee conduct.

Technical:

(1) Encourage cruise line organisations to seek or develop certifications/standards.
(2) Try to ensure that more social and economic benefits accrue for local people.
(3) Educate staff and tourists about the importance of conservation in cruise tourism.

(4) Cooperate with other cruise lines for the betterment of the industry.
(5) Try to operate or support regional tourism planning schemes as much as possible.
(6) Encourage trip/programme evaluation to continually modify existing social and environmental measures.

Transition II: Socioecological change

The transition to the principled level of organisational culture involves what was termed by Malloy and Fennell (1998b) as a socioecological change. Where the first transition is indicative of the leadership's or membership's dissatisfaction with its current practice of free market capitalism, the second transition is a function of dissatisfaction with tourism driven by a social agenda as opposed to an ecological one. Pragmatically, this second transition is made possible when relative economic stability exists within the organisation and the host country. When economic viability is secured, ecological awareness is realistically conceivable. In this transition, a broadened concern exists for the wellbeing of internal (i.e. organisational members) and external (i.e. clientele and members of the host society) stakeholders. In addition, there may be an expanded view of the impact of tourism upon the resource base generally. The underlying theme for this transition is a movement towards a cosmopolitan, as opposed to a local, theme (see Gouldner, 1964). The structural inequalities that are inherent in the cruise line industry fully prevent such a transition at present. If it were to occur, however, it may include the following aspects:

Political:

(1) Encourage the participation of a range of stakeholders in decision-making, including those with global and regional interests (e.g. Greenpeace or the Sierra Club).
(2) When making decisions include not only economic and social but also ecological variables in the process, in balance with:
   (i) ecological principles and values to be considered;
   (ii) the relationship of cruise tourism with other types of land use;
   (iii) development of policy to control waste management, noise pollution, carrying capacity, etc.; and
   (iv) how cruise tourism fits into the context of a broader ecosystem management plan.
(3) Where possible and appropriate, attempt to decentralise or distribute meaningful authority and responsibility to staff as well as to those whose interests lie external to the organisation (e.g. development of partnerships in order to involve as many stakeholders in the industry as possible).

Cultural:

(1) Through consultation with relevant stakeholders, develop an organisational vision that reflects a socially responsible, economically viable and ecologically conscious perspective.
(2) Communicate this vision through the development, promotion and implementation of organisational mission and value statements as well as enforced codes of ethics.
(3) Foster, within the context of job training and organisational socialisation, a sense that economically, socially and ecologically responsible behaviour is a basic assumption of all members of the organisation.
(4) Develop a system of promotion and tenure that demonstrates that efficient and effective performance and socially and ecologically responsible conduct is fundamental to all employee conduct.

Technical:

(1) Require cruise operators to have special training regarding local and international regulations of the sea.
(2) Attempt to facilitate community development through social, economic and ecological benefits, while in the process ensuring the integrity of the resource base.
(3) Educate staff/tourists about conservation of the setting in the context of broader ecological issues.
(4) Seek out international cooperation to stimulate training and technology, and to monitor the impacts that cruise liners have on the environment, in the interests of all.

Conclusion

This chapter sought to exercise part of the gamut of moral decision-making frameworks as identified in previous chapters. Although some measure was taken to find complementarity between an issue and a particular model, the true test of a model is its applicability across the board, which is the case with the models included here. Although it was beyond the scope of the book to go into great detail on issues such as sex tourism, cruise tourism, all-inclusives and so on, it was only necessary to identify a few salient issues and apply the models accordingly. Even so, with a limited amount of information on each scenario, researchers may eventually come up with different conclusions on the application of the models to the scenarios. With this in mind, it becomes essential that those endeavouring to apply decision-making models to tourism situations: (1) have as complete a base of knowledge on the situation as a possible; (2) understand how the model works and how it can apply to the tourism case; and (3) be prepared to work with colleagues (brainstorming) in fleshing out as many interpretations of the situation as possible to avoid bias.

12 A Moral Tourism Industry?

Teaching and imparting of knowledge make sense in an unchanging environment.... But if there is one truth about modern man it is that he lives in an environment that is continually changing. The only man who is educated is the man who has learned how to learn ... how to adapt and change... who has learned that no knowledge is secure, that only the process of seeking knowledge gives a basis for security

Carl Rogers (2004)

Introduction

This chapter focuses on integrating the main themes of this book into a conceptual framework. These themes include the base of existing tourism knowledge, macro interactions (broad factors that dictate the evolution of the tourism industry) and micro interactions (relationships between tourists, hosts and service providers), as well as the importance of the humanities and biosciences in creating an enhanced base of knowledge for tourism ethics (adapted from Fennell, 2003b). This framework is anchored by the belief that tourism research, more than ever, needs to be interdisciplinary in its focus; that tourism issues are inherently complex, and will remain as such; and that the knowledge base in tourism needs to be extended in more effectively grasping the true underlying nature of tourism problems. A series of outcomes are identified through this framework, one of which, education, is discussed briefly. The chapter concludes with a review of some of the main ideas outlined in the book, including specific emphasis on how tourism ought to be part of, not excluded from, the global expansion of the circle of morality.

Interdisciplinarity

The generation of knowledge through what we view as science today has followed a long and established tradition that can be traced back to the scientific revolution of the 17th century. During this period, scholars such as Newton, Bacon and Descartes advocated a scientific and technical approach to knowledge generation over traditional approaches adopted by the humanities (the dichotomy between approaches was termed the 'Ancients' vs. the Moderns, or the 'Battle of the Books'). Moderns felt that belief systems that were wholly supported by Greek, Roman and theist doctrines failed to substantiate 'facts' because they were devoid of scientific proof, which could only come about through the discovery of natural laws and their submission to rigorous scientific methods (e.g. Darwin's natural selection). The fundamental schism between these two protagonists was further intensified through reductionism, which has continued to act as a cornerstone to science, enabling the development of innumerable achievements through specialised fields of inquiry. Although one cannot argue against the power of reductionism in science over the years, it has slowly and steadily forced us to look inwards at the specific in our efforts to expand the frontiers of knowledge. In doing so, however, there has been an associated erosion of the connectedness between fields. This specialisation has been most notable at the broadest level, where a divisiveness between the humanities and sciences, as noted above, was forcefully articulated in C.P. Snow's 1959 Rede Lecture Address at Cambridge University, where he stated that:

> I believe the intellectual life of the whole western society is increasingly being split into two polar groups: ... At one pole we have the literary intellectuals, who incidentally while no one was looking took to referring to themselves as 'intellectual' as though there were no others ... at the other, scientists, and as the most representative, the physical scientists. Between the two a gulf of mutual incomprehension – sometimes (particularly among the young) hostility and dislike, but most of all lack of understanding. They have a curious distorted image of each other. Their attitudes are so different that, even on the level of emotion, they can't find much common ground. (Snow, as cited by Gould, 2003: 90–91)

While the distancing has become readily apparent at the broadest level, it has continued within major disciplinary domains such as social science and the natural sciences. A classic example of this 'siloing effect' lies in resource management, where natural and social scientists have, in the manner discussed above, traditionally held very little in common. Social scientists, if even considered at all, are often left off research teams in favour of natural resource-based scientists who are seen to be more in touch with the issues in question (see Freudenburg & Gramling, 2002). Attwell and Cotterill (2000), for example, note that African conservation science is not open to deconstruction, but rather is essential in its form as an absolute science in isolating phenomena for inquiry for the purpose of making knowledge available to politicians for a balanced approach to decision-making. Social science, therefore, has a role to play in this process through the exploration of taboos, the underlying reasons for high fertility, the use of aid in funding and cultural biases that exist in the adoption of certain practices. Together, social and biological sciences can make inroads into very complex problems in African conservation as well as other areas. Unhappily, however, social science has been caught in the middle of this scholarly tug-of-war between the humanities and biology, because of what some scholars view as a limited foundational body of knowledge (Pinker, 2002), and where scholars are said to labour under the burden of a false science, where experiments prove that no measurable progress can be made towards our understanding of human nature (Saul, 1995).

This sentiment was also maintained by Claude Lévi-Strauss, who said the following regarding the place and role of social sciences, well before the present time:

> Therefore, the social sciences can claim only a formal, not a substantial, homology with the study of the physical world and living nature. It is precisely when they try to come closer to the ideal of scientific knowledge that it becomes most obvious that they offer no more than a prefiguration, on the walls of the cave, of operations that will have to be validated later by other sciences, which will deal with the real objects of which we are examining the reflections. (Lévi-Strauss, 1981: 643)

For this reason it becomes essential that scholars avoid the examination of different areas of research in isolation, especially with regard to phenomena such as conservation and sustainability, which beg for an interdisciplinary approach (Kauffmann & Cleveland, 1995; see also Echtner & Jamal, 1997). This was recognised in the mid-1960s by Wagar (1964) and separately by Lucas (1964), in their research on recreational carrying capacity, which relies both on natural and social science research. The move towards the integration of ideas is taking place once again in geography, where, after three decades of a move away from the regional science perspective, some geographers have begun to advocate the importance of examining geographical phenomena from a more diverse intellectual perspective (Markuson, 2002). The same has taken place in other areas of study, where researchers from many different fields have collaborated on projects such as plate tectonics, DNA and the human genome, because of the inherent complexity of these dynamic systems of study (see, for example, Brainard, 2002). It is therefore the problem itself, rather than the disciplinary orientation, which should be our central concern.

Interdisciplinary research is based on the integration of ideas from across fields and directed towards a common goal. In this regard it is essential that those involved have a fundamental understanding of the core concepts of the area, its research traditions or themes and the basic questions under consideration (Mitchell, 1995). Figure 12.1 illustrates five

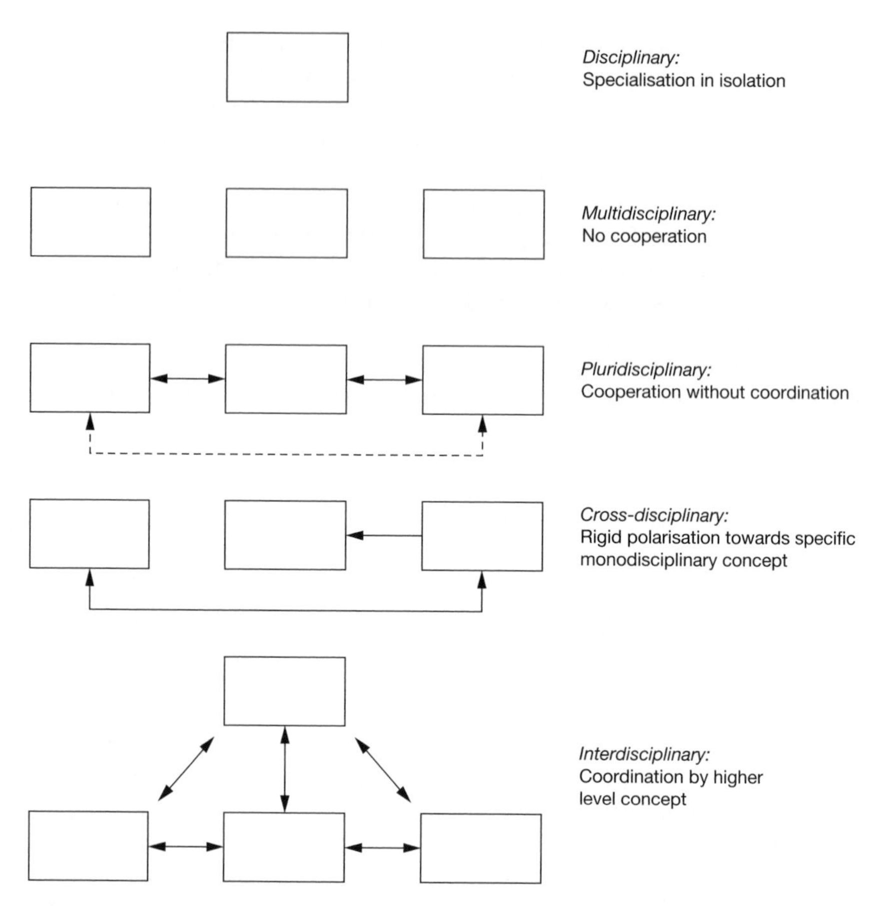

Figure 12.1 Increasing cooperation and coordination in research
Source: Jantsch (1972).

different disciplinary perspectives, the first four of which are said to be subsumed in the interdisciplinary approach (after Jantsch, 1972). The first of these, the disciplinary approach, relates to specialisation in isolation, while the multidisciplinary approach – one that is often confused with interdisciplinarity – generates little or no cooperation between areas. Pluridisciplinary approaches include those where cooperation exists but without coordination; while the cross-disciplinary perspective is premised on the notion of rigid polarisation towards a specific monodisciplinary concept. Coordination by higher-level concept is the chief characteristic of interdisciplinary science, providing a common link across several fields. For example, the study of rural tourism might best be accomplished through the work of geographers, agriculturalists, planners and environmental scientists. The arrows between disciplines in Figure 12.1 can be viewed as integrating mechanisms in emphasising that there may be a myriad of different ways, techniques, research methodologies, stakeholder groups and so on, through which to make connections between fields in finding optimal solutions.

While interdisciplinarity in principle holds great potential for research, it is not without its detractors. Disciplinary jealousies and territorialities – an *odium literarium* – are difficult to overcome, as well as self-interest that goes along with individual recognition over group recognition. Added to these problems are interpersonal conflicts, credit for teamwork, existing institutional structure and a series of logistical dilemmas that are often not anticipated at the beginning of a project. Academics come by this individualistic approach to research honestly (perhaps more so in some fields), as they are conditioned to work in isolation by virtue of their undergraduate and graduate school training.

The discussion on interdisciplinarity has been examined by E.O. Wilson (1999) in his work on consilience – the melding or jumping together of the humanities and the natural sciences. Wilson feels that it is no longer acceptable to 'silo' the major magesteria, because the most pressing questions on human nature cannot be absolved without the application of the humanities and the natural sciences together. In Wilson's words:

> To grasp human nature objectively, to explore it to the depths scientifically, and to comprehend its ramifications by cause-and-effect explanations leading from biology into culture, would be to approach if not attain the grail of scholarship, and to fulfill the dreams of the Enlightenment. (Wilson, 2000: vii)

This theme was taken up by Stephen J. Gould (2003; see also Daily & Ehrlich, 1999; Pinker, 2002) in his second last book, entitled *The Hedgehog, the Fox, and the Magister's Pox*. In it Gould, who claims ownership over the usage of the word 'consilience' over his intellectual adversary, Wilson (in reference also to *odium literarium*), based on the foundational work of William Whewell during the mid-1880s, argues that, while foxes have many ways of attaining a meal (consilience and interdisciplinarity), the hedgehog has only one way of protecting itself (i.e. to curl up in a ball for the purpose of self-preservation, relating to reductionism). The suggestion here is that scientists need to become rather more fox-like in realising that there are many different ways to solve very complex problems. Although specialisation has served us well in the academy, an interdisciplinary focus, i.e. bridging across fields, is in our best interests (Note: I have also argued in Fennell (2013c) that more specialisation is needed in how we *organise* our field, especially our journals, in an effort to become less amorphous and to move away from the audit culture that has tourism researchers, programmes and universities competing against each other.)

The need to cross over into other fields for theoretical and conceptual guidance is especially important in tourism, which can claim little theory of its own. This should be abundantly clear at the conclusion of this book. Although the multidisciplinary foundation of tourism has been instrumental in its development as a field, true interdisciplinarity is perhaps far less evident. As stated many times previously, the most obvious implication of this failed union is the many chronic issues that continue to plague our field. What is most ironic about this is that tourism fails to actualise the very basis upon which it has evolved. We are truly multicultural, in the disciplinary sense of the word, but how has this contributed to getting the job done on the ground? We are told that tourism is important by the many world leaders who continue to push their regions, especially after catastrophic events. But it is further ironic that, although tourism continues to be so important as a foreign exchange earner, its credibility as a legitimate science continues to be questioned (or ignored) in relation to other disciplines that are seen to have much more practical and theoretical relevance. For example, a US team from the University of Cincinnati has undertaken interdisciplinary research on the Greek island of Santorini, for the purpose of softening the impact that 3 million tourists create per year. At only 14 km long, with a population of 12,000, the team of planners, architects and biologists is trying to contain the sprawl, curb the sewage problems, reduce the reliance on cars and protect the sensitive cliff environments that make the region so attractive to tourists (Spears, 2004). In view of the growing complex array of problems in tourism at all scales, a move towards interdisciplinary research needs to take place if we are to create new opportunities for research and understanding for the purpose of filling many long-standing voids in knowledge both in and between fields of study.

Complexity

Unhappily, tourism scholars have come to the realisation that our problems are as chronic and multidimensional as they have ever been.

For example, in Chapter 3, I summarised Strassberg's (2003) work on the producers and consumers of ethics, as part of an ethical system. Moral systems can be less resilient in the face of chaos and complexity, Strassberg notes, as evidenced in the 11th September, 2001 attack on the USA, where producers and transmitters of ethics failed to inform society of critical knowledge a priori, which may have saved hundreds of lives. We can only assume that those who were charged with the responsibility for human safety and security at this time failed to do so, not out of ignorance, but rather self-interest.

In the absence of a firm grasp of an understanding of change, pressure and impacts through conventional models, researchers have begun to embrace new concepts and theories, such as complexity (Russell & Faulkner, 1999). This new science evolved during the 1970s in large part due to the work of two scholars in two very different fields (as described by Reed & Harvey, 1992). Bhaskar (1978) developed the philosophical ontology for complexity from the notion that the experimental, reductionist approach to science, and the strict laws inherent in this, served only to produce limited and idealised knowledge. Bhaskar argued that science's proper object of enquiry is one that is open to a collection of interacting entities that exist in the real world. Because of the diversity inherent in this openness and complexity, prediction through conventional scientific methods may produce uncertain results. One of his most convincing arguments lay in the inability of laboratory experiments, with too many uncontrollable and intervening variables, to replicate the real world:

> Bhaskar's philosophical ontology depicts the world as being a complex, emergently-structured and multi-layered universe of discrete entities and mechanisms. The world is historically open and loosely integrated. It consists of a manifold set of interacting powers and their overdeterminations. This interactive complexity and openness creates not only a basis for structural slippages and accidents, but gives nature itself, at all levels, an inherent historicity, one that is replete with evolutionary possibilities. (Reed & Harvey, 1992: 358)

The scientific proof of Bhaskar's work came at the hands of Prigogine through his work on dissipative structures and the laws of thermodynamics (Prigogine *et al.*, 1972). As described by Reed and Harvey (1992), conventional wisdom on systems dynamics postulated that each system typically moves towards a state of disorder called thermodynamic equilibrium, where energy is distributed rather evenly throughout the system (a disordered state referred to as positive entropy). What Prigogine discovered was that systems will also demonstrate a negative entropy, where the system exists in a state of thermal disequilibrium and is further characterised by evolutionary growth and internal complexity. Prigogine demonstrated that thermodynamic phenomena can be organised in three ways: (1) as equilibrated systems, which refers to the conventional systems theory perspective; (2) as near-to-equilibrated systems; and (3) as far-from-equilibrated systems. The state that is most relevant to our discussion here is the far-from-equilibrated system, described as follows:

> Far-from-equilibrated systems differ from their near-to-equilibrated counterparts by being evolving entities. In addition to preserving a state of minimal entropy production, they can also increase in complexity, and hence, increase their negative entropy production. Instead of being homeostatically constrained, their evolution is irreversible in time. As such, they

> move naturally from one equilibrated state to another, often radically different, reference point. Furthermore, unlike conservative systems, far-from-equilibrium configurations are subject to spontaneous internal fluctuations. These fluctuations constantly prove and push the system beyond its boundaries. ... For this reason there is seldom any way of using our knowledge of a system's past behavior to predict its future sequential unfolding or evolutionary end-point. (Reed & Harvey, 1992: 363)

The proof behind the far-from-equilibrated perspective opened the door for an entirely new approach to science based on complexity – defined as a system where there are a 'large number of simple elements, or "intelligent" agents, interacting with each other and the environment' (Lam, 2000: 71). In reiterating the initial basis of the complexity idea, Lam suggests that the behaviour of the system cannot be learned by reduction; that is, a knowledge of the parts is not sufficient to allow an understanding of the whole system. Systems can appear to be complex because we have insufficient knowledge to understand them (perhaps because of the application of reductionist means). Once this knowledge is in place, Lam notes, the system becomes simple. Complexity is the antithesis of the traditional deterministic view of science, which was based on the notion that, if we know all the initial components of a system, then we might be able to predict the future state of that system. Those who subscribe to the complexity approach to science will consider simple deterministic models as archaic because complex systems have emergent properties that cannot be controlled for in deterministic experimentation (Williams, 2000). The number of variables in such complex environments, which grow exponentially, might actually demand computing environments larger than the environment under study. And because complexity makes it difficult to understand systems and how they change, we need to learn how to live within systems rather than control them (Williams, 2000). What this implies for the progression of science is that complexity has set the foundation for the development of an epistemological bridge between traditionally disparate sciences, not only *within* humanities and sciences, but also *between*, as a notion quite contrary to the age-old tradition (see Byrne, 1999; Eve *et al.*, 1997; Perrings *et al.*, 1995; Williams, 2000). But although complexity represents a fresh alternative to a moribund and exhausted systems theory in the social sciences (Walker *et al.*, 2002), there is also the recognition that systems can be much more difficult to study (Stewart, 2001). This means that, instead of looking downwards in the system (as natural scientists do as in the case of the molecular level) or upwards (as social scientists often do), scientists will need to look laterally in discovering how other disciplinary approaches might address common problems at different scales and through the use of different methodologies (Katz & Kahn, 1978).

The open nature of tourism systems – the 'reality' component as discussed earlier by Lam – means that problems may be less structured and less predictable, or 'messy' (Checkland & Scholes, 1990). In such cases there might not be any one rigid solution to problems that occur within the tourism industry, but one or more optimal or best-fit solutions for a particular point in time. In one of a very few number of papers on tourism and complexity, Milne and Ateljevic (2001) write that it is vital that we transcend traditional boundaries of economic and cultural consumption if we are to more effectively understand commodification beyond the monetary exchange. Their discussion is hinged on the belief that complexity must be 'embraced if we are really to gain an understanding of the links that exist between tourism and the broader processes of development' (Milne & Ateljevic, 2001: 379). These are attainable, the authors note, through a number of mechanisms, including mutual dependency, adaptation, discussion and negotiation, honesty, long-term commitment, quality control and shared knowledge. An understanding of different forms and systems of knowledge is essential in coming to grips with complex problems under conditions of scientific uncertainty. Because the knowledge of the system we deal with is always incomplete, there is the need to integrate multiple expertises – scientific and lay public – in making appropriate decisions about systems (Holling, 1993, 2001; Scoones, 1999).

While complexity must surely provide us with the ability to look at phenomena differently, and perhaps even more effectively, we should still be careful about accepting it outright. Through complexity we see that destinations suddenly get discovered for no other reason than reasons we cannot seem to explain. But surely we can ascribe certain characteristics to the onset and evolution of tourism in an area (see Butler, 1980). But by explaining these developments away to complexity and chaos, do we not lose perspective on the human influences that have played a part, time and time again, in the processes of destination rise and fall? Surely the marketers have discovered these destinations too during the early stages of their development and applied their timeworn practices to induce visitation. I am not sure we should or can develop effective models of chaos and complexity to *replace* the older deterministic and linear models, but rather as heuristic devices to *complement* what we have learned thus far. So, while swinging the pendulum in the other direction is always a good idea, the avoidance of a 'complexity' or 'chaos' determinism is just as important as other forms of determinism.

Knowledge

What has made us so successful in cultural and evolutionary terms, at least from the perspective of Clark, Bronowski and others, is the need for knowledge. However, as described in the previous section, our ability to achieve more knowledge has been problematic because of the shifting sands of knowledge based on intra- and interdisciplinary variances (see

Lévi-Strauss, above). Part of the problem according to some scholars is that in pursuing knowledge we have tended to side with facts alone, drawn through instrumental means. But facts do not necessarily allow the lantern to burn on into the night, as much as ideas and imagination, which provide the real fuel for the revealing of knowledge. This has prompted Saul to note that we in society do not *understand* most of what we *know*. Citing the Scottish philosopher Thomas Reid, who said 'when a man is conscious of pain, he is certain of its existence', Saul observes that the recognition of the existence of pain does not mean we understand it. To have knowledge is to understand the relationship between what people know and do. Our actions are based on very narrow bands of quite specialised information, with little understanding of the broader picture. Shared knowledge is important, Saul (2001: 32) writes, because it is 'a manifestation of our collective unconscious'. The contractual obligations that pervade society serve only to destroy the ability of the citizenry to use this collective unconscious. Society is thus legitimised through the expression of this collective unconscious. As such, when shared knowledge is taken away, we reduce our capacity to solve problems. We let contractual obligations take precedence over citizen rights.

Wolfe and Fuller (1983) provide a good example of the utility of a shared knowledge perspective in examining the relationship between the generation of knowledge and how such knowledge can be applied to social utility and policy. These authors wrote on the importance of knowledge transfer between the Western ideologies (exogenous knowledge generated by experts through empirical observation, experimentation and theory building), and traditional knowledge (endogenous knowledge derived from members of aboriginal communities), in attempts to find common ground between people, institutions and ideas (cf. Chapter 9 and the discussion on Western and traditional codes of ethics).

While Saul (1995) would argue that shared knowledge lies at the core of any successful society, the same argument holds true for organisations and businesses. In business, knowledge management has become a key tool for the private sector and an important determinant of economic growth for governments at the macro scale, but also at the micro scale (i.e. for service providers) (Beijerse, 1999). Knowledge management entails 'a process by which information is transformed into capabilities for effective action and used to reduce the uncertainty of decision making' (Cooper, 2002: 375). Information is any content that can be communicated, whereas knowledge includes a context that makes the information both meaningful and useful. The intent is to allow entities like organisations to act as intelligently as possible through the elimination of chaos (Wiig, 1997).

One model of knowledge management that has generated enthusiasm among scholars recently is the communities of practice approach, which focuses on the analysis of groups of people who share common concerns, shared problems and passion about the problems/area, and who wish to broaden their knowledge of the problems/area by interacting on a regular basis (Wenger *et al.*, 2002). The authors suggest that a community of practice has three main elements, including a domain (the area of knowledge involved), community (people) and practice (set of shared ideas, frameworks and so on from which group members work); what Bourdieu has referred to as habitus. From the travel and tourism industry perspective, it can be suggested that the concept of community of practice as an ethical strategy might be broadened to entail the following:

(1) The domain of knowledge in tourism must be broadened, or the domain in other aligning areas must include tourism research, through interdisciplinarity, consilience or other means.
(2) The people who should be involved in this area must also reflect the different interdisciplinary approaches that are required to fix tourism problems equally and without prejudice.
(3) Increasingly, the types of methods and approaches required to fix such problems need to reflect an interdisciplinary perspective. Consequently, tourism problems need the operationalisation of more than just tourism approaches.

This is very close to Saul's (2001) notion that shared knowledge involves relationships between humans, the relationship of these humans to place, and the joining together of shared knowledge, which may be based on different assumptions.

In linking with the previous discussion on interdisciplinarity, knowledge may be of a practical nature or of a theoretical nature. Both are important, and just as in the earlier discussion on theoretical and applied ethics, both have a role to play in society. This is the message in the work of Blyth (2003), who developed what is referred to as 'Blyth's Bridge', in illustrating the relationship between practical knowledge and research-oriented knowledge. In his conceptualisation, Blyth identifies two main supports of his bridge. The first, *research*, deals with existing knowledge through a variety of research-oriented techniques designed to foster improved knowledge. The other main support, *practice*, seeks to move from current practices towards those that are designed to have an improved impact on society. The conceptualisation is framed by a series of support mechanisms, including strategic educational supports (e.g. higher learning academies and training programmes), as well as policy-oriented supports, which enable the connection between practice and research through research-inspired reflection and user-inspired research. Important in this work is that these main pillars and the various support mechanisms in place allow for applied studies, which can only evolve through the sharing of knowledge between research and practice.

Sensitivity to both the practical and research ends of tourism is especially important, because tourism is one of three prime agents (the other two being

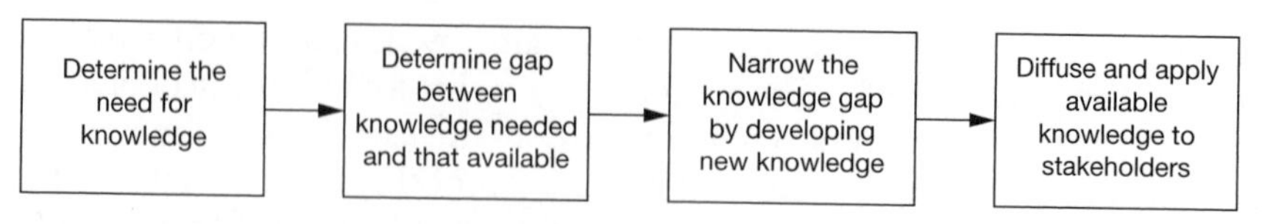

Figure 12.2 The knowledge value chain
Source: Weggeman, as cited in Cooper (2002).

media and the internet) responsible for steadily eroding national boundaries, raising consciousness of global realities and expanding the knowledge horizon (Hall, 2001). To this we may add that, while this 'shrinking of the world' is a phenomenon that will continue to unfold, the repercussions of this trend have largely been ignored. The framework in Figure 12.2 illustrates the importance of eliminating knowledge gaps through the generation of new knowledge, and how such knowledge might be used by an organisation. More specifically, Figure 12.2 illustrates that the flow of knowledge can be viewed along a chain. Initially there is a recognition of the need for knowledge. This may surface as a result of any number of social or ecological tourism problems within a region, and the implications for the region itself and perhaps beyond. Gorilla tourism in Central Africa is a case in point, where researchers have suggested that the health of mountain gorillas is in danger because of exposure to volume and frequency of tourists. Studies of gorilla faeces, for example, indicate the presence of human parasites that have persisted since the introduction of tourism. Beyond the initial stages of the conceptualisation there is the need to determine the knowledge gap between the level of knowledge needed and that which is available. If the level of knowledge available is not sufficient to ameliorate the problem, the knowledge gap must be narrowed through the development of new knowledge. In the case of mountain gorillas, it might mean determining the spatial and temporal frequency and level of interaction that might occur between tourist and animal. It would also entail the extent to which wild gorilla populations are susceptible to human exposure, perhaps in comparison to the susceptibility of captive groups. Once new knowledge has been generated, the final link in the value chain involves its diffusion and application to stakeholders. Although knowledge management is being used widely in a business context, the application of this model can directly impact investigations that are geared towards the development of new research and its applicability to theory and practice in other areas.

Synthesis

Figure 12.3 is an attempt to reflect on the previous discussion in this book in searching for a way forward in tourism ethics. While no attempt is

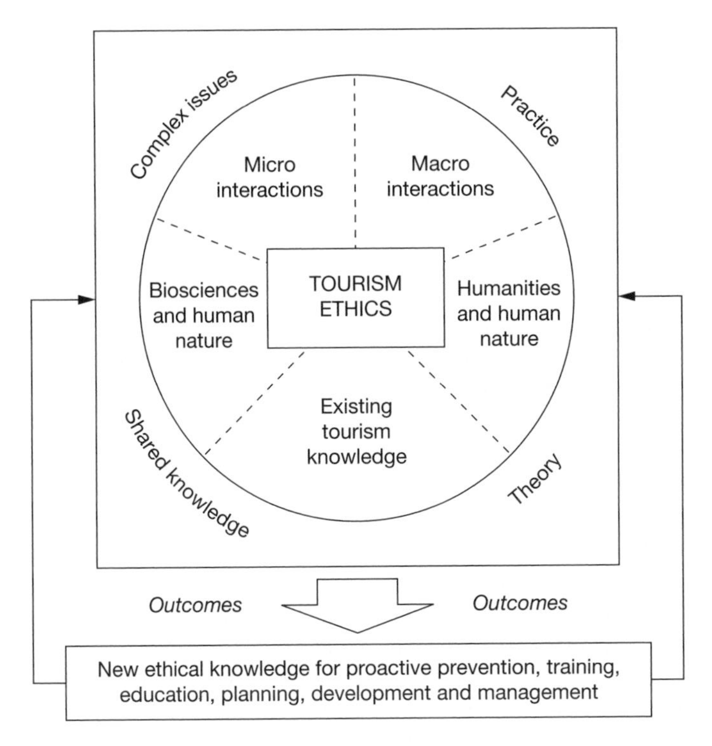

Figure 12.3 Framework for tourism ethics

made to organise the material in the figure according to scale (i.e. certain aspects of the figure as preceding other aspects, and in recognising that the small, local operator may be challenged differently from tourism operators at the largest scale), there are two main components of the illustration that are distinct. The outer realm, or the square, illustrates that tourism problems are the result of many complex issues, which are diverse and multifaceted given the social, economic and ecological basis of tourism. We understand these to be complex because there are often few ways to ameliorate or mitigate them, which has led to various long-term impacts. 'Practice' is included in the figure to support the practical and multifaceted aspects of the industry. For example, ecotourism, as one small market in tourism, can be practised in many different ways, some of which are ethically based and some of which are not. Theory is included, not as the antithesis of practice, but to suggest that, although these two aspects are often treated as such, there needs to be a strong bond between the two in our attempts to move forward ethically. This is further represented by the 'Shared knowledge' realm, which underscores the need to share knowledge not only in regard to theory and practice from the standpoint of research, but also from the perspective of the many different stakeholder groups who may be able to provide a clearer understanding of the issues that are most salient – for both researchers and stakeholders. The circular realm contains five main areas, which are said to cover, generally, the main aspects of ethics and tourism from both the industry and tourism research standpoints. The dashed lines in Figure 12.3, between the five sectors, illustrate that all of these components must be linked through the recognition of common problems brought on by the temporary movement of people across space. Each of these will be discussed in more detail below.

(1) *Micro interactions*. This aspect of the model takes into consideration the myriad tourist, host and service provider relationships, which create a huge array of different levels of interaction. Tourists, for example, travel for countless reasons, including to learn, relax and be self-absorbed, self-actualised and hedonistic. In catering to tourists, however, we should be wise to the words of Raymond Williams, who observed that for well over 200 years there has been a theme suggesting that the outsider has come to be morally superior to the insider (cited in De Botton, 2002). The present-day manifestation of this is the proclivity of tourists to feel that they are a superior breed over the local people who serve them, and therefore richly deserving of the experiences they seek. Host individuals, which may include a wide variety of different stakeholders, may be involved directly (e.g. through jobs) or indirectly (e.g. by having to wait in line behind groups of tourists). Depending on the extent of tourism within the locality, they will have to compete with tourists for their fair share of local resources. Tourism service providers may be motivated by a number of different reasons to be involved in the tourism industry. A primary motivator, however, must be commercial success, which will temper most interactions with hosts and tourists. This is not to say that interactions cannot be positive, but rather that relationships are struck with other stakeholders because of a commercial orientation. Furthermore, because all of these groups are involved in a wide array of different interactions on the basis of time, setting and circumstance, all of these are subject to consideration on moral grounds. For example, in destinations that are more mature (consolidated or stagnated, for example, in Butler's (1980) terms), and depending on the type of traveller in questions, tourists are often disadvantaged by where they come from and what they represent. What else can a host of a lesser developed country conclude about the wealthy people they encounter? This has prompted scholars to note that it is cultures, not liberal democracies, which are in conflict with one another, because of differing value sets and associated insecurities. While we can generalise about the feelings that might be shared between hosts and visitors, and these cultural differences, we should also be careful to suggest, or rather to hope, that the prejudice or the preconception can be stripped away through personal

exchanges of one form or another. These can be commercial, a smile, a conversation, an act or a gesture. In this we are presented with an opportunity to break down the preconceived notions about the industry, service and expectation, through genuine acts of kindness that allow us to bring meaning into our experiences and lives. Before such encounters tourists are simply part of an industry. Through these interactions we have the ability to emerge as individuals who are framed in circumstance only, and not in class or status. So, although we are blessed with the privilege of existence (Bryson, 2003), we have the singular ability to appreciate it and make it even better through any number of different ways – including travel (see Gibson, 2010, who argues that ethics should be about the multisensory dimensions of encounters, without being a slave to ethical essentialism, e.g. codes of ethics).

(2) *Macro interactions*. This component of a larger scale than (1) above contains a cross-section of broadly based ecological, social, economic, technical and political aspects regarding the movement of tourists from specific origins to specific destinations. Ethical consideration must be afforded to issues such as SARS, the sociopolitical climate of destinations, health risks to travellers, ecological impacts, risks to host populations, development inequities and so on. When regions get forced into tourism development we create a whole host of problems for the people who live in these places and the places themselves. For example, Niagara-on-the-Lake (population about 15,000) is a beautiful Victorian-style community about 25 minutes by car from where I live. The community plays host to the Shaw Festival of the Arts, attracting many hundreds of thousands of people to the region each year. I visit this place two or three times a year for various functions, but I could not live there because all I see is tour buses, no vacancy signs, manicured lawns and other accoutrements of the tourism trade. If I lived there it would not be my community. It would be owned temporarily by the hordes (stemming from the Turkish word *ordū*, which originally meant a camp, but which has further come to mean a disorderly swarm, as described by Boorstin, 1985), who visit this small place every year and who seldom interact with anything or anyone beyond service providers and the theatre. The competition for resources in time and space in Niagara-on-the-Lake reminds us that we should be sensitive to changing histories and changing values. I recall having a conversation with an elderly man in Cancun who, when asked about his feelings regarding change in his town, commented that the generation of memories seemed to cease when development took place – the happy ones anyway. We often forget that the places we frequent as travellers were once stable communities that had changed very little over many years. The commons issues that were discussed in Chapter 8 shed light on the struggle for resources that are unevenly distributed in time and space. Tourism is said to be a

commons issue because all do not share evenly in the profits, which many would argue need to be equalised across the board (common pool equity). This rather utopian vision might only come as a result of placing other more intrinsic values squarely in the middle of resource development and management.

Expected outcomes that are thought to come forth from the framework (Figure 12.3), include the emergence of new ethical knowledge for the creation of a safer, healthier and more responsible travel industry (prevention); the training of graduate students who should be better able to appreciate the magnitude of travel-related issues from the perspective of ethics; the education of service providers about how to balance financial success with ecological stewardship; the ability to plan with more than just economics as the only significant bottom line; development that chooses not to marginalise local populations; and management that enshrines the rights of employees. These outcomes must also allow us to learn about what has worked and what has not (i.e. what is ethical and what is not), and so are illustrated to be fed back into the model to better inform those working and researching in the field.

(3) *Biosciences and human nature*. Efforts to link ethics and tourism more effectively will be hastened through the recognition that many of the most important theoretical contributions to human nature – the root of ethics and of tourism – are to be found in biology (see Fennell, 2006). As noted by Pinker and others, the theory of reciprocal altruism provides one of the few foundational explanations for human relationships based on self-interest, altruism and cooperation, which form the cornerstone for our understanding of why people interact in business and other ventures, and what they wish to achieve from these interactions (Pinker, 2002). Bats, it was said, have up to 18 years to work on their reciprocal relationships. Tourists have rather less than 18 hours. Because of the lack of formal and long-standing relationships between tourists and hosts, the evolutionary biologist would suggest that it is beneficial for tourists or hosts to be amoral, or to take without giving back (i.e. in an effort to maximise benefits and minimise costs). The long-term and cumulative effects of this on the industry would obviously be catastrophic. As long as there is justice in value for money, than reciprocity unfolds. It is when fair exchanges do not take place that we can say that reciprocal altruism will be constrained, leading to an associated erosion of cooperation and confidence. Frank's (1988) theory of commitment must also be part of this learning in theoretical attempts to balance passion with reason, and to move away from a calculus based on commerce alone in forging virtuous, valued relationships.

Similarly, the theory of inclusive fitness also clearly articulates why we favour those who are closest to us genetically, and are less altruistic to those who are not related. The fact that these theoretical approaches have been built into psychology and anthropology lends weight to the notion that tourism can also benefit from this theory, as well as others that have emerged from evolutionary biology. For example, scientists have found a way to genetically detect if someone is prone to unethical behaviour. In a study of young men from Dunedin, New Zealand, who were maltreated in their youth and of little variation in class or wealth, researchers tested subjects for difference in the gene monoamine oxidase A (MAOA), and compared these differences to their upbringing (Caspi *et al.*, 2002). Subjects were further divided on the basis of high-activity MAOA and low-activity MAOA. Those found to have high-activity MAOA genes did not get into trouble as they got older even if maltreated, while those who had low-activity MAOA were much more likely to get into trouble after being maltreated, committing four times the number of nasty crimes. This study illustrates that one not only has to experience maltreatment, but also have a specific low-active gene. The whole scenario can be recreated in mice by knocking the MAOA gene out altogether and, in the process, creating an aggressive mouse. The reverse holds true by reinstating the gene. While other factors may have played a role in the behaviour of these individuals, it is startlingly clear the role that genes – and the environment in this case – play in the behaviour of people. The implications for the behaviour of tourists are indeed savoury.

In addition, brain scanning has recently proven to be a fertile ground on which to measure the cognitive aspects of morality (Goldberg, 2004). Scientists measure the blood flow in the brain when exposing subjects to different ethical dilemmas. In wartime, for example, it was common for people to hide in their basements to avoid enemy soldiers. Those with young children found it difficult to quiet crying babies. If you cover its mouth you risk killing the baby, but at the same time, if you are found, there is the risk that the whole family dies. What do you do? In the first case, neuroscientists say the more primitive, emotional parts of the brain activate the 'protect your kin' cognitions, where the thought of killing one's own child is repellent. However, the more advanced reasoning aspects of the brain, which are responsible for higher-level thinking (the pre-frontal cortex), are later activated, which enables the agent to rationalise how saving the baby would benefit the whole family, especially if the baby is to be killed anyway. This hybrid morality theory – partly emotional and partly rational – suggests that one of these alone is not enough to provide an effective balance in making morally good decisions. But there are cases where one system overrides another. Take this classic problem: waiting for a trolley while on vacation, you notice that the trolley is out of control and about to hit a group of people. Is it morally acceptable to flip a switch diverting the trolley so that it hits and kills only one person? Or is it acceptable to push someone in front of the trolley thus derailing it and saving the other five? Most people say it is morally acceptable to do the former but not the latter. In the latter case, scientists have found that the situation is too up close and personal because it activates the brain's more primitive moral system (see Greene *et al.*, 2001). Scenarios involving sex tourism, cruise line tourism and so on, in the manner discussed above, may prove to be fertile grounds for unlocking the ethical tendencies of certain groups of tourists.

(4) *Tourism and social science knowledge*. Any inroads we make in elevating ethics to a position of greater relevance in tourism can only be founded on the rich base of existing tourism knowledge. This includes the many disciplinary contributions that have been made in the past, especially in recognition of geography, economics, political science, anthropology and psychology. These will continue to be the main vehicle by which to educate students and stimulate research in our field, but they cannot provide the basis for a strong and sustained focus on ethics. For this, tourism researchers will need to seek expertise from other magisteria in creating a path to the future – the main idea behind this book – and in reference to the intellectual time warp described by McKercher at the outset. In exploring these other pathways we begin the progressive step towards moving away from a focus on the same issues in the same ways, which has likely unfolded both from the infancy of tourism studies and, perhaps more importantly, from the evolution of studies based on a common ancestry. Researchers in tourism have emerged from a common pool of mentors, who themselves have either been taught the discipline from even fewer mentors, or been instrumental in defining the field itself, leading to a certain degree of homogeneity. This is certainly not to knock the wonderful efforts of so many excellent researchers, but merely to suggest that, with time and new students, unexamined perspectives will need to be cast over the field in order to achieve further progress. This has compelled theorists in other fields to note that science progresses one funeral at a time (Ehrlich, 2000), which is a morbid way of suggesting that ideas change slowly in a discipline. The foregoing serves to suggest that we must be vigilant in detecting the gaps that exist in our field. I hesitate to state, but nevertheless feel it important to note, that our contribution to a greater understanding of human nature is marginal at best. This is because of our narrowly prescribed view of the world, which is almost exclusively based on social science. In this we have been *les savants ne sont pas curieux*, as to the integration of theory from other disciplines, despite our 'interdisciplinary' roots and mandate.

(5) *Humanities and human nature*. In allowing ethics into tourism, we open the door to philosophy and the humanities. Saul noted that those who are transfixed by reductionism founded only through instrumental ends, will sadly venture no closer to the truth. In this he supports the

belief that we need to be able to think more clearly to see the bigger conceptual picture, which might only come about through the rich history of human nature via the humanities. This has been echoed by leading evolutionary scholars such as Gould (2003) and Wilson (1999), above, who argue for consilience as an essential ingredient in allowing researchers to attain higher meaning. Although self-interest can be understood from a biological standpoint, through reciprocal altruism and inclusive fitness, it is a well-documented theme in philosophy. For example, in her book *The Virtue of Selfishness: A New Concept of Egoism*, Rand (1964) takes forward her theory of objectivism, which is a philosophy of ethics that upholds rational selfishness. Citing the dictionary definition of selfishness as 'concern with one's own interests', Rand notes that there is no moral evaluation inherent in this definition; that is, there is no guidance telling us whether this concern for the self is good or evil. Altruism, she notes, has falsely protracted selfishness as evil, by lumping together two questions into one: (1) What are values? (2) Who should be the beneficiary of values? It is the beneficiary of an action that has emerged as the only criterion of moral value. As long as the beneficiary is anyone other than oneself, there is not a problem. Rand notes that in redeeming both humans and morality it is the concept of selfishness that must be addressed, and the realisation that the individual has a right to a moral existence – a moral code – in the fulfilment of one's own life. This means that humans must be able to benefit from their own moral actions. The basic premise behind Rand's Objectivist ethics is as follows:

> The Objectivist ethics holds that the actor must always be the beneficiary of his action and that man must act for his own *rational* self-interest. But his right to do so is derived from his nature as man and from the function of moral values in human life – and, therefore, is applicable *only* in the context of a rational, objectively demonstrated and validated code of moral principles which define and determine his actual self-interest. It is not a license to 'do as he pleases' and it is not applicable to the altruists' image of a 'selfish' brute nor to any man motivated by irrational emotions, feelings, urges, wishes or whims. (Rand, 1964: ix)

Foot, too, engages self-interest in her attempts to balance reason and emotion. The rationality of self-preservation must be weighed against other selfless acts like helping neighbours, she notes, in our efforts to equilibrate self-interest. This is central to her concept of practical rationality, which must be derived from the 'goodness of the will' (Foot, 2003: 11).

Unfortunately, however, we have been taught to view self-interest as anything but good. In families of young children, the demands on parents (e.g. sports games and practices, dance lessons, music lessons, school, jobs, housework, shopping, appointments with healthcare specialists, birthday parties and so on) can be extreme. These can be overwhelming to the point where parents become fatigued or ill. Psychiatrists are now recommending that parents not feel guilty or the least bit selfish by focusing on themselves in the face of so many demands. Their rationale for this is that, by taking care of one's self and keeping the body physically and mentally at peak levels, they are in a better condition to take care of the overall needs of the family.

In an attempt to forge a link between the main theoretical and conceptual constructs discussed in this chapter, i.e. the work of Jantsch (1972) on interdisciplinarity, Weggeman's (1997) conceptualisation of the knowledge value chain and the framework for tourism ethics, an integrated model has been developed in an effort to forge a synthesis (Figure 12.4). At the outset of the chapter, interdisciplinarity was discussed as a means by which to address some of the most pressing issues in tourism. This is set out in Figure 12.4 and includes research and practical considerations in our efforts to become more ethical in tourism. This forms the central part of the Figure 12.4 (identified as the 'Interdisciplinary realm'). The first two steps of Weggeman's knowledge–value chain have been adapted at the top part of the figure to demonstrate the need for new ethical knowledge as a result of the severity of ethical issues in the field, as well as the critical step of determining the gap between what is needed and what is available regarding ethics. The third step in Weggeman's model, which is designed to narrow the knowledge gap by developing new knowledge, is replaced by the interdisciplinary realm. Hall (2001) observes that knowledge creates knowledge when it is shared. Underpinning such an emphasis, he notes, are relationships. These relationships are in turn based on language, which in turn embed human values as the core of who we are and what we do. It is therefore essential that these collaborative relationships be based on shared values (research, practice and so on) and that the knowledge that is to be diffused and applied be representative of the same shared values as outcomes of the research. The basis from which to share knowledge through interdisciplinarity rests on the integration of ethical behaviour of all stakeholder groups in the industry at different scales, knowledge transfer within the service sector, and the culmination of many different approaches in tourism studies, bioscience research and the humanities. What emerges from this is the creation of new ethical knowledge, and the diffusion and application of this new knowledge among stakeholders in an integrated fashion. Figure 12.4 also incorporates a feedback loop to demonstrate that knowledge development should enable the system to 'learn' through continuous iterations of research and/or application. As such, the approach to knowledge accumulation should be premised on a system that encourages the continuous gathering of knowledge.

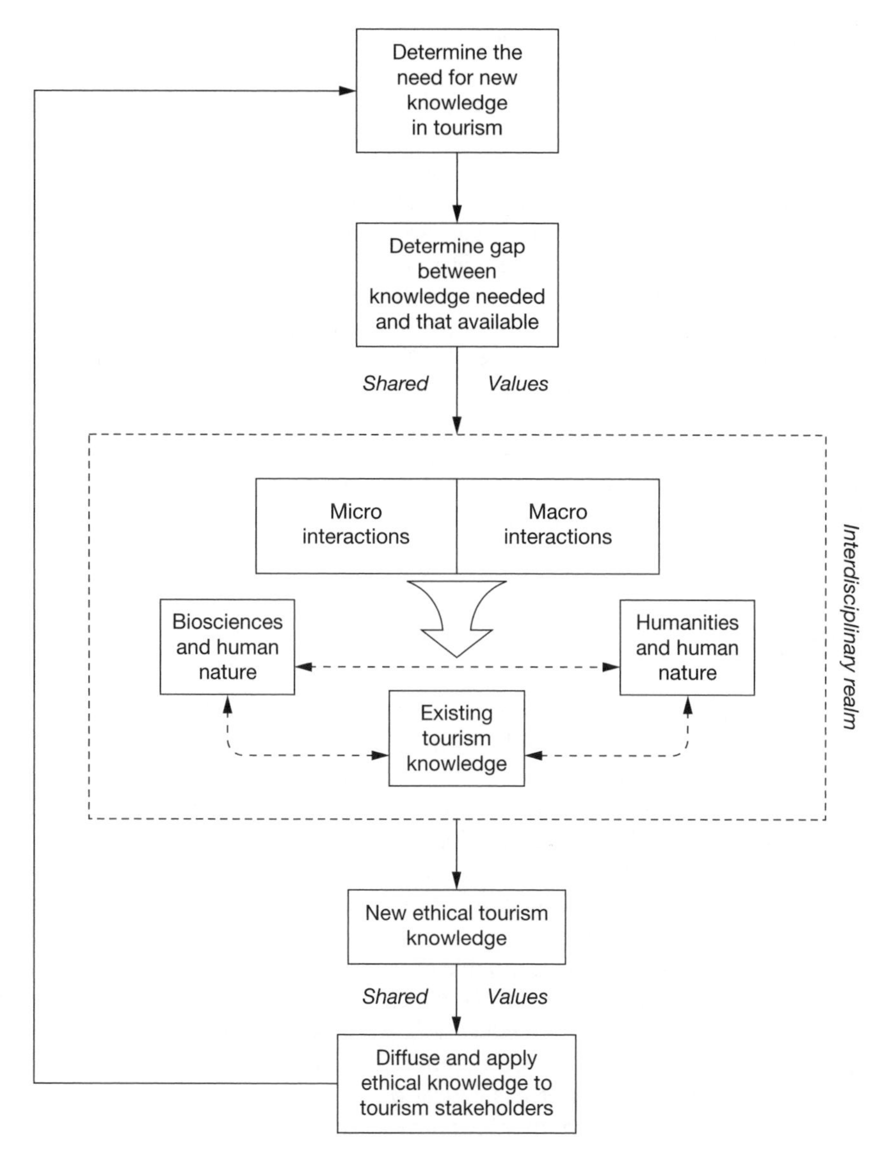

Figure 12.4 Framework for the interdisciplinary exchange and diffusion of ethical knowledge in tourism
Source: Adapted from Fennell (2004).

Tourism scholars have also begun to ask questions about the role of ethics in the process of conducting research. Moscardo (2010) argues that such is important because the field of tourism does not have an existing code of ethics at present to lead the field in this area. Moscordo contends that moving across cultures and between governments, funding agencies and community members opens up several opportunities for unethical behaviour on the part of the scholar. Ryan (2005) urges scholars to be reflective in their research because of the range of competing responsibilities that researchers have towards respondents, ethics committees, other researchers, and readers of our books and journals, in placing our best research foot forward.

An ethical issue that has emerged over many years is the audit culture that pervades the academic system at many different scales and geographic regions. Simply stated, professors, departments, universities, journals, and so on, are now in direct competition with one another for supremacy in academic research, because more, especially in the best journals, is better. Fennell (2013c) illustrates that such amounts to more instrumental reason because structure and form are more important than content. Where we publish, is now more important than what we publish because of the demands that the system places on scholars, with extrinsic motivators (more money, prestige, competition), surfacing as more important than the intrinsic nature of research.

Education

One of the main outcomes of the preceding model is the generation of new ethical knowledge that will aid in the further academic and practical education of students and practitioners in tourism. It has been said that civilisation owes its existence to education as the continual acquisition of mental nourishment for further development (Bronowski, 1981; Clark, 1969). In consideration of the current state of affairs in tourism education, and keeping in mind that education is akin to mental nourishment, how is it that we have been nourished regarding the difference between what is right and wrong in tourism? If we have trouble answering this question, we need to be wise to the words of Oakeshott (1962: 62), who noted that 'Every form of the moral life ... depends upon education. And the character of each form is reflected in the kind of education required to nurture and maintain it'. So it is not only about a lack of ethics education in tourism, but also about the direction and quality of the instruction necessary to go along with it. Given the frequency and severity of ethical transgressions in tourism, we can ill afford to continue to ignore this in our instruction. By allowing ethics in the door, we open it wider to the possibility of learning that much more about the subtle and not-so-subtle ethical nuances of a very complex industry.

Although I cannot be certain of this, opening the following up to further study, if tourism students get ethical instruction at all, it is most likely as an elective or as part of a humanities credit, which is required to fulfil what passes these days as a liberal arts degree. It may also be that a class in ethics may be acquired as a component part of a business degree – in which more and more tourism courses seem to be subsumed these days. In this regard, ethics in business has become more a problem than a solution. Although the public wants it, the philosophers who should teach it are suspicious of the free enterprise system, so avoid involvement altogether. Coupled with this is the notion that students who enter business clamour for courses that give them what they want and are less favourable towards those that venture outside the central theme. These complications have prompted scholars to suggest that business schools have no place in the university at all – 'university' meaning universal thought and liberal arts (Saul, 2001). While this is most draconian, it is perhaps more hopeful to strive for less prescription in business studies and more of an openness to liberal arts, which would have the benefit of exposing students to a wider view of the world. The same argument can be made in tourism, where curricula tend towards a higher degree of prescription in accordance with the professional school mentality.

To theorists such as Behrman (1981), there is the belief that the university cannot teach values, wisdom or judgement in the same manner we teach our other various subjects. Values, it is suggested, are learned before the first grade at home and tested continuously throughout one's career. These are acquired through experience and observation, and exist as the most important ingredient in decision-making. Values are what we live *by* and die *for*, Behrman notes, and unfortunately not what we seem to *live* for. We would rather strive for more power, status, profit and goods. In response to such criticisms (the moral character of students already formed when they get to university, and the university using the wrong methods and approaches), Sharp Paine (1991) argues that individuals not only have the power to be ethical themselves, but also to influence the conduct of others. She argues for strength in character, which can take place through reinforcements put into place in curricula that enhance students' integrity and the integrity of those they work with. Instruction should include logical analysis, the history of organisations, reflection on works of fiction, assessment of ethical decision-making models and training in sound reasoning. While it is open to debate as to whether or not courses in ethics will change the moral fabric of the individual, as we have seen here (and as observed by Aristotle so many years ago), schooling in ethics may provide the needed tension by which to allow people to make more informed choices on how to act. This observation is substantiated by other researchers, who observe that ethics in the classroom can help students think more clearly in the day-to-day issues that they will continue to confront (Vallen & Casado, 2000). This has been reinforced through research that compared the responses of hotel managers with those of college students on a number of ethical scenarios. The investigators found that, although all situations were viewed as unethical in a relative sense by both groups, students were less likely to view certain situations as unethical than their managerial counterparts (Stevens & Fleckenstein, 1999). This has led scholars to the conclusion that unless ethics becomes more formally institutionalised, there is danger that the whole profession may grow weary of this 'well doing' as newer events and issues surface, leading to the ultimate end of irrelevance (Rohr, 1978).

The differences between business scholars and scholars in the humanities also surface through the content and manner of teaching ethics in the classroom, as noted above. While business instructors rely heavily on the case study approach to learning as well as ethical theory and decision-making models, some theorists in the humanities, including Klonoski (2003), have called for more of a focus on discussions of central issues in classical philosophy, which include those central to the foundation of contemporary ethical issues in all the professions. Here it is argued that an 'unapplied' approach is required, which allows for a deep appreciation of the historical lineage of philosophical problems, and also the ability to think not only in a practical sense but also in a theoretical sense regarding the perennial issues of classical antiquity. Klonoski uses the example of advertising as a link to the past. Advertising is often said to be premised on the avoidance of truth in reality. That is, the influence that advertisements have on consumers is such that we may be induced to purchase goods that are neither necessary nor useful for living well. But the issue around differences in reality and appearance is, in fact, an old one, represented in Plato's *Republic*. In the dialogue between the Sophist Thrasymachus and Socrates, the former took the position that:

> there was really no such thing as truth, that the appearance of justice or moral righteousness was valuable as a front for achieving one's own selfish interests. In the end, this Sophist's view was in part that the clever creation of images or appearances of goodness, whether in speech or practice, with no commitment to goodness itself, is the best way to live. (Klonoski, 2003: 25)

Klonoski suggests that such a position represents an early delineation of advertising in that truth is supposed to dwell in the eyes of the beholder, and so the reality of a product is said to lie in the image that is presented. Whether consumers are able to distinguish between images and originals or reality and appearance opens up a series of epistemological issues, such as: how does a consumer gain access to reality if only presented with images? What does it mean to have knowledge of things as they are in themselves? By what means might consumers distinguish truth from fantasy?

Tribe (2002) suggests that we are in need of an educational approach in tourism that allows our students to demonstrate ethical competence in their professional lives through a strong sense of critical knowing, in the context of morality. This is because rules are often overlooked or ignored in tourism, with a failure to improve outcomes. As noted previously in this book, rules and laws can just as easily be unethical as ethical. What is needed is a foundation of ethics that guides all of our actions in tourism. Without this, we are bound to continue to make the same mistakes.

In bridging the gap, Tribe proposes a model of ethical tourism action from three perspectives. The first, 'action to follow reflection', is grounded in disciplinary thinking, where there is a linear flow from reflection (on what the discipline says may be right or wrong) to action. The second, 'reflection-in-ethical-tourism-action' is less strongly tied to disciplinary routines, where actions are instigated by the problem itself rather than the discipline (a theme strongly supported in this chapter). This perspective acknowledges the inherent difficulties in trying to solve real-world complexities from theory alone. Instead, there is a dynamic that needs to exist between action and reflection in the search for the 'good'. Third, the 'knowing-in-ethical-tourism-action' perspective emerges when reflection and action are integrated for the purpose of striving for what is good in tourism: 'This is where theories of ethical action are embedded in the action itself. These theories can be distilled out by reflection-in-action, but the point is that the artistry implied suggests a much more fluid exchange between action and reflection' (Tribe, 2002: 313).

Ethical tourism action, according to Tribe, takes place by embracing Aristotle's notion of *phronesis*, or 'practical wisdom', which is defined as 'knowledge of how to secure the ends of human life', and 'a true and reasoned state of capacity to act with regard to the things that are good and bad for men' (Aristotle, 1980: 142, as cited in Tribe, 2002: 314). The important aspects of *phronesis* are said to be knowledge; 'the good'; actions, practice and experience; and disposition. This means, among other things, that knowledge alone is not sufficient, knowledge must have intrinsic purpose in striving for the good, practical judgement is a temperament towards 'good' rather than what is deemed 'correct', and repeated experience or practice is important in acquiring *phronesis*.

We can operationalise phronesis, Tribe notes, by understanding that reflection produces strong theories, but is less applicable; while action, although practically relevant, is intellectually indefensible. In pulling theory and practice closer together, tourism educators are urged to use any of three activities in bolstering ethics in the tourism curriculum, including case studies, role play and work placement (see work by Jamal (2004) that frames Tribe's work in the context of sustainable tourism).

Another ethics-based model of education is structured by Hay and Foley (1998), and is based on the notion that being a responsible citizen means going beyond subscribing to ethical prescriptions alone, and moving towards a method that entails: (1) stimulating the moral imagination; (2) recognising ethical issues; (3) developing analytical skills; (4) eliciting a sense of moral obligation and personal responsibility; and (5) tolerating – and resisting – disagreement and ambiguity. The best way in which to accomplish these ends, according to the authors, is through the case method approach. While case *studies* refer to examples of ethical situations used in lectures, the case *method* approach

> refers to a student-centred teaching-and-learning approach in which the teacher's role is to facilitate reflection and focused inquiry, and to promote discussion about some real-world phenomenon whose characteristics are outlined in an article, book, video-tape or other intellectually provocative material. (Hastings Centre, 1979: 174)

Hay and Foley recommend the use of a jurisprudential inquiry model for the case method, which employs real-world ethical dilemmas in promoting thought-provoking discussion and argument. It does so with the intent of allowing students to gain an appreciation of the moral stance of other classmates as well as to critically evaluate their own ethical beliefs on the topic. There are six phases in this process as follows (Hay & Foley, 1998: 175; adapted from Joyce *et al.*, 1996).

Phase 1. *Orientation to the case*
Teacher introduces material.

Phase 2. *Identifying the issues*
Students synthesise facts into an ethics issue.
Students select one issue for discussion.
Students identify values and value conflicts.
Students recognise underlying factual and definitional questions.

Phase 3. *Taking positions*
Students articulate a position.
Students state basis of position (e.g. social, economic or ecological consequences).

Phase 4. *Exploring the stance patterns of argumentation*
Establish the point at which value is violated (factual).
Prove the desirable or undesirable consequences of a position (factual).
Clarify the value conflict with analogies.
Set priorities. Assert priority of one value over another and demonstrate lack of gross violation of second value.

Phase 5. *Refining and qualifying the positions*
Students state positions and reasons for positions, and examine similar situations.
Students qualify positions.

Phase 6. *Testing factual assumptions behind qualified assumptions*
Identify factual assumptions and determine whether they are relevant.
Determine the predicted consequences and examine their factual validity (will they actually occur?).

The proposed method is said to allow students to more thoroughly identify and analyse ethical questions, practice forceful dialogue, and appreciate the advantage of the pluralistic approach to ethics and dialogue, which allows students to respect the opinions of others. It also has advantages over other teaching methods based on simplicity, ease of application and cost-effectiveness (Hay & Foley, 1998).

In tourism, researchers and practitioners are not always on the same page regarding the value of education. For the latter group, education is quite often seen as existing outside their reach and responsibility. This was the case with regard to comments made by a representative of the German tourism industry at the Bad Boll Conference in Germany (1986) on Third World tourism. In representing the world of the tourism operator, he said that: 'We also have to develop products according to the expectations of our clients ... the major motivation of tourists is recreation and holiday on the beach. ... We are in no position to educate our clients' (D'Sa, 1999: 65). Surely such a position is unrealistic at the dawn of the new millennium? Is it not the responsibility of all who are involved in the provision of services for tourism to educate travellers, either directly or indirectly, however meagrely? Ecotourism comes to mind as one of those forms of tourism where there is a heavy mandate to educate tourists about natural history and local customs. This has been the finding of researchers in tourism, who conclude that, for ecotourism to become effective, structured education programmes based on sound methodologies can induce changes in tourist behaviour (Lück, 2003; Orams, 1997). Indeed, in many cases ecotourists were found wanting more information on specific environments, which has the benefit not only of informing tourists but also providing a means by which to safeguard the natural world. The implications for ecotourism management are not hard to see: more education is required in order to help reach the lofty goals that define it as different from other forms of tourism. But it is not only what we view as educative or ethical at any point in time, but also how such values change. In a longitudinal study of outdoor recreationists, researchers discovered that there was a general improvement in the values that visitors placed on the wilderness over time, with a corresponding positive change in what was considered appropriate behaviour in a wilderness setting. The authors attributed this to higher levels of education, increased memberships in conservation-oriented organisations and manager-initiated education programmes, as well as the impact of society as a whole in bringing the public's attention to ethical issues in parks (Watson *et al.*, 1996). Better facilitation of visitor *experiences* in visitor management strategies, as a soft management approach, is felt to be essential in leading to better conduct of tourists with a reduction in negative visitor impacts (Mason, 2005).

Such results appear to indicate that we should be receptive to changing patterns, changing times and changing philosophies about the role of ethics in tourism both in the university and within society. But receptivity has been slow and relatively unnoticed. In hospitality studies, two non-profit institutions continue to be dedicated to the advancement of ethics in the hospitality industry. These are the International Institute for Quality and Service in Ethics and Tourism (IIQUEST), as noted in Chapter 1, and the Marion Isbell Endowment for Hospitality Ethics Centre (see Vallen & Casado, 2000). These centres are involved in the study and promotion of hospitality ethics and in surveys of ethics in hospitality and organise contests for student articles and essays from hospitality schools around the world. The International Hotel and Restaurant Association (2001) has recognised the dearth of knowledge in their field by developing an environmental teaching resource pack for the hospitality industry that includes a wealth of information needed to develop and expand environmental education in hotel schools. With the understanding that such education is needed in the hospitality sector (indeed, theorists in the hospitality sector have been much more active in the examination of ethics in education than their tourism studies counterparts (see Enghagen, 1990; Hagarty, 1990; O'Halloran, 1991), the report argues that tourism and hospitality need to play a leading role in environmental responsibility in the day-to-day operations of those working in the field. But are these enough? The argument here is that they are not, and will continue not to be, unless there is a heightened level of industry and academic support for centres of tourism ethics that would work in an integrated fashion with related centres in other fields.

A final model of education that is relevant to an ethics approach is through the work by the Tourism Education Futures Initiative (TEFI), which is a collection of tourism scholars interested in changing the research paradigm that currently exists in our field. TEFI supports a number of core values in helping to frame this change, including stewardship, knowledge, professionalism, mutuality and ethics. Sheldon *et al.* (2011) illustrate that the teaching of ethics in curricula needs to include defining ethics, exposing students to ethical traditions and principles, showing students how to achieve reconciliation through ethics, making connections to power and politics evoking good actions through multiple means. It is encouraging to see that scholars have applied ethical theories, models and scenarios to different student groups in increasing awareness. Examples include cross-cultural differences (Litvin *et al.*, 2004), students' fears of moral situations when they enter the workforce (Marnburg, 2006), ethical beliefs of hospitality and tourism students towards their school life along the lines of selfishness, unfair advantage, plagiarism and the violation of school rules (Yeung *et al.*, 2002). Research has also sought to investigate practitioners on the basis of ethics. Yeung (2004) asked industry professionals about the type of ethical content that should shape hospitality curricula, while Hudson and Miller (2006) studied the degree to which tourism industry practitioners know the difference between right and wrong based on their responses to scenarios.

A comprehensive ethical programme

Bridging back into the realm of business, ethical education and training within the organisation is essential, especially in light of the many ethical transgressions identified in earlier chapters. In Chapter 9, a process was developed for the creation of a code of ethics. But, as McDonald (1999) observes, the establishment of a code of ethics is only the initial step in a broader overall ethical programme. Other aspects include an activation of the role of management, a code of conduct, ethical training and evaluation. These general aspects are elaborated upon in this section for the purpose of demonstrating the extent to which organisations may go in their attempts to be ethical. The nature of this comprehensive ethical programme, in general, includes the following:

A. *Activate the role of upper management.* Seen as essential because of the leadership role of these individuals within the firm.

B. *Create the code of ethics.* This may include sections on employment practice, ethics in general, conflict of interest, bribery and gifts, safety and occupational health, customer relations, relations with suppliers, environment, responsibilities to shareholders, monitoring and compliance and the means of enforcement.

C. *Conduct ethical training.* (1) Establish programme objectives. These may include: the fostering of awareness of the ethical components of managerial decision-making; the legitimisation of ethical components; the provision of conceptual frameworks for analysing the ethical components and to help individuals become confident in their use; the helping of participants to apply ethical analysis to day-to-day business activities. (2) Design the programme. This includes a detailed training manual for facilitators or trainers, who will then be charged with the responsibility of training employees and developing a workshop. The manual includes sections on developing videos for training purposes; why standards of conduct are important; and why the code of conduct is so important. (3) Facilitate the programme. This stage of implementation includes the actual training of trainers and later employees through a variety of methods. This may include: (1) the use of biographies of morally laudable people; (2) videos on business ethics; (3) the use of role plays that stem from examples of ethical dilemmas; (4) the use of current newspaper reports on situations of ethical transgressions, or cases of superior ethical behaviour; (5) the use of ethical board games, including ethical scenarios; (6) the use of a conference format which, included specialists in the area of ethics; and (7) following up the conference with dilemma workshops to further emphasise the importance of ethics in the workplace.

D. *Evaluate ethical programmes.* The evaluation for the course can include questions related to train-the-trainer sessions on the training material itself, the extent to which employees had been aware of the existing code of conduct, and the extent to which employees knew of the importance of business ethics in general. This focus on evaluation in the context of programme design serves to underscore the importance of a strategic approach to programme planning, which has been so useful in other fields, but not in tourism (Fennell, 2002).

The future activities of the organisation could also focus on a number of related aspects. These include:

Ethical audit. This is deemed important as a means by which to bring the organisation's core values and ethics into focus. Although often seen as a control mechanism, the audit should be viewed as an opportunity to heighten ethical awareness. This would include open discussion and dialogue from facilitators (see Gray, 1996, for a discussion of ethics and auditing).

Appointment of an ethics ombudsperson/ethics officer. This role would include investigation of accounts of ethical violations, counselling, and advising. Such individuals may be asked to find resolutions or act on certain issues. McDonald (1999) notes that employees who are experienced, and have plateaued on the pay scale or are near retirement, are best suited to these positions.

Ethical committees. Such committees have membership that rotates among all employees for the purpose of distributing accountability throughout the organisation. The decisions made provide clear guidelines for action, along with functions geared towards policy development.

Open channels of communication. This approach is typified by ethics 'hotlines', where employees can make phone calls or send e-mails to those who can offer advice. Employees can ask questions that guide their own conduct, or report ethical violations that take place in the organisation.

The development of this framework demands a significant outlay of resources in the form of time, money and people. While it will have application to smaller firms, it may be more useful to larger firms where there is the chance for a greater frequency of unethical behaviour, and where there are likely to be more resources available to implement such a system (Murphy *et al.*, 1992; see Cohen, 2010, who argues that the teaching of ethics should be based on an ethics of responsibility).

Concluding Thoughts

Throughout this book I have been decidedly critical of tourism, especially from the industry standpoint. Ever mindful of this, the focus of the criticism is a protraction of what appears in the literature on tourism. For example, Ladkin and Martinez Bertramini (2002: 71) write that, despite the overall importance of tourism as an engine of growth, 'there are increasing criticisms of the industry by scholars, development institutions and governments, for the negative economic, environmental and sociocultural impacts evident in host destinations' (see also various authors in Singh, 2015). Although researchers continue to recognise tourism's shortcomings, the aim here was to examine these in light of a body of research (ethics) that has not generally been emphasised in tourism studies. Notwithstanding, I would be remiss by failing to mention that there are many positive examples of tourism development and tourism interactions in the world (see, for example, Pigram, 2004, in the context of Australia). It is, however, noteworthy that these fail to appear in the academic literature as frequently as they ought to, which might therefore provide scope for a continuation of the work found here within, but cast in a different light.

Perhaps one of the most important conclusions of this book is the difficulty in finding what might be viewed as 'perfect' solutions in addressing tourism problems and issues – a point not lost on Smith and Duffy (2003). Unfortunately, we cannot look into our microscopes, manipulate the fine adjustment knobs and fully bring into focus an understanding of the parameters of our tourism-oriented problems. Taking the analogy further, our limits of understanding are subject more to the coarse adjustment of a microscope – knowing that they exist and seeing the outline of a subject under scrutiny, but often never fully able to sharpen the image. It is perhaps overstated, but worthwhile remembering, that knowledge is never absolute (Bronowski, 1981), and that the only certainty is uncertainty, as noted in the quote leading into this chapter. Werner Heisenberg realised this in the 1920s in his attempts to record the direction and speed of electrons, which physicists had yet to do. The imprecision and imperfection of knowledge has led physicists – arguably the most precise of all sciences – to view physics not as certainty, but rather as philosophy. This leap is an important one because it demonstrates that philosophy helps us understand our place in the physical universe and gives us the opportunity to reflect upon knowledge, authority, technology, responsibility, kinship and group dynamics. I find it intriguing that whenever we are confronted with difficult problems, it is to philosophy that we look in judging whether or not we have taken, or are about to take, a right or wrong step. Ethics, as a fundamental part of philosophy, is not some exotic beast to be used only by heroes and saints, and unattainable to the average mortal, but rather something that is part of ourselves and of society in the most normal of ways (Saul, 2001). This is no coincidence. The need to be good is part of our genetic and cultural fabric, a part that we will continue to need and feed if we are to be successful in addressing the many challenging times that surely lie ahead.

But ethics is subject to misrepresentation in the most horrific ways imaginable. Reason provided the grounds for early explorers to claim superiority over indigenous peoples. It was reason that convinced them they were in fact better, measured by some cost-benefit calculation of relativistic values (e.g. technology). This provided clear guidance towards the right to their lands, women, water and possessions. But when we reason independent of passion, do we not also open ourselves up to self-interest and the erosion of our ability to imagine the other? So, while reason has often been argued to sit atop our human qualities, it has not been enough to allow us to find the truth behind human nature. This was demonstrated by Marx, Nietzsche and Freud:

> Marx had shown that the capitalist economic system is governed by laws of its own which cannot be controlled by reason and which create the potential for an imminent crisis. Nietzsche had unmasked the belief of the Enlightenment in human self-determination as nothing more than the product of the desire for power. Finally Freud proceeded to cast doubt on man's rational ability to control his own inner self, his emotions and instincts, by describing them as the forms taken by a ubiquitous sexual drive. All these ideas radically questioned the power of reason: in central areas of man's self-understanding it had turned out that it was not the self-assured subject but rather blind forces that were in control. (Delius *et al.*, 2000: 96)

The inability to control our destiny through reason alone is testament to the need for an equilibrium based on reason, passion and morality. And yet the world has changed so very much since the birth of Christ, since the Renaissance and since the Enlightenment. We continue to face new problems and new pressures, e.g. increasing population, urbanisation and genetically modified foods, which seem to magnify the problems that humanity has endured over time. Never before, one could argue, has there been the need to test how ethics can be used to address the pressures within a world of increasing globalisation.

At the same time, I also find optimism in the research of Singer and others, who have made reference to the widening of the circle of morality, which has had a dramatic effect on who we are in time and space (and perhaps in keeping pace with the pressures identified above). We know that people in all of the world's cultures (remember Morton who said that we all descended from a common culture) can determine right from wrong (Durant, 1935/1963); children are able to determine if people are in a state of distress and will reach out to them (Zahn-Waxler *et al.*, 1992); and we continue to expand the circle of morality through exercising the sentiments and passions that we were endowed with (Frank, 1988). If, as a species, we are evolving to accept morality to a greater degree in our social and individual interactions and transactions, then what about tourism? To some tourism researchers, however, the act of travel exempts individuals from inclusion as moral beings by virtue of their status in a club that bases its ethics on being free to do and say as one pleases simply because one has invested precious time and money in the process: I paid for it, so I deserve it (Oh yes, those halcyon days!).

But if we can accept that morality is a condition of humanity, that we are continually modifying our ethical precepts, and that we are becoming increasingly moral, then should not tourism jump aboard too? To do otherwise seems counter-intuitive. If this means that tourism needs to better play by the rules, then so be it. (Having said this, there will always be places like Ibiza where tourists can more freely exercise their hedonistic rights, i.e. the self-interested side of human nature over the cooperative one, as noted in Chapter 2.) In this way tourism is seen as a vehicle by which to unleash this less virtuous side of who we are, outside the boundaries of our everyday existence – away from those we coexist with as family, as friends and at work. But doesn't this mean that there are benefits to the tourist and costs borne by others? If so, is this something we are prepared to live with, collectively?

But if we can accept that the freedom to do as one pleases as a tourist is no longer as acceptable, there must be an explanation. First, we cannot pretend that tourism today is the same as tourism in the 1960s, where the free-wheeling nature of the beast was highly acceptable. Second, the sheer volume of tourism in this day and age constrains tourists from hanging on to some of these older 'practices' (e.g. littering or photographing people without their permission). Finally, we are a great deal wiser about the effects of tourism on the economy, on other people and the natural world. What this means is that, instead of stigmatising a more ethical approach to tourism, we should embrace it. In support of this mindset, we ought to continue to search for approaches that can be effective in balancing commerce (profit), people and the environment in our efforts to ensure that morality does not remain a dirty word.

The aforementioned suggests that morality cannot be externalised from decision-making, and research for that matter, but rather must exist as its central guiding force. In this, there is one question that must be at the front of our minds when we attempt to make decisions in tourism: is it ethical? This is especially difficult to accomplish when we are more prone to identifying and rewarding what is wrong rather than what is right. What do I mean by this? Too easily do we forget about the many transgressions that we more frequently see in the news regarding some of the most prominent people in society. We return politicians 'on the take' to multiple terms in office; we accept that our favourite athletes need to take 'medicinal supplements' in their pursuit of gold medals; we encourage our corporations to seek recognition – whether good or bad – in their efforts to sell more goods, because even bad recognition is 'good for business'; and we seek relationships with people who are wealthy, because more money is more virtuous. All of this flies in the face of attempts to bring us back on line through the adoption of ethics as a central guiding force in our lives. We are ethical by nature, but are we not also competitive and self-interested, as noted above?

At the outset of the book it was proposed that ethics has real potential to emerge as a new platform in tourism studies (Macbeth, 2005). In promoting it as such, we open the door to a broad base of literature, which historically has provided guidance and knowledge about a range of human behaviours, including who we are and how we ought to behave in time, space and in full view of circumstance. In their efforts to include ethics and philosophy in their work, tourism researchers and practitioners can expect to make proactive and timely contributions not only to the field of tourism – especially in addressing the tourism impact dilemma – but also towards an understanding of human nature in general.

Appendix: WTO Global Code of Ethics for Tourism

ARTICLE 1: Tourism's contribution to mutual understanding and respect between peoples and societies

(1) The understanding and promotion of the ethical values common to humanity, with an attitude of tolerance and respect for the diversity of religious, philosophical and moral beliefs, are both the foundation and the consequence of responsible tourism, stakeholders in tourism development and tourists themselves should observe the social and cultural traditions and practices of all peoples, including those of minorities and indigenous peoples and to recognize their worth;

(2) Tourism activities should be conducted in harmony with the attributes and traditions of the host regions and countries and in respect for their laws, practices and customs;

(3) The host communities, on the one hand, and local professionals, on the other, should acquaint themselves with and respect the tourists who visit them and find out about their lifestyles, tastes and expectations; the education and training imparted to professionals contribute to a hospitable welcome;

(4) It is the task of the public authorities to provide protection for tourists and visitors and their belongings; they must pay particular attention to the safety of foreign tourists owing to the particular vulnerability they may have; they should facilitate the introduction of specific means of information, prevention, security, insurance and assistance consistent with their needs; any attacks, assaults, kidnappings or threats against tourists or workers in the tourism industry, as well as the wilful destruction of tourism facilities or of elements of cultural or natural heritage should be severely condemned and punished in accordance with their respective national laws;

(5) When travelling, tourists and visitors should not commit any criminal act or any act considered criminal by the laws of the country visited and

abstain from any conduct felt to be offensive or injurious by the local populations, or likely to damage the local environment; they should refrain from all trafficking in illicit drugs, arms, antiques, protected species and products and substances that are dangerous or prohibited by national regulations;

(6) Tourists and visitors have the responsibility to acquaint themselves, even before their departure, with the characteristics of the countries they are preparing to visit; they must be aware of the health and security risks inherent in any travel outside their usual environment and behave in such a way as to minimize those risks.

ARTICLE 2: Tourism as a vehicle for individual and collective fulfilment

(1) Tourism, the activity most frequently associated with rest and relaxation, sport and access to culture and nature, should be planned and practised as a privileged means of individual and collective fulfilment; when practised with a sufficiently open mind, it is an irreplaceable factor of self-education, mutual tolerance and for learning about the legitimate differences between peoples and cultures and their diversity;

(2) Tourism activities should respect the equality of men and women; they should promote human rights and, more particularly, the individual rights of the most vulnerable groups, notably children, the elderly, the handicapped, ethnic minorities and indigenous peoples;

(3) The exploitation of human beings in any form, particularly sexual, especially when applied to children, conflicts with the fundamental aims of tourism and is the negation of tourism; as such, in accordance with international law, it should be energetically combatted with the cooperation of all the States concerned and penalized without concession by the national legislation of both the countries visited and the countries of the perpetrators of these acts, even when they are carried out abroad;

(4) Travel for purposes of religion, health, education and cultural or linguistic exchanges are particularly beneficial forms of tourism, which deserve encouragement;

(5) The introduction into curricula of education about the value of tourist exchanges, their economic, social and cultural benefits, and also their risks, should be encouraged.

ARTICLE 3: Tourism, a factor of sustainable development

(1) All the stakeholders in tourism development should safeguard the natural environment with a view to achieving sound, continuous and sustainable economic growth geared to satisfying equitably the needs and aspirations of present and future generations;

(2) All forms of tourism development that are conducive to saving rare and precious resources, in particular water and energy, as well as avoiding so far as possible waste production, should be given priority and encouraged by national, regional and local public authorities;
(3) The staggering in time and space of tourist and visitor flows, particularly those resulting from paid leave and school holidays, and a more even distribution of holidays should be sought so as to reduce the pressure of tourism activity on the environment and enhance its beneficial impact on the tourism industry and the local economy;
(4) Tourism infrastructure should be designed and tourism activities programmed in such a way as to protect the natural heritage composed of ecosystems and biodiversity and to preserve endangered species of wildlife; the stakeholders in tourism development, and especially professionals, should agree to the imposition of limitations or constraints on their activities when these are exercised in particularly sensitive areas: desert, polar or high mountain regions, coastal areas, tropical forests or wetlands, propitious to the creation of nature reserves or protected areas;
(5) Nature tourism and ecotourism are recognized as being particularly conducive to enriching and enhancing the standing of tourism, provided they respect the natural heritage and local populations and are in keeping with the carrying capacity of the sites.

ARTICLE 4: Tourism, a user of the cultural heritage of mankind and contributor to its enhancement

(1) Tourism resources belong to the common heritage of mankind; the communities in whose territories they are situated have particular rights and obligations to them;
(2) Tourism policies and activities should be conducted with respect for the artistic, archaeological and cultural heritage, which they should protect and pass on to future generations; particular care should be devoted to preserving and upgrading monuments, shrines and museums as well as archaeological and historic sites which must be widely open to tourist visits; encouragement should be given to public access to privately-owned cultural property and monuments, with respect for the rights of their owners, as well as to religious buildings, without prejudice to normal needs of worship;
(3) Financial resources derived from visits to cultural sites and monuments should, at least in part, be used for the upkeep, safeguard, development and embellishment of this heritage;
(4) Tourism activity should be planned in such a way as to allow traditional cultural products, crafts and folklore to survive and flourish, rather than causing them to degenerate and become standardized.

ARTICLE 5: Tourism, a beneficial activity for host countries and communities

(1) Local populations should be associated with tourism activities and share equitably in the economic, social and cultural benefits they generate, and particularly in the creation of direct and indirect jobs resulting from them;
(2) Tourism policies should be applied in such a way as to help to raise the standard of living of the populations of the regions visited and meet their needs; the planning and architectural approach to and operation of tourism resorts and accommodation should aim to integrate them, to the extent possible, in the local economic and social fabric; where skills are equal, priority should be given to local manpower;
(3) Special attention should be paid to the specific problems of coastal areas and island territories and to vulnerable rural or mountain regions, for which tourism often represents a rare opportunity for development in the face of the decline of traditional economic activities;
(4) Tourism professionals, particularly investors, governed by the regulations laid down by the public authorities, should carry out studies of the impact of their development projects on the environment and natural surroundings; they should also deliver, with the greatest transparency and objectivity, information on their future programmes and their foreseeable repercussions and foster dialogue on their contents with the populations concerned.

ARTICLE 6: Obligations of stakeholders in tourism development

(1) Tourism professionals have an obligation to provide tourists with objective and honest information on their places of destination and on the conditions of travel, hospitality and stays; they should ensure that the contractual clauses proposed to their customers are readily understandable as to the nature, price and quality of the services they commit themselves to providing and the financial compensation payable by them in the event of a unilateral breach of contract on their part;
(2) Tourism professionals, insofar as it depends on them, should show concern, in co-operation with the public authorities, for the security and safety, accident prevention, health protection and food safety of those who seek their services; likewise, they should ensure the existence of suitable systems of insurance and assistance; they should accept the reporting obligations prescribed by national regulations and pay fair compensation in the event of failure to observe their contractual obligations;
(3) Tourism professionals, so far as this depends on them, should contribute to the cultural and spiritual fulfilment of tourists and allow them, during their travels, to practise their religions;

(4) The public authorities of the generating States and the host countries, in cooperation with the professionals concerned and their associations, should ensure that the necessary mechanisms are in place for the repatriation of tourists in the event of the bankruptcy of the enterprise that organized their travel;

(5) Governments have the right and the duty – especially in a crisis, to inform their nationals of the difficult circumstances, or even the dangers they may encounter during their travels abroad; it is their responsibility however to issue such information without prejudicing in an unjustified or exaggerated manner the tourism industry of the host countries and the interests of their own operators; the contents of travel advisories should therefore be discussed beforehand with the authorities of the host countries and the professionals concerned; recommendations formulated should be strictly proportionate to the gravity of the situations encountered and confined to the geographical areas where the insecurity has arisen; such advisories should be qualified or cancelled as soon as a return to normality permits;

(6) The press, and particularly the specialized travel press and the other media, including modern means of electronic communication, should issue honest and balanced information on events and situations that could influence the flow of tourists; they should also provide accurate and reliable information to the consumers of tourism services; the new communication and electronic commerce technologies should also be developed and used for this purpose; as is the case for the media, they should not in any way promote sex tourism.

ARTICLE 7: Right to tourism

(1) The prospect of direct and personal access to the discovery and enjoyment of the planets resources constitutes a right equally open to all the worlds inhabitants; the increasingly extensive participation in national and international tourism should be regarded as one of the best possible expressions of the sustained growth of free time, and obstacles should not be placed in its way;

(2) The universal right to tourism must be regarded as the corollary of the right to rest and leisure, including reasonable limitation of working hours and periodic holidays with pay, guaranteed by Article 24 of the Universal Declaration of Human Rights and Article 7.d of the International Covenant on Economic, Social and Cultural Rights;

(3) Social tourism, and in particular associative tourism, which facilitates widespread access to leisure, travel and holidays, should be developed with the support of the public authorities;

(4) Family, youth, student and senior tourism and tourism for people with disabilities, should be encouraged and facilitated.

ARTICLE 8: Liberty of tourist movements

(1) Tourists and visitors should benefit, in compliance with international law and national legislation, from the liberty to move within their countries and from one State to another, in accordance with Article 13 of the Universal Declaration of Human Rights; they should have access to places of transit and stay and to tourism and cultural sites without being subject to excessive formalities or discrimination;
(2) Tourists and visitors should have access to all available forms of communication, internal or external; they should benefit from prompt and easy access to local administrative, legal and health services; they should be free to contact the consular representatives of their countries of origin in compliance with the diplomatic conventions in force;
(3) Tourists and visitors should benefit from the same rights as the citizens of the country visited concerning the confidentiality of the personal data and information concerning them, especially when these are stored electronically;
(4) Administrative procedures relating to border crossings whether they fall within the competence of States or result from international agreements, such as visas or health and customs formalities, should be adapted, so far as possible, so as to facilitate to the maximum freedom of travel and widespread access to international tourism; agreements between groups of countries to harmonize and simplify these procedures should be encouraged; specific taxes and levies penalizing the tourism industry and undermining its competitiveness should be gradually phased out or corrected;
(5) So far as the economic situation of the countries from which they come permits, travellers should have access to allowances of convertible currencies needed for their travels.

ARTICLE 9: Rights of the workers and entrepreneurs in the tourism industry

(1) The fundamental rights of salaried and self-employed workers in the tourism industry and related activities, should be guaranteed under the supervision of the national and local administrations, both of their States of origin and of the host countries with particular care, given the specific constraints linked in particular to the seasonality of their activity, the global dimension of their industry and the flexibility often required of them by the nature of their work;
(2) Salaried and self-employed workers in the tourism industry and related activities have the right and the duty to acquire appropriate initial and continuous training; they should be given adequate social protection; job insecurity should be limited so far as possible; and a specific status,

with particular regard to their social welfare, should be offered to seasonal workers in the sector;

(3) Any natural or legal person, provided he, she or it has the necessary abilities and skills, should be entitled to develop a professional activity in the field of tourism under existing national laws; entrepreneurs and investors – especially in the area of small and medium-sized enterprises – should be entitled to free access to the tourism sector with a minimum of legal or administrative restrictions;

(4) Exchanges of experience offered to executives and workers, whether salaried or not, from different countries, contributes to foster the development of the world tourism industry; these movements should be facilitated so far as possible in compliance with the applicable national laws and international conventions;

(5) As an irreplaceable factor of solidarity in the development and dynamic growth of international exchanges, multinational enterprises of the tourism industry should not exploit the dominant positions they sometimes occupy; they should avoid becoming the vehicles of cultural and social models artificially imposed on the host communities; in exchange for their freedom to invest and trade which should be fully recognized, they should involve themselves in local development, avoiding, by the excessive repatriation of their profits or their induced imports, a reduction of their contribution to the economies in which they are established;

(6) Partnership and the establishment of balanced relations between enterprises of generating and receiving countries contribute to the sustainable development of tourism and an equitable distribution of the benefits of its growth.

ARTICLE 10: Implementation of the principles of the Global Code of Ethics for Tourism

(1) The public and private stakeholders in tourism development should cooperate in the implementation of these principles and monitor their effective application;

(2) The stakeholders in tourism development should recognize the role of international institutions, among which the World Tourism Organization ranks first, and non-governmental organizations with competence in the field of tourism promotion and development, the protection of human rights, the environment or health, with due respect for the general principles of international law;

(3) The same stakeholders should demonstrate their intention to refer any disputes concerning the application or interpretation of the Global Code of Ethics for Tourism for conciliation to an impartial third body known as the World Committee on Tourism Ethics.

References

Adams, J.S., Tashchian, A. and Shore, T.H. (2001) Codes of ethics as signals for ethical behavior. *Journal of Business Ethics* 29, 199–211.
Ahmed, Z.U., Krohn, F.B. and Heller, V.L. (1994) International tourism ethics as a way to world understanding. *The Journal of Tourism Studies* 5 (2), 36–44.
Alcock, J. (2001) *The Triumph of Sociobiology*. New York: Oxford University Press.
Alexander, R.D. (1979) *Darwinism and Human Affairs*. Seattle, WA: University of Washington Press.
Alexander, R.D. (1987) *The Biology of Moral Systems*. New York: Aldine de Gruyter.
Alvard, M. (1994) Conservation by native peoples: Prey choice in a depleted habitat. *Human Nature* 5, 127–154.
Ammon, R. (1997) Risk management process. In D.J. Cotten and T.J. Wilde (eds) *Sport Law for Sport Managers*. Dubuque, I: Kendall/Hunt.
Andersen, J.E. (1994) *Public Policymaking: An Introduction*. Dallas: Houghton Mifflin.
Anon. (1999) Forest fears. *The Economist* 351 (8125), 54–55.
Ardrey, R. (1970) *The Social Contract*. New York: Dell.
Aristotle (1980) *The Nicomachean Ethics* (D. Ross, trans.). Oxford: Oxford University Press.
Aristotle (1998) *The Nicomachean Ethics* (D. Ross, trans.). Oxford: Oxford University Press.
Aronsson, L. (2000) *The Development of Sustainable Tourism*. London: Continuum.
Aspinall, R. (2017) Will turtles and tourism always be at loggerheads? See https://www.theguardian.com/science/blog/2017/feb/03/will-turtles-and-tourism-always-be-at-loggerheads-zakynthos (accessed July 27).
Attwell, C. and Cotterill, F. (2000) Postmodernism and African conservation science. *Biodiversity and Conservation* 9, 559–577.
Axelrod, R. (1984) *The Evolution of Cooperation*. New York: Basic Books.
Axelrod, R. and Hamilton, W.D. (1981) The evolution of cooperation. *Science* 211, 1390–1396.
Ayala, F.J. (1987) The biological roots of morality. *Biology and Philosophy* 2, 235–252.
Backoff, J.F. and Martin, C.L. (1991) Historical perspectives: Development of codes of ethics in the legal, medical, and accounting professions. *Journal of Business Ethics* 10, 99–110.
Baptista, J.A. (2012) The virtuous tourist: Consumption, development, and nongovernmental governance in a Mozambican village. *American Anthropologist* 114 (4), 639–651.
Baranoff, T.I. (1918) *On the Question of the Biological Basis of Fisheries* (W.E. Ricker, trans.). Nanaimo, BC: Fisheries Research Board of Canada.
Barash, D.P. (1977) *Sociobiology and Behavior*. New York: Elsevier.
Barnes, J. (1990) *The Toils of Scepticism*. Cambridge: Cambridge University Press.
Barnett, T., Bass, K. and Brown, G. (1996) Religiosity, ethical ideology and intentions to report a peer's wrongdoing. *Journal of Business Ethics* 15, 1161–1174.

Baron-Cohen, S. (1995) *Mindblindness: An Essay on Autism and Theory of Mind.* Cambridge, MA: MIT Press.
Baron-Cohen, S. (2002) The extreme male brain theory of autism. *TRENDS in Cognitive Sciences* 6 (6), 248–254.
Barnhart, R.K. (1988) *The Barnhart Dictionary of Etymology.* New York: The H.W. Wilson Co.
Barr, C.W. (1995) Thais target world trade in child sex. *Christian Science Monitor* 87 (98), 1–5.
Batson, C.D. (1990) How social an animal? The human capacity for caring. *American Psychologist* 45 (3), 336–346.
Batson, C.D., Ahmad, N., Yin, J., Bedell, S.J., Johnson, J.W., Templin, C.M. and Whiteside, A. (1999) Two threats to the common good: Self-Interested egoism and empathy-induced altruism. *Personality and Social Psychology Bulletin* 25 (1), 3–16.
Batten, J., Hettihewa, S. and Mellor, R. (1999) Factors affecting ethical management: Comparing a developed and developing economy. *Journal of Business Ethics* 19, 51–59.
Bauer, T. and Dowling, R. (2003) Ecotourism policies and issues in Antarctica. In D.A. Fennell and R.K. Dowling (eds) *Ecotourism Policy and Planning* (pp. 309–329). Wallingford: CABI.
Bauman, Z. (1993) *Postmodern Ethics.* Oxford: Blackwell.
Baumhart, R. (1968) *Ethics in Business.* New York: Holt, Rinehart and Winston.
Bayles, M.D. (1981) *Professional Ethics.* Belmont, CA: Wadsworth.
Beach, F.A. (1964) Biological bases for reproductive behavior. In W. Etkin (ed.) *Social Behavior and Organization Among Vertebrates* (pp. 117–142). Chicago, IL: University of Chicago Press.
Beauchamp, T.L. and Bowie, N.E. (eds) (1979) *Ethical Theory and Business.* Englewood Cliffs, NJ: Prentice-Hall.
Becker, L.C. (1977) *Property Rights: Philosophic Foundations.* London: Routledge & Kegan Paul.
Beddoe, C. (2003) Ending child sex tourism: A vision for the future. In B. McKercher and T. Bauer (eds) *Sex and Tourism: Journeys of Romance, Love, and Lust* (pp. 197–207). New York: The Haworth Hospitality Press.
Behrman, J.N. (1981) *Discourses on Ethics and Business.* Cambridge, MA: Oelgeschlager, Gunn & Hain.
Beijerse, R.P. (1999) Questions in knowledge management: Defining and conceptualising a phenomenon. *Journal of Knowledge Management* 3 (2), 94–109.
Beim, H. (1998) Ethics. In R.R. Dupont, T.E. Baxter and L. Theodore (eds) *Environmental Management: Problems and Solutions* (pp. 255–266). Boca Raton, FL: Lewis Publishers.
Belhassen, Y. and Caton, K. (2011) On the need for critical pedagogy in tourism education. *Tourism Management* 32, 1389–1396.
Benedict, R. (1934) *Patterns of Culture.* Boston, MA: Houghton-Mifflin.
Benjamin, J., Li, L., Patterson, C., Greenberg, B., Murphy, D. and Hamer, D. (1996) Population and familial association between the D4 dopamine receptor gene and measures of novelty seeking. *Nature Genetics* 12, 81–84.
Benjamin, M. (1985) Ethics and animal consciousness. In M. Velasquez and C. Rostankowski (eds) *Ethics: Theory and Practice* (pp. 491–499). Englewood Cliffs, NJ: Prentice-Hall.
Benn, P. (1998) *Ethics.* Montreal: McGill-Queen's University Press.
Berkes, F. and Folke, C. (2000) *Linking Social and Ecological Systems: Management Practices and Social Mechanisms for Building Resilience.* Cambridge: Cambridge University Press.

Bermuda National Trust (1999) *Report on the Potential Impacts of Cruise Ships on Bermuda's Environment*. Bermuda: Bermuda National Trust.

Bermuda National Trust (2001) The things they leave behind: Black water, did you say? [Electronic version] (19 May). *Economist* 359 (8222), 31–32.

Bernasconi, R. (1999) The third party: Levinas on the intersection of the ethical and the political. *Journal of the British Society for Phenomenology* 30 (1), 76–87.

Berry, J.F. and Shine, R. (1980) Sexual size dimorphism and sexual selection in turtles (order Testudines). *Oecologia* 44, 185–191.

Bersoff, D.M. (1999) Why good people sometimes do bad things: Motivated reasoning and unethical behavior. *Personality and Social Psychology Bulletin* 25 (1), 28–39.

Bertella, G. (2013) Ethical content of pictures of animals in tourism promotion. *Tourism Recreation Research* 38 (3), 281–294.

Bertella, G. (2016) Experiencing nature in animal-based tourism. *Journal of Outdoor Recreation and Tourism* 14, 22–26.

Beversluis, E.H. (1987) Is there 'no such thing as business ethics'? *Journal of Business Ethics* 6, 81–88.

Bertella, G. (2013). Ethical content of pictures of animals in tourism promotion. *Tourism Recreation Research* 38(3): 281-294.

Bertella, G. (2016). Experiencing nature in animal-based tourism. *Journal of Outdoor Recreation and Tourism* 14: 22-26.

Bhaskar, R. (1978) *A Realist Theory of Science*. Brighton, Sussex: The Harvester Press.

Bhide, A. and Stevenson, H.H. (1990) Why be honest if honesty doesn't pay? *Harvard Business Review* 68, 121–129.

Bigelow, H. and Schroeder, W. (1953) *Fishes of the Gulf of Maine. Fishery Bulletin 53*. Washington, DC: US Government Printing Office.

Birch, C. (1993) *Regaining Compassion for Humanity and Nature*. Kensington, NSW: New South Wales University Press.

Black, J.S., Stern, P.C. and Elworth, J.T. (1985) Personal and contextual influences on household energy adaptations. *Journal of Applied Psychology* 70, 3–21.

Blackburn, S. (2001) *Being Good: A Short Introduction to Ethics*. Oxford: Oxford University Press.

Blackstone, W.T. (1985) Ethics and ecology. In M. Velasquez and C. Rostankowski (eds) *Ethics: Theory and Practice* (pp. 451–453). Englewood Cliffs, NJ: Prentice-Hall.

Blackwell, T. (2003) Show extolling Thai sex trade got tax credit from Ottawa. *National Post*, 22 November, A5.

Blyth, D.A. (2003) Bridging research and practice: Challenges and opportunities. Paper presented at University of Tucson, Tucson, Arizona, 11 February 2003.

Boehmer-Christiansen, S. (1994) The precautionary principle in Germany: Enabling government. In T. O'Riordan and J. Cameron (eds) *Interpreting the Precautionary Principle* (pp. 31–60). London: Earthscan.

Boorstin, D.J. (1985) *The Discoverers: A History of Man's Search to Know his World and Himself*. New York: Vintage Books.

Bouazza Ariño, O. (2002) Sustainable tourism and taxes: An insight into the Balearic eco-tax law. *European Environmental Law Review* 11 (6), 169–174.

Bowie, N.E. (1978) Business codes of ethics: Window dressing or legitimate alternative to government regulation? In T.L. Beauchamp and N.E. Bowie (eds) *Ethical Theory and Business* (pp. 234–239). Englewood Cliffs, NJ: Prentice-Hall.

Bowie, N.E. (2001) New directions in corporate social responsibility. In W.M. Hoffman, R.E. Frederick and M.S. Schwartz (eds) *Business Ethics: Readings and Cases in Corporate Morality* (4th edn; pp. 178–188). Boston: McGraw Hill.

Boyd, R. (1992) The evolution of reciprocity when conditions vary. In A.H. Harcourt and F.B.M. de Waal (eds) *Coalitions and Alliances in Humans and other Animals*. Oxford: Oxford University Press.

Boyd, R. and Richerson, P. (1985) *Culture and the Evolutionary Process*. Chicago, IL: University of Chicago Press.

Boyd, S. (1999) Tourism: Searching for ethics under the sun. *Latinamerica Press* 31 (33), 1–2, 10.

Boyd, S.W. (1991) Towards a typology of tourism: Setting and experience. Paper presented at the Annual Meeting of the Association of American Geographers, Ohio State University, Youngstown, Ohio, 1–2 November.

Brainard, J. (2002) US agencies look to interdisciplinary science. *Chronicle of Higher Education* 48 (40), 20–23.

Bramwell, B. and Lane, B. (1993) Sustainable tourism: An evolving global approach. *Journal of Sustainable Tourism* 1 (1), 1–5.

Brandt, R. (1959) *Ethical Theory: The Problems of Normative and Critical Ethics*. Englewood Cliffs, NJ: Prentice-Hall.

Brewer, R. (1979) *Principles of Ecology*. Philadelphia, PA: Saunders College.

Briassoulis, H. (2002) Sustainable tourism and the question of the commons. *Annals of Tourism Research* 29 (4), 1065–1085.

British Columbia Ministry of Development, Industry and Trade (1991) *Developing Code of Ethics: British Columbia's Tourism Industry*. Victoria, BC: Ministry of Development, Trade and Tourism.

Britton, R.A. (1977) Making tourism more supportive of small-state development: The case of St Vincent. *Annals of Tourism Research* 4 (5), 268–278.

Britton, S.G. (1982) The political economy of tourism in the Third World. *Annals of Tourism Research* 9 (3), 331–358.

Brody, B. (1983) *Ethics and its Applications*. New York: Harcourt Brace Jovanovich Inc.

Bronowski, J. (1981) *The Ascent of Man*. London: British Broadcasting Corporation.

Brookes, L.J. (1991) Codes of conduct for business: Are they effective, or just window-dressing? *Canadian Public Administration* 34 (1), 171–176.

Brookfield, H. (1975) *Interdependent Development*. London: Methuen.

Broom, A. (1987) *The Closing of the American Mind*. New York: Simon & Schuster.

Broome, J. (1991) *Weighing Goods: Equality, Uncertainty and Time*. Cambridge, MA: Basil Blackwell.

Brown, D.E. (1991) *Human Universals*. Philadelphia, PA: Temple University Press.

Brown, L. (2013) Tourism: A catalyst for existential authenticity. *Annals of Tourism Research* 40, 176–190.

Brundtland Report. See http://www.un-documents.net/our-common-future.pdf

Bryson, B. (2003) *A Short History of Nearly Everything*. Toronto, ON: Doubleday.

Buchanan, J.M. and Tullock, G. (1982) *Towards a Theory of the Rent-Seeking Society*. College Station, Texas: A&M Press.

Buckley, R. (2002) Tourism ecolabels. *Annals of Tourism Research* 29 (1), 183–208.

Budowski, G. (1976) Tourism and environmental conservation: Conflict, coexistence, or symbiosis. *Environmental Conservation* 3 (1), 27–31.

Bull, A.O. (1996) The economics of cruising: An application to the short ocean cruise market. *The Journal of Tourism Studies* 7 (2), 28–35.

Burkenroad, M.D. (1953) Theory and practice of marine fishery management. *Journal du conseil permanent international pour l'exploration de la mer* XVIII (3).

Burns, G.L., MacBeth, J. and Moore, S. (2011) Should dingoes die? Principles for engaging ecocentric ethics in wildlife tourism management. *Journal of Ecotourism* 10 (3), 179–196.

Buss, D.M. (1995) Evolutionary psychology: A new paradigm for psychological science. *Psychological Inquiry* 6 (1), 1–30.

Buss, D.M. (2000) *The Dangerous Passion: Why Jealousy Is as Necessary as Love and Sex*. New York: The Free Press.
Buss, L.W. (1987) *The Evolution of Individuality*. Princeton, NJ: Princeton University Press.
Butcher, J. (2003) *The Moralisation of Tourism: Sun, Sand … and Saving the World?* London: Routledge.
Butler, J. (1950) *Five Sermons*. New York: Liberal Arts Press.
Butler, R.W. (1980) The concept of a tourist area cycle of evolution: Implications for management of resources. *Canadian Geographer* 24 (1), 5–12.
Butler, R.W. (1990) Alternative tourism: Pious hope or Trojan horse? *Journal of Travel Research* 28 (3), 40–45.
Butler, R.W., Fennell, D.A. and Boyd, S.W. (1992) *The POLAR Model: A System for Managing the Recreational Capacity of Canadian Heritage Rivers*. Ottawa, ON: Environment Canada.
Byrne, D. (1999) *Complexity Theory and the Social Sciences: An Introduction*. London: Routledge.
Calamai, P. (n.d.) Arctic lake pollution goes back centuries. Unknown.
Callicott, J.B. (1984) Non-anthropocentric value theory and environmental ethics. *American Philosophical Quarterly* 21, 299–309.
Cantaloops, A.S. (2004) Policies supporting sustainable tourism development in the Balearic Islands. *The Ecotax. Anatolia* 15 (1), 39–56.
Carlson, D.S. and Perrewe, P.L. (1995) Institutionalization of organizational ethics through transformational leadership. *Journal of Business Ethics* 14, 829–838.
Carneiro, R.L. (1970) The theory of the origin of the state. *Science* 169, 733–738.
Carriere, J. (1991) The crisis in Costa Rica: An ecological perspective. In D. Goodman and M.R. Redclift (eds) *Environment and Development in Latin America; The Politics of Sustainability: Issues in Environmental Politics* (pp. 184–204). Manchester: Manchester University Press.
Carson, R. (1962/1987) *Silent Spring*. Boston: Houghton Mifflin.
Caspi, A., McClay, J., Moffitt, T., Mill, J., Martin, J., Craig, I., Taylor, A. and Poulton, R. (2002) Role of genotype in the cycle of violence in maltreated children. *Science* 297, 851–854.
Castañeda, Q. (2012) The neoliberal imperative of tourism: Rights and legitimization in the UNWTO Global Code of Ethics for Tourism. *Society for Applied Anthropology* 34 (3), 47–51.
Caton, K. (2012) Taking the moral turn in tourism studies. *Annals of Tourism Research* 39 (4), 1906–1928.
Caton, K. (2014) Humanism and tourism: A moral encounter. In M. Mostafanezhad and K. Hannam (eds) *Moral Encounters in Tourism* (pp. 185–195). London: Routledge.
Caton, K. and Santos, C.A. (2009) Selling study abroad in a postcolonial world. *Journal of Travel Research* 48 (2), 191–204.
Ceasar, M. (1999) The myth of 'ecotourism'. *Latinamerica Press* 31 (33), 3.
Cerruti, J. (1964) The two Acapulcos. *National Geographic Magazine* 126 (6), 848–878.
Chagnon, N. (1983) *Yanomamo, the Fierce People* (3rd edn). New York: Holt, Rinehart & Winston.
Checkland, P. and Scholes, J. (1990) *Soft Systems Methodology in Action*. Chichester: John Wiley & Sons.
Chester, G. (2002) *Mohonk Agreement*. The Institute for Policy Studies. See http://www.rainforest-alliance.org/business/tourism/documents/mohonk.pdf (accessed 7 May 2012).
Chichilnisky, G. (1996) The economic value of the Earth's resources. *Trends in Ecology and Evolution* 11 (3), 135–140.
Clark, C.W. (1973) Profit maximization and the extinction of animal species. *Journal of Political Economy* 81 (4), 950–961.

Clark, K. (1969) *Civilisation*. New York: Harper & Row.

Clark, J. and Dawson, L. (1996) Personal religiousness and ethical judgements: An empirical analysis. *Journal of Business Ethics* 15, 359–372.

Clawson, M. and Knetsch, J.L. (1966) *Economics of Outdoor Recreation*. Baltimore, MD: Johns Hopkins University Press.

Cleverdon, R. and Kalisch, A. (2000) Fair trade in tourism. *International Journal of Tourism Research* 2, 171–187.

Cochrane, D. (2017) Liberals to announce marijuana will be legal by July 1, 2018. http://www.cbc.ca/news/politics/liberal-legal-marijuana-pot-1.4041902 (accessed 29 March 2017).

Cohen, D.V. (1995) Creating ethical work climates: A socioeconomic perspective. *The Journal of Socio-Economics* 24, 317–343.

Cohen, E. (1987) Alternative tourism: A critique. *Tourism Recreation Research* 12 (2), 13–18.

Cohen, E. (1978) The impact of tourism on the physical environment. *Annals of Tourism Research* 5 (2), 215–237.

Cohen, E. (1972) Toward a sociology of tourism. *Social Research* 39 (1), 164–182.

Cohen, E. (2010) Towards an ethics of responsibility in tourism education. *Tourism Recreation Research* 35 (3), 302–303.

Cohen, E. (2013) 'Buddhist compassion' and 'animal abuse' in Thailand's Tiger Temple. *Society & Animals* 21, 266–283.

Cohen, J., Plant, L. and Sharp, D. (1993) A validation and extension of a multidimensional ethics scale. *Journal of Business Ethics* 12, 13–26.

Cole, S. (2007) Implementing and evaluating a code of conduct for visitors. *Tourism Management* 28, 443–451.

Coles, T., Fenclova, E. and Dinan, C. (2013) Tourism and corporate social responsibility: A critical review and research agenda. *Tourism Management Perspectives* 6, 122–141.

Colish, M. (1978) Cicero's *De Officiis* and Machiavelli's *Prince*. *Sixteenth Century Journal* 9, 91–94.

Collinson, D. (1987) *Fifty Major Philosophers: A Reference Guide*. London: Croom Helm.

Collins, J.C. and Porras, J.I. (1996) Building your company's vision. *Harvard Business Review* 74 (4), 65–77.

Concise Oxford Dictionary (1976) (6th edn) (J.B. Sykes, ed.). Oxford: Clarendon Press.

Connell, J. (2006) Medical tourism: Sea, sun, sand and…surgery. *Tourism Management* 27, 1093–1100.

Connellan, H., Baron-Cohen, S., Wheelwright, S., Batki, A. and Ahluwalia, J. (2000) Sex difference in neonatal social perception. *Infant Behavior and Development* 23, 113–118.

Convenvention on Biological Diversity (CBD) (2001) Biological Diversity and Tourism: International Guidelines for Sustainable Tourism. See http://biodiv.unwto.org/content/applying-guidelines-biological-diversity-and-tourism-development

Cooper, C. (2002) Knowledge management. *Current Issues in Tourism* 5 (2), 375–377.

Cooper, T.L. (1987) Hierarchy, virtue, and the practice of public administration: A perspective for normative ethics. *Public Administration Review* 47, 320–328.

Cosmides, L., Tooby, J. and Barkow, J.H. (1992) Introduction: Evolutionary psychology and conceptual integration. In J.H. Barkow, L. Cosmides and J. Tooby (eds) *The Adapted Mind: Evolutionary Psychology and the Generation of Culture* (pp. 3–15). New York: Oxford.

Costanza, R. (1987) Social traps and environmental policy. *BioScience* 37 (6), 407–412.

Costanza, R., d'Arge, R., de Groot, R., Farber, S., Grasso, M., Hannon, B., Limburg, K., Naeem, S., O'Neill, R.V., Paruelo, J., Raskin, R.G., Sutton, P. and van den Belt, M. (1997) The value of the world's ecosystem services and natural capital. *Nature* 387, 253–260.

Cottingham, J. (1998) *Philosophy and the Good Life: Reason and the Passions in Greek, Cartesian and Psychoanalytic Ethics*. Cambridge: Cambridge University Press.
Cummings, B. (2012) Benefit corporations: How to enforce a mandate to promote the public interest. *Columbia Law Review* 112 (3), 578–627.
Curry, P. (2011) *Ecological Ethics: An Introduction* (2nd edn). Cambridge: Polity Press.
Daily, G. (ed.) (1997) *Nature's Services: Societal Dependence on Natural Ecosystems*. Washington, DC: Island Press.
Daily, G.C. and Ehrlich, P.R. (1999) Managing earth's ecosystems: An interdisciplinary challenge. *Ecosystems* 2, 277–280.
Daly, M. and Wilson, M. (1988) *Homicide*. Hawthorne, NY: Aldine de Gruyter.
Daly, M. and Wilson, M. (1999) *The Truth about Cinderella: A Darwinian View of Parental Love*. New Haven, CT: Yale University Press.
D'Amore, L.J. (1993) A code of ethics and guidelines for socially and environmentally responsible tourism. *Journal of Travel Research* 31 (3), 64–66.
Darwin, C. (1859/2007) *ON the Origin of Species by means of Natural Selection*. New York: Cosimo Classics.
Davies, T. and Cahill, S. (2000) *Environmental Implications of the Tourism Industry*. Washington, DC: Resources for the Future.
Davis, P.B. (1998) Beyond guidelines: A model for Antarctic tourism. *Annals of Tourism Research* 23 (3), 546–533.
Dawkins, R. (1999) *The Selfish Gene*. Oxford: Oxford University Press.
De Botton, A. (2002) *The Art of Travel*. London: Penguin Group.
De George, R.T. (1987) The status of business ethics: Past and future. *Journal of Business Ethics* 6, 201–211.
Degler, C.N. (1991) *In Search of Human Nature: The Decline and Revival of Darwinism in American Social Thought*. New York: Oxford University Press.
De Grosbois, D. and Fennell, D.A. (2011) Carbon footprint of the global hotel companies: Comparison of methodologies and results. *Tourism Recreation Research* 36 (3), 231–245.
Delius, C., Gatzemeier, M., Sertcan, D. and Wunscher, K. (2000) *The Story of Philosophy: From Antiquity to Present*. Cologne: Konemann.
Demont, J. (2003) Finding the higher ground in Saint John. *MacLean's* 116 (41), 65, October 13.
Denhardt, K.G. (1989) The management of ideals: A political perspective on ethics. *Public Administration Review* 48, 187–193.
Denhardt, K.G. (1991) Unearthing the moral foundation of public administration: Honor, benevolence, and justice. In J.S. Brown (ed.) *Ethical Frontiers in Public Management: Seeking New Strategies for Resolving Ethical Dilemmas* (pp. 91–113). San Francisco, CA: Jossey-Bass.
Dennett, D.C. (1984) *Darwin's Dangerous Idea: Evolution and the Meanings of Life*. New York: Touchstone.
Dernoi, L.A. (1981) Alternative tourism: Towards a new style in North–South relations. *Tourism Management* 2, 253–264.
Descartes, R. (1637/1998) *Discourse on Method* (D.A. Cress, trans.) (3rd edn). Cambridge: Hackett Co.
Descartes (1641) *Meditations on First Philosophy* (J. Cottingham, trans.). Cambridge: Cambridge University Press, 1996.
Devall, B. and Sessions, G. (1985) *Deep Ecology: Living as if Nature Mattered*. Salt Lake City, UT: Peregrine Smith.
de Waal, F. (1989a) Food sharing and reciprocal obligations among chimpanzees. *Journal of Human Evolution* 18, 433–459.
de Waal, F. (1989b) *Chimpanzee Politics: Power and Sex Among the Apes*. Baltimore, MD: Johns Hopkins University Press.

Diamond, C. (2001) Injustice and animals. In C. Elliott (ed.) *Slow Cures and Bad Philosophers: Essays on Wittgenstein, Medicine, and Bioethics* (pp. 118–148), Durham, NC: Duke University Press.

Diamond, J. (1993) New Guineans and their natural world. In S.R. Kellert and E.O. Wilson (eds) *The Biophilia Hypothesis* (pp. 251–271). Washington, DC: Island Press.

Dickson, B. (1999) The precautionary principle in CITES: A critical assessment. *Natural Resources Journal* 39, 211–228.

Dieke, P. (2001) Kenya and South Africa. In D.B. Weaver (ed.) *The Encyclopedia of Ecotourism* (pp. 89–105). London: CABI Publishing.

Dobson, J. (2011) Towards a utilitarian ethic for marine wildlife tourism. *Tourism in Marine Environments* 7 (3–4), 213–222.

Dobson, S., Gill, A. and Baird, S. (2001) *A Primer on the Canadian Pacific Cruise Ship Industry*. (Draft). See http://www.pac.dfo-mpo.gc.ca/oceans/Policy/cruiseship_e.htm (accessed 15 October 2003).

Donovan, P.J. (1986) Do different religions share common ground? *Religious Studies* 22 (3–4), 367–375.

Dowd, D. (2000) *Capitalism and its Economics: A Critical History*. London: Pluto Press.

Dower, N. (1989) What is environmental ethics? In N. Dower (ed.) *Ethics and Environmental Responsibility* (pp. 11–37). Aldershot: Avebury.

Dowling, R.K. (1992) *The Ecoethics of Tourism: Guidelines for Developers, Operators and Tourists*. Canberra: Bureau of Tourism Research.

Drengson, A. and Inoue, Y. (eds) (1995) *The Deep Ecology Movement: An Introductory Anthology*. Berkeley, CA: North Atlantic Books.

Driscoll, D.-M. and Hoffman, W.M. (1999) Gaining the ethical edge: Procedures for delivering values-driven management. In W.M. Hoffman, R.E. Frederick and M.S. Schwartz (eds) *Business Ethics: Readings and Cases in Corporate Morality* (4th edn; pp. 573–584). Boston, MA: McGraw Hill.

Drummond, A. (1998) The Padaung 'human zoo'. *In Focus* Autumn (29), 8–9.

D'Sa, E. (1999) Wanted: Tourists with a social conscious. *International Journal of Contemporary Hospitality Management* 11 (2/3), 64–68.

Dubos, R.J. (1973) Humanizing the earth. *Science* 179, 769–772.

Dunfee, T.W. and Black, B.M. (1996) Ethical issues confronting travel agents. *Journal of Business Ethics* 15, 207–217.

Duenkel, N. and Scott, H. (1994) Ecotourism's hidden potential – altering perceptions of reality. *Journal of Physical Education, Recreation and Dance* October, 40–44.

Duffy, R. and Moore, L. (2010) Neoliberalising nature? Elephant-back tourism in Thailand and Botswana. *Antipode* 42 (3), 742–766.

Dunfee, T.W. and Werhane, P. (1997) Report on business ethics in North America. *Journal of Business Ethics* 16, 1589–1595.

Dunlap, R.E. and Van Liere, K.D. (1978) The 'New Environmental Paradigm'. *The Journal of Environmental Education* 9 (4), 10–19.

Durant, W. (1935/1963) *The Story of Civilization I: Our Oriental Heritage*. New York: Simon & Schuster.

Dwyer, L. and Forsyth, P. (1997) Economic significance of cruise tourism. *Annals of Tourism Research* 25 (2), 393–415.

Dyson, P. (2012) Slum tourism: Representing and interpreting 'reality' in Dharavi, Mumbai. *Tourism Geographies: An International Journal of Tourism Space, Place and Environment*, doi:10.1080/14616688.2011.609900.

Eades, J.S. (2009) Moving bodies: The intersections of sex, work, and tourism. In *Economic Development, Integration, and Morality in Asia and the Americas* (pp. 225–253). See https://doi.org/10.1108/S0190-1281(2009)0000029011 (accessed March 8, 2015).

Echtner, C.M. and Jamal, T.B. (1997) The disciplinary dilemma of tourism studies. *Annals of Tourism Research* 24 (4), 868–883.

Echtner, C.M. and Prasad, P. (2003) The context of Third world tourism marketing. *Annals of Tourism Research* 30 (3), 660–682.

Edel, A. (1964) *Ethical Judgement: The Use of Science in Ethics*. New York: Free Press of Glencoe.

Edgell, D.L. (1999) *Tourism Policy: The Next Millennium*. Champaign, IL: Sagamore.

Edney, J.J. (1980) The commons problem: Alternative perspectives. *American Psychologist* 35, 131–150.

Ehrenfeld, D. (1981) *The Arrogance of Humanism*. Oxford: Oxford University Press.

Ehrlich, P.R. (2000) *Human Natures: Genes, Cultures, and the Human Prospect*. New York: Penguin.

Ellis, C.J. (2001) Common pool equities: An arbitrage based non-cooperative solution to the common pool resource problem. *Journal of Environmental Economics and Management* 42, 140–155.

Ellis, J. (2000) The precautionary principle: From paradigm to rule of law. *International Law* 2, 127–129.

Ellul, J. (1965) *The Technological Society* (J. Wilkinson, trans.). New York: Alfred A. Knopf.

Elm, D.R. and Nichols, M.L. (1993) An investigation of the moral reasoning of managers. *Journal of Business Ethics* 12, 817–833.

Enderle, G. (1987) Some perspectives of managerial ethical leadership. *Journal of Business Ethics* 6, 657–663.

Enghagen, L.K. (1990) Ethics in hospitality and tourism education: A survey. *Hospitality Research Journal* 14, 113–118.

Erisman, H.M. (1983) Tourism and cultural dependency in the West Indies. *Annals of Tourism Research* 10 (3), 337–361.

Eve, R.A., Horsfall, S. and Lee, M. (eds) (1997) *Chaos, Complexity and Sociology: Myths, Models and Theories*. Thousand Oaks, CA: Sage.

Fainstein, S.S. (2010) *The Just City*. Ithaca, NY: Cornell University Press.

Fairweather, J.R., Maslin, C. and Simmons, D.G. (2005) Environmental values and response to ecolabels among international visitors to New Zealand. *Journal of Sustainable Tourism* 13 (1), 82–98.

Farrell, B.J., Cobbin, D.M. and Farrell, H.M. (2002) Can codes of ethics really produce consistent behaviors? *Journal of Managerial Psychology* 17 (6), 468–490.

Farrington, P. (1999) All-inclusives: A new apartheid. *In Focus* 33, 10–11, 17.

Feinberg, J. (1985) The rights of animals and unborn generations. In M. Velasquez and C. Rostankowski (eds) *Ethics: Theory and Practice* (pp. 466–476). Englewood Cliffs, NJ: Prentice-Hall.

Fennell, D.A. (2000a) Ecotourism on trial: The case of billfishing as ecotourism. *Journal of Sustainable Tourism* 8 (4), 341–345.

Fennell, D.A. (2000b) Tourism and applied ethics. *Tourism Recreation Research* 25 (1), 59–70.

Fennell, D.A. (2002) *Ecotourism Programme Planning*. Wallingford: CABI.

Fennell, D.A. (2003a) *Ecotourism: An Introduction* (2nd edn). London: Routledge.

Fennell, D.A. (2003b) Towards interdisciplinarity in tourism: Making a case through shared knowledge and complexity. *Recent Advances and Research Updates* 4 (2), 221–232.

Fennell, D.A. (2004) Deep ecotourism: seeking reverence in theory and practice. In T.V. Singh (ed.) *Novelty Tourism: Strange Experiences and Stranger Practices* (pp. 109–120). London: Routledge.

Fennell, D.A. (2006) Evolution in tourism: Implications of the theory of reciprocal altruism. *Current Issues in Tourism* 9 (2), 105–124.
Fennell, D.A. (2008) Responsible tourism: A Kierkegaardian interpretation. *Tourism Recreation Research* 33 (1), 3–12.
Fennell, D.A. (2009) The nature of pleasure in pleasure travel. *Tourism Recreation Research* 34 (2), 123–134.
Fennell, D.A. (2012a) *Tourism and Animal Ethics*. London: Routledge.
Fennell, D.A. (2012b) Tourism and animal rights. *Tourism Recreation Research* 37 (2), 157–166.
Fennell, D.A. (2012c) Tourism, animals and utilitarianism. *Tourism Recreation Research* 37 (3), 239–249.
Fennell, D.A. (2013a) Ecotourism, animals and ecocentrism: A reexamination of the billfish debate. *Tourism Recreation Research* 38 (2), 3–13.
Fennell, D.A. (2013b) Tourism and animal welfare. *Tourism Recreation Research* 38 (3), 325–340.
Fennell, D.A. (2013c) The ethics of excellence in tourism research. *Journal of Travel Research* 52 (4), 417–425.
Fennell, D.A. (2014) Exploring the boundaries of a new moral order for tourism's global code of ethics: An opinion piece on the position of animals in the tourism industry. *Journal of Sustainable Tourism* 22 (7), 983–996.
Fennell, D.A. (2015a) Akrasia and tourism: Why we sometimes act against our better judgement? *Tourism Recreation Research* 40 (1), 95–106.
Fennell, D.A. (2015b) *Ecotourism*. London: Routledge.
Fennell, D.A. (2016) (ed.) *Tourism Ethics*. London: Routledge.
Fennell, D.A. and Ebert, K. (2004) Tourism and the precautionary principle. *Journal of Sustainable Tourism* 12 (6), 461–479.
Fennell, D.A. and Malloy, D.C. (1995) Ethics and ecotourism: A comprehensive ethical model. *Journal of Applied Recreation Research* 20 (3), 163–183.
Fennell, D.A. and Malloy, D.C. (1999) Measuring the ethical nature of tourism operators. *Annals of Tourism Research* 26 (4), 928–943.
Fennell, D.A. and Malloy, D.C. (2007) *Codes of Ethics in Tourism: Practice, Theory, Synthesis*. Clevedon: Channel View Publications.
Fennell, D.A. and Przeclawski, K. (2003) Generating goodwill in tourism through ethical stakeholder interaction. In S. Singh, D.J. Timothy and R.K. Dowling (eds) *Tourism in Destination Communities* (pp. 135–152) Wallingford: CABI.
Fennell, D.A. and Weaver, D.B. (2005) The ecotourium concept and tourism-conservation symbiosis. *Journal of Sustainable Tourism* 13 (4), 373–390.
Ferrell, O.C. and Skinner, S.J. (1988) Ethical behavior and bureaucratic structure in marketing research organizations. *Journal of Marketing Research* 25, 103–109.
Finnis, J. (1980) *Natural Law and Natural Rights*. New York: Oxford University Press.
Fishbein, M. and Ajzen, I. (1975) *Belief, Attitude, Intention and Behaviour: An Introduction to Theory and Research*. Reading, MA: Addison-Wesley.
Fiske, A.P. (1992) The four elementary forms of sociality: Framework for a unified theory of social relations. *Psychological Review* 99 (4), 689–723.
Fleckenstein, M.P. and Huebsch, P. (1999) Ethics in tourism: Reality or hallucination? *Journal of Business Ethics* 19, 137–142.
Florman, S.C. (1994) *The Existential Pleasures of Engineering* (2nd edn). New York: St Martin's Griffin.
Font, X. (2002) Environmental certification in tourism and hospitality: Progress, process and prospects. *Tourism Management* 23, 197–205.
Font, X. and Tribe, J. (2001) Promoting green tourism: The future of environmental awards. *International Journal of Tourism Research* 3, 9–21.

Foot, P. (1959) Moral beliefs. *Proceedings of the Aristotelian Society* 59, 83–104.
Foot, P. (1978) *Virtues and Vices and Other Essays in Moral Philosophy*. Berkeley, CA: University of California Press.
Foot, P. (2003) *Natural Goodness*. Oxford: Clarendon Press.
Ford, P. and Blanchard, J. (1993) *Leadership and Administration of Outdoor Pursuits*. State College, PA: Venture Publishing.
Fox, R. (1989) *The Search for Society: Quest for a Biosocial Science and Morality*. New Brunswick, NJ: Rutgers University Press.
Fox, R. (1994) *The Challenge of Anthropology: Old Encounters and New Excursions*. New Brunswick, NJ: Transaction.
Fox, R. (1997) *Conjectures & Confrontations: Science, Evolution, Social Concern*. London: Transaction.
Fox, R.M. and DeMarco, J.P. (1986) The challenge of applied ethics. In J.P. DeMarco and R.M. Fox (eds) *New Directions in Ethics: The Challenge of Applied Ethics* (pp. 1–18). New York: Routledge & Kegan Paul.
Fraedrich, J.P. (1993) The ethical behavior of retail managers. *Journal of Business Ethics* 12 (3), 207–218.
Frank, R. (1988) *Passions within Reason: The Strategic Role of the Emotions*. New York: W.W. Norton & Co.
Frankena, W.K. (1963) *Ethics*. Englewood Cliffs, NJ: Prentice-Hall.
Frankl, V. (1985) *Man's Search for Meaning*. New York: Washington Square Press.
Franklin, A. (2001) Neo-Darwinian leisures, the body and nature: Hunting and angling in modernity. *Body & Society* 7 (4), 57–76.
Frantz, D. (1999) Gaps in sea laws shield pollution by cruise lines. *The New York Times* (3 January) CXCVIII (51, 391), 19–20.
Fountain, H. (2017) With more ships in the Arctic, fears of disaster rise. The New York Times (July 23). See https://www.nytimes.com/2017/07/23/climate/ships-in-the-arctic.html
Freeman, R.E. (ed.) (1990) *Business Ethics: The State of the Art*. New York: Oxford University Press.
Freeman, R.E. (1999) Divergent stakeholder theory. *Academy of Management Review* 24, 233–236.
Freeman, R.E. and Gilbert, D.R. (1988) *Corporate Strategy and the Search for Ethics*. Englewood Cliffs, NJ: Prentice-Hall.
Freestone, D. and Hey, E. (1996) Origins and development of the precautionary principle. In D. Freestone and E. Hey (eds) *The Precautionary Principle and Environmental Law: The Challenge of Implementation* (pp. 3–12). The Hague: Kluwer Law International.
Freidmann, J. and Alonso, W. (eds) (1974) *Regional Development and Planning: A Reader*. Cambridge, MA: MIT Press.
Friedman, M. (1970) The social responsibility of business is to increase its profits. In W.M. Hoffman, R.E. Frederick and M.S. Schwartz (eds) *Business Ethics: Readings and Cases in Corporate Morality* (4th edn; pp. 156–160). Boston, MA: McGraw Hill.
Frenzel, F., Koens, F., and Steinbrink, M. (eds) (2012) *Slum Tourism: Poverty, Power and Ethics*. London: Routledge.
Freudenburg, W.R. and Gramling, R. (2002) Scientific expertise and natural resource decisions: Social science participation on interdisciplinary scientific committees. *Social Science Quarterly* 83 (1), 7–17.
Fry, D.P. (2000) Conflict management in cross-cultural perspective. In F. Aureli and F. de Waal (eds) *Natural Conflict Resolution* (pp. 334–351). Berkeley, CA: University of California Press.

Fukuyama, F. (1995) *Trust: The Social Virtues and the Creation of Prosperity*. New York: Free Press Paperbacks.
Furedi, F. (1997) *Population and Development: A Critical Introduction*. New York: St Martin's Press.
Gadgil, M. (1991) Conserving India's biodiversity: The societal context. *Evolutionary Trends in Plants* 5 (1), 3–8.
Gadgil, M., Berkes, F. and Folke, C. (1993) Indigenous knowledge for biodiversity conservation. *Ambio* 22 (2–3), 151–156.
Gago, A., Labandeira, X., Picos, F. and Rodríguez, M. (2009) Specific and general taxation of tourism activities. Evidence from Spain. *Tourism Management*, 30 (3), 381–392.
Garely, E. (2004) Child sex tourism persists. See *eTurbo News.com* (accessed 11 May 2004).
Garofalo, C. and Geuras, D. (1999) *Ethics in the Public Service: The Moral Mind at Work*. Washington, DC: Georgetown University Press.
Garrod, B. and Fennell, D.A. (2004) An analysis of whalewatching codes of conduct. *Annals of Tourism Research* 31 (2), 334–352.
Genot, H. (1995) Voluntary environmental codes of conduct in the tourism sector. *Journal of Sustainable Tourism* 3 (3), 166–172.
Ghimire, K.B. and Pimbert, M.P. (eds) (1997) *Social Change and Conservation: Environmental Politics and Impacts of National Parks and Protected Areas*. London: Earthscan.
Gibson, C. (2010) Geographies of tourism: (Un)ethical encounters. *Progress in Human Geography* 34 (4), 521–527.
Gibbs, J.C. (1977) Kohlberg's stages of moral development: A constructive critique. *Harvard Educational Review* 47, 43–59.
Gjerdalen, G. and Williams, P.W. (2000) An evaluation of the utility of a whale watching code of conduct. *Tourism Recreation Research* 25 (2), 27–37.
Gilligan, C. (1982) *In a Different Voice: Psychological Theory and Women's Development*. Cambridge, MA: Harvard University Press.
Gioia, D.A. (1999) Practicability, paradigms, and problems in stakeholder theorizing. *Academy of Management Review* 24, 228–232.
Glasbergen, P. (1998) The question of environmental governance. In P. Glasbergen (ed.) *Co-operative Environmental Governance* (pp. 1–18). London: Kluwer Academic Publishers.
Godoi Trigo, L.G. (2003) The old problems of Brazilian tourism. *Tourism Review* 58 (1), 19–24.
Goldberg, C. (2004) Mind's morality play. *National Post*, 14 June, C12.
González Bernáldez, F. (1994) Tourism and the environment. *All of Us* 15, 2.
Goodpaster, K.E. and Matthews, J.B. (1982) Can a corporation have a conscience? In W.M. Hoffman, R.E. Frederick and M.S. Schwartz (eds) *Business Ethics: Readings and Cases in Corporate Morality* (4th edn; pp. 147–156). Boston: McGraw Hill.
Goodwin, H. (2003) Ethical and responsible tourism: Consumer trends in the UK. *Journal of Vacation Marketing* 9 (3), 271–284.
Goodwin, H. and Roe, D. (2001) Tourism, livelihoods and protected areas: Opportunities for fair trade tourism in and around national parks. *International Journal of Tourism Research* 3, 377–391.
Gordon, H.S. (1954) The economic theory of a common-property resource: The fishery. *Journal of Political Economy* 62, 124–142.
Gosling, J. (1969) *Pleasure and Desire: The Case of Hedonism Reviewed.* Oxford: Clarendon Press.
Gould, S.J. (2003) *The Hedgehog, the Fox, and the Magister's Pox*. New York: Harmony Books.
Gouldner, A. (1964) *Patterns of Industrial Bureaucracy*. New York: Free Press.
Government of Canada (1998) *Voluntary Codes: A Guide for their Development and Use*. A joint initiative of the Office of Consumer Affairs, Industry Canada, and the

Regulatory Affairs Division, Treasury Board Secretariat. Ottawa, ON: Distribution Services Communications Branch.
Government of Canada. (2001) *A Canadian Perspective on the Precautionary Principle/ Approach.* See http://cppa/HTML/pamphlet_e.htm.
Gowans, C.W. (ed.) (1987) *Moral Dilemmas*. New York: Oxford University Press.
Grafton, A. (1999) Introduction. In N. Machiavelli (ed.) *The Prince* (pp. vx–xxix). London: Penguin Books.
Gray, S.T. (1996) Auditing your ethics. *Association Management* 48 (9), 188.
Greene, J., Sommerville, R.B.C., Nystrom, L.E., Dorley, J.M. and Cohen, J.D. (2001) An FMRI investigation of emotional engagement in moral judgement. *Science* 293 (5537), 2105–2108.
Grenier, A.A. (1998) *Ship-based Polar Tourism in the Northwest Passage: A Case Study.* Roveniemi, Finland: University of Lapland.
Green Globe 21 (2002) The Green Globe 21 Path. See https://greenglobe.com/about/ (Accessed 3 June 2002).
Griffin, J. (1997) *Value Judgement: Improving Our Ethical Beliefs*. Oxford: Clarendon Press.
Grimwood, B.S.R. (2013) Illuminating traces: Enactments of responsibility in practices of Arctic river tourists and inhabitants. *Journal of Ecotourism* 12 (2), 53–74.
Grimwood, B.S.R. (2014) Advancing tourism's moral morphology: Relational metaphors for just and sustainable Arctic tourism. *Tourist Studies* 15 (1), 3–26.
Grossman, B. (1988) *Corporate Loyalty: A Trust Betrayed*. Markham: Penguin.
Gullison, R.E., Rice, R.E. and Blundell, A.G. (2000) 'Marketing' species conservation. *Nature* 404, 923–924.
Guy, M.E. (1990) *Ethical Decision Making in Everyday Work Situations*. Westport, CT: Greenwood Press.
Habermas, J. (1979) *Communication and the Evolution of Society*. Boston: Beacon Press.
Habermas, J. (1984) *Theory of Communicative Action, Vol. I: Reason and the Rationalization of Society* (trans. Thomas McCarthy) Boston: Beacon Press.
Habermas, J. (1993) *Justification and Application: Remarks on Discourse Ethics* (C.P. Cronin, trans.). Cambridge, MA: MIT Press.
Hagarty, J.A. (1990) Ethics in hospitality education. *International Journal of Hospitality Management* 9 (2), 106–109.
Hagedorn, R. (1981) *Essentials of Sociology*. Toronto, ON: Holt, Rinehart & Winston.
Haidt, J. (2002) The moral emotions. In. R.J. Davidson, K.R. Scherer and H.H. Goldsmith (eds) *Handbook of Affective Sciences* (pp. 852–870). New York: Oxford University Press.
Haidt, J. (2001) The emotional dog and its rational tail: A social intuitionist approach to moral judgement. *Psychological Review* 108 (4), 814–832.
Haig, D. (1993) Genetic conflict in human pregnancy. *The Quarterly Review of Biology* 68 (4), 495–532.
Hall, B. (2001) Values development and learning organizations. *Journal of Knowledge Management* 5 (1), 19–32.
Hall, C.M. (1994) *Tourism and Politics: Policy, Power and Place*. Chichester: John Wiley & Sons.
Hall, C.M. and Wouters, M. (1995) Issues in Antarctic tourism. In C.M. Hall and M.E. Johnston (eds) *Polar Tourism: Tourism in the Arctic and Antarctic Regions* (pp. 147–166). Chichester: John Wiley & Sons.
Hall, D. and Brown, F. (2006) *Tourism and Welfare: Ethics, Responsibility and Sustained Wellbeing*. Wallingford: CABI.
Hall, S. (ed.) (1993) *Ethics in Hospitality Management: A Book of Readings*. East Lansing, MI: Educational Institute of the American Hotel and Motel Association.
Hames, R. (1990) Game conservation or efficient hunting? In B.J. McKay and J.M. Acheson (eds) *The Question of the Commons: The Culture and Ecology of Communal Resources* (pp. 92–107). Tucson, AZ: The University of Arizona Press.

Hamilton, C. (2003) *Growth Fetish*. Crows Nest, NSW: Allen and Unwin.
Hamilton, W.D. (1964) The genetical evolution of social behaviour (I and II). *Journal of Theoretical Biology* 7, 1–52.
Hamilton, W.D. (1971) Geometry of the selfish herd. *Journal of Theoretical Biology* 31, 295–311.
Hammond, A. (1998) *Which World? Scenarios for the 21st Century: Global Destinies, Regional Choices*. Washington, DC: Island Press.
Hampton, G. (1999) Environmental equity and public participation. *Policy Sciences* 32, 163–174.
Hansen, R. (1992) A multidimensional ethics scale for measuring business ethics: A purification and refinement. *Journal of Business Ethics* 11, 523–534.
Hardin, G. (1968) The tragedy of the commons. *Science* 162, 1243–1248.
Harding, K. (2003) Thirst for ethics unslacked. *The Globe and Mail*, 3 May, C1.
Hargrove, E.C. (1993) The ontological argument for the preservation of nature. In S.J. Armstrong and R.G. Botzler (eds) *Environmental Ethics: Divergence and Convergence* (pp. 158–163). New York: McGraw-Hill.
Harrington, I. (1971) The trouble with tourism unlimited. *New Statesman* 82, 176.
Harris, J.R. (1998) *The Nurture Assumption: Why Children Turn Out the Way they Do*. New York: Free Press.
Harris, M. (1989) *Our Kind: Who We Are, Where We Came From, Where We Are Going*. New York: Harper and Row.
Harris, R. and Jago, L. (2001) Professional accreditation in the Australian tourism industry; an uncertain future. *Tourism Management* 22, 383–390.
Harrison, D. (2008) Pro-poor tourism: A critique. *Third World Quarterly* 29 (5), 851–868.
Harrison, P. (1997) Do animals feel pain? In E. Soifer (ed.) *Ethical Issues: Perspectives for Canadians* (pp. 40–51). Peterborough, ON: Broadview Press.
Hart, D.K. (1984) The virtuous citizen, the honorable bureaucrat, and 'Public Administration'. *Public Administration Review* 44, 111–120.
Harvey, D. (2005) *A Brief History of Neoliberalism*. Oxford: Oxford University press.
Hastings Center (Institute of Society, Ethics and the Life Sciences) (1979) *The Teaching of Ethics in Higher Education*. New York: Hastings Center, Hastings-on-Hudson.
Hawkes, K. (1993) Why hunter-gatherers work. *Current Anthropology* 34, 341–361.
Hay, I. and Foley, P. (1998) Ethics, geography and responsible citizenship. *Journal of Geography in Higher Education* 22 (2), 169–183.
Hayek, F. (1962) *The Road to Serfdom*. Chicago, IL: University of Chicago Press.
Haywood, K.M. (1993) Sustainable development for tourism: A commentary with an organizational perspective. In J.G. Nelson and R.W. Butler (eds) *Tourism and Sustainable Development: Monitoring, Planning, Managing* (pp. 233–241). Waterloo, ON: Heritage Resources Centre Joint Publication Number 1, University of Waterloo.
Heaton, H. (1939) *A History of Trade and Commerce – With Special Reference to Canada* (rev. edn). Toronto: Thomas Nelson and Sons.
Hegarty, W. and Sims, H. (1979) Organizational philosophy, policies, and objectives related to unethical decision behavior: A laboratory experiment. *Journal of Applied Psychology* 64 (3), 331–338.
Heidegger, M. (1966) *Discourse on Thinking*. New York: Harper Torchbooks.
Heidegger, M. (1962) Being and time. New York: Harper & Row.
Heintzman, P. (1995) Leisure, ethics, and the Golden Rule. *Journal of Applied Recreation Research* 20 (3), 203–222.
Held, V. (2004) Care and justice in the global context. *Ratio Juris* 17 (2), 141–155.
Hewitt de Alcantara, C. (1998) *Uses and Abuses of the Concept of Governance*. UNESCO. Oxford: Blackwell Publishers.

Hick, J. (1992) The universality of the Golden Rule. In J. Runzo (ed.) *Ethics, Religion and the Good Society: New Directions in a Pluralistic World* (pp. 155–166). Louisville, KY: Westminster/John Knox.

Higgins-Desboilles, F. (2006) More than an 'industry': The forgotten power of tourism as a social force. *Tourism Management* 27, 1192–1208.

Higgins-Desboilles, F. (2008) Justice tourism and alternative globalisation. *Journal of Sustainable Tourism* 16 (3), 345–364.

Higgins-Desbiolles, F. (2009) Indigenous ecotourism's role in transforming ecological consciousness. *Journal of Ecotourism* 8 (2), 144–160.

Hill, K. and Kaplan, H. (1985) Food sharing among Ache foragers: Tests of explanatory hypotheses. *Current Anthropology* 26, 223–245.

Hill, K. and Kaplan, H. (1993) On why male foragers hunt and share food. *Current Anthropology* 34, 701–710.

Hills, T. and Lundgren, J. (1977) The impact of tourism in the Caribbean: A methodological study. *Annals of Tourism Research* 4 (5), 248–267.

Hjalager, A.-M. (2002) Repairing innovation defectiveness in tourism. *Tourism Management* 23, 465–474.

Hobbes, T. (1651/1957) *Leviathan; or the Matter, Form, and Power of a Commonwealth Ecclesiastical and Civil*. Oxford: Blackwell.

Hobson, J.S.P. (1993) Analysis of the US cruise line industry. *Tourism Management* 14, 453–462.

Hodgkinson, C. (1991) *Educational Leadership: The Moral Art*. Syracuse, NY: State University of New York Press.

Hoese, K. (1999) The Gwaii Haanas Watchman Program: Sustainable tourism development in a protected area. In P. Williams and I. Budke (eds) *On Route to Sustainability: Best Practices in Canadian Tourism* (pp. 102–108). Ottawa, ON: Canadian Tourism Commission Distribution Centre.

Hoffman, W.M. (1991) Business and environmental ethics. In W.M. Hoffman, R.E. Frederick and M.S. Schwartz (eds) *Business Ethics: Readings and Cases in Corporate Morality* (4th edn; pp. 434–443). Boston: McGraw Hill.

Hoffman, W.M., Frederick, R.E. and Schwartz, M.S. (2001) Introduction. In W.M. Hoffman, R.E. Frederick and M.S. Schwartz (eds) *Business Ethics: Readings and Cases in Corporate Morality* (4th edn; pp. 1–43). Boston: McGraw Hill.

Holcomb, J.L., Upchurch, R.S. and Okumus, F. (2007) Corporate social responsibility: What are top hotel companies reporting? *International Journal of Contemporary Hospitality Management* 19 (6), 461–475.

Holden, A. (2003) In need of a new environmental ethics for tourism? *Annals of Tourism Research* 30 (1), 95–108.

Holden, A. (2005) Achieving a sustainable relationship between common pool resources and tourism: The role of environmental ethics. *Journal of Sustainable Tourism* 13 (4), 339–352.

Holden, A. (2009) The environment-tourism nexus: influence of market ethics. *Annals of Tourism Research* 36 (3), 373–389.

Holland, J. (1997/8) Does Jamaica have a choice? *In Focus* 26, 10–11.

Holland, S.M., Ditton, R.B. and Graefe, A.R. (1998) An ecotourism perspective on billfish fisheries. *Journal of Sustainable Tourism* 6 (2), 97–116.

Holland, S.M., Ditton, R.B. and Graefe, A.R. (2000) A response to 'Ecotourism on Trial': The case of billfish angling as ecotourism. *Journal of Sustainable Tourism* 8 (4), 346–351.

Hölldobler, B. and Wilson, E.O. (1990) *The Ants*. Cambridge, MA: Harvard University Press.

Holling, C.S. (1993) Investing in research for sustainability. *Ecological Applications* 34, 552–555.

Holling, C.S. (2001) Understanding the complexity of economic, ecological, and social systems. *Ecosystems* 4, 390–405.
Honderich, T. (ed.) (1995) *The Oxford Companion to Philosophy*. Oxford: Oxford University Press.
Honey, M. and Rome, A. (2001) *Certification and Ecolabelling*. Washington, DC: Institute for Policy Studies.
Hope, K.R. (1980) The Caribbean tourism sector: Recent performance and trends. *Tourism Management* 1 (3), 175–183.
Hornsby, R. (nd). What Heidegger means by Being-in-the-world. http://royby.com/philosophy/pages/dasein.html (accessed 20 March 2017).
Hudson, S. (2007) To go or not to go? Ethical perspectives on tourism in an 'Outpost of Tyranny'. *Journal of Business Ethics* 76 (4), 385–396.
Hudson, S. and Miller, G. (2005) The responsible marketing of tourism: The case of Canadian Mountain Holidays. *Tourism Management* 26 (2), 133–142.
Hudson, S. and Miller, G. (2006) Knowing the difference between right and wrong: The response of tourism students to ethical dilemmas. *Journal of Teaching in Travel & Tourism* 6 (2), 41–59.
Hughes, G. (1995) The cultural construction of sustainable tourism. *Tourism Management* 16 (1), 49–59.
Hughes, P. (2001) Animals, values and tourism – structural shifts in UK dolphin tourism provision. *Tourism Management* 22 (4), 321–329.
Hultsman, J. (1995) Just tourism: An ethical framework. *Annals of Tourism Research* 22 (3), 553–567.
Hume, D. (1739/1978) *A Treatise of Human Nature*. Oxford: Clarendon Press.
Humphrey, N. (1992) *A History of the Mind: Evolution and the Birth of Consciousness*. London: Chatto and Windus.
Hunt, S.D. and Chonko, L.B. (1984) Marketing and Machiavellianism. *Journal of Marketing* 48 (3), 30–42.
Hunt, S.D. and Vasquez-Parraga, A.Z. (1993) Organizational consequences, marketing ethics, and salesforce satisfaction. *Journal of Marketing Research* 21, 78–90.
Hunt, S.D. and Vitell, S. (1986) A general theory of marketing ethics. *Journal of Macromarketing* 1, 5–16.
Hunt, S.D., Wood, V.R. and Chonko, L.B. (1989) Corporate ethical values and organizational commitment in marketing. *Journal of Marketing* 53, 79–90.
Huntsman, A.G. (1944) Fishery depletion. *Science* XCIX, 534.
Huxley, T.H. (1894/1968) *Evolution and Ethics and Other Essays*. New York: Greenwood Press.
International Hotel and Restaurant Association (2001) Preparing for a Green Future. See http://www.iftta.org/content/international-hotel-and-restaurant-association (accessed 21 May 2002).
Issaverdis, J.-P. (2001) The pursuit of excellence: Benchmarking, accreditation, best practice and auditing. In D.B. Weaver (ed.) *The Encyclopedia of Ecotourism* (pp. 579–594). Wallingford: CABI.
Jacobs, J. (2004) *Dark Age Ahead*. Toronto, ON: Random House.
Jamal, T. (2004) Virtue ethics and sustainable tourism pedagogy: Phronesis, principles and practice. *Journal of Sustainable Tourism* 12 (6), 530–545.
Jamal, T. and Alejandra Camargo, B. (2013) Sustainable tourism, justice and an ethic of care: Toward the Just Destination. *Journal of Sustainable Tourism*, DOI: 10.1080/09669582.2013.786084.
Jantsch, E. (1972) *Technological Planning and Social Futures*. New York: Halsted Press.
Jeng, M. (1997/8) Denied the crumbs: The hosts' perspective. *In Focus* 26, 12.
Jenkins, C.L. (1982) The effects of scale in tourism projects in developing countries. *Annals of Tourism Research* 9 (2), 229–249.

Litvin, S.W., Tan, P.S.K., Tay, P.F.J. and Aplin, K. (2004) Cross-cultural differences: An influence on tourism ethics? *Tourism* 52 (1), 39–50.
Locke, J. (1690/1979) An essay concerning humane understanding. (P.H. Nidditch (ed.)). Oxford: Oxford University Press.
John, N. (1950) Equilibrium points in n-person games. *Proceedings of the National Academy of Sciences* 36 (1), 48–49.
Johnson, D.B. (1998) Green businesses: Perspectives from management and business ethics. *Society & Natural Resources* 11 (3), 259–266.
Johnson, E. (1984) Treating the dirt. In T. Regan (ed.) *Earthbound: New Introductory Essays in Environmental Ethics* (pp. 346–360). New York: Random House.
Jones, H. (1972) Gozo: The living showpiece. *Geographical Magazine* 45 (1), 53–57.
Jos, P.H. (1988) Moral autonomy and the modern organization. *Polity* 21, 321–343.
Jose, A. and Thibodeaux, M.S. (1999) Institutionalization of ethics: The perspective of managers. *Journal of Business Ethics* 22, 133–143.
Josephides, N. (1997/8) The invasion of the body-snatchers. *In Focus* 26, 6.
Joyce, B., Weil, M. and Showers, B. (1996) *Models of Teaching* (4th edn). Boston, MA: Allyn & Bacon.
Kagan, J. (1971) *Change and Continuity in Infants*. New York: Wiley.
Kagan, J. (1984) *The Nature of the Child*. New York: Basic Books.
Kagan, J. (1998) *Three Seductive Ideas*. Cambridge, MA: Harvard University Press.
Kahlenborn, W. and Dominé, A. (2001) The future belongs to international ecolabelling schemes. In X. Font and R. Buckley (eds) *Tourism Ecolabelling: Certification and Promotion of Sustainable Management* (pp. 247–258). Wallingford: CABI.
Kalisch, A. (2000) Corporate social responsibility in the tourism industry. *Tourism Concern Bulletin* 2 (Autumn), 2.
Kant, I. (1781) *Critique of Pure Reason* (P. Guyer and A.W. Wood, trans. eds). Cambridge: Cambridge University Press.
Kant, I. (1788/1977) *Critique of Practical Reason* (L.W. Beck, trans.) Indianapolis, IN: Bobbs-Merrill.
Kaplan, J. (1973) *Criminal Justice: Introductory Cases and Materials*. Mineola, NY: The Foundation Press.
Kasser, T. (2003) *The High Price of Materialism*. Cambridge, MA: The MIT Press.
Karwacki, J. and Boyd, C. (1995) Ethics and ecotourism. *A European Review* 4, 225–232.
Katz, D. and Kahn, R. (1978) *The Social Psychology of Organizations*. New York: John Wiley & Sons.
Kauffman, W. (1995) *No Turning Back: Dismantling the Fantasies of Environmental Thinking*. New York: Basic Books.
Kaufmann, R.K. and Cleveland, C.J. (1995) Measuring sustainability: An interdisciplinary approach to an interdisciplinary concept. *Ecological Economics* 15, 109–112.
Kavallinis, I. and Pizam, A. (1994) The environmental impacts of tourism: Whose responsibility is it anyway? The case study of Mykonos. *Journal of Travel Research* 33 (2), 26–32.
Kay, J. (2004a) Ethical dilemmas. *National Post Business* March, 78–87.
Kay, J. (2004b) The best way to get from A to Z. *National Post*, 24 January, Section RB: 1, 4–5.
Keating, B. (2009) Managing ethics in the tourism supply chain: The case of Chinese travel to Australia. *International Journal of Tourism Research* 11, 403–408.
Keefe, J. and Wheat, S. (1998) *Tourism and Human Rights*. London: Tourism Concern.
Keenan, J.P. (2002) Whistleblowing: A study of managerial differences. *Employee Responsibilities and Rights Journal* 14 (1), 17–32.
Keiffer, G.H. (1979) *Bioethics: A Textbook of Issues*. New York: Addison-Wesley.
Keith, A. (1947) *Evolution and Ethics*. New York: Putnam.

Kimber, R. (1981) Collective action and the fallacy of the liberal fallacy. *World Politics* 178–196.
Kinnier, R.T., Kernes, J.L. and Dautheribes, T.M. (2000) A short list of universal moral values. *Counseling and Values* 45 (1), 4–17.
Kitcher, P. (1993) The evolution of human altruism. *The Journal of Philosophy* XC (10), 497–516.
Klein, R. (2002) *Cruise Ship Blues: The Underside of the Cruise Industry*. Gabriola Island, BC: New Society Publishers.
Klingberg, H. (2001) *When Life Calls Out to Us: The Love, and Life Work of Victor and Ely Frankl*. New York: Doubleday.
Klonoski, R.J. (2003) Unapplied ethics: On the need for classical philosophy in professional ethics education. *Teaching Business Ethics* 7, 21–35.
Koehn, D. (1999) What can Eastern philosophy teach us about business ethics? *Journal of Business Ethics* 19, 71–79.
Kohlberg, L. (1969) Stage and sequence: The cognitive-developmental approach to socialization. In D.A. Gosslin (ed.) *Handbook of Socialization Theory and Research* (pp. 347–480). Chicago, IL: Rand McNally.
Kohlberg, L. (1981) *The Philosophy of Moral Development*. New York: Harper & Row.
Kohlberg, L. (1984) *Philosophy of Moral Development*. New York: Harper & Row.
Krippendorf, J. (1977) *Les Devoreurs des paysages*. Lausanne: 24 Heures.
Krippendorf, J. (1987) Ecological approach to tourism marketing. *Tourism Management* 8 (2), 174–176.
Kropotkin, P. (1902/unknown) *Mutual Aid: A Factor in Evolution*. Boston, MA: Extending Horizons Books.
Kruuk, H. (1976) The biological function of gulls' attraction towards predators. *Animal Behavior* 24, 146–153.
Kurzban, R. (2003) Biological foundations of reciprocity. In E. Ostrom and J. Walker (eds) *Trust and Reciprocity: Interdisciplinary Lessons from Experimental Research* (pp. 105–127). New York: Russell Sage Foundation.
Kurzban, R., Tooby, J. and Cosmides, L. (2001) Can race be erased? Coalitional computation and social categorization. *Proceedings of the National Academy of Sciences* 98 (26), 15387–15392.
Kutay, K. (1989) The new ethic in adventure travel. *Buzzworm* 1, 31–36.
Labich, K. (1992) The new crisis in business ethics. *Fortune* 25, 167–176.
Ladkin, A. and Martinez Bertramini, A. (2002) Collaborative tourism planning: A case study of Cusco, Peru. *Current Issues in Tourism* 5 (2), 71–93.
Lam, L. (2000) How nature self-organizes: Active walks in complex systems. *Skeptic* 8 (3), 71–77.
Lane, D.E. and Stephenson, R.L. (2000) Institutional arrangements for fisheries: Alternative structures and impediments to change. *Marine Policy* 24, 385–393.
Latour, B. (2014) Another way to compose the common world. *Journal of Ethnographic Theory* 4 (1), 301–307.
Lea, J.P. (1993) Tourism development ethics in the Third World. *Annals of Tourism Research* 20, 701–715.
Lee, S. and Jamal, T. (2008) Environmental justice and environmental equity in tourism: Missing links to sustainability. *Journal of Sustainable Tourism* 7 (1), 44–67.
Leheny, D. (1994) A political economy of Asian sex tourism. *Annals of Tourism Research* 22 (2), 367–384.
Leopold, A. (1949/1966) *A Sand County Almanac*. New York: Ballantine Books.
Lessnoff, M. (1986) *Social Contract*. Atlantic Highlands, NJ: Humanities Press International.
Lévi-Strauss, C. (1981) *The Naked Man* (J. and D. Weightman, trans.). New York: Harper & Row.

Levinas, E. (1985) *Ethics and Infinity*. R.A. Cohen (trans.). Pittsburgh, PA: Duquesne University Press.
Levitt, T. (1979) The dangers of social responsibility. In T.L. Beauchamp and N.E. Bowie (eds) *Ethical Theory and Business* (pp. 138–141). Englewood Cliffs, NJ: Prentice-Hall.
Lewontin, R.C. (1961) Evolution and the theory of games. *Journal of Theoretical Biology* 1, 382–403.
Lewontin, R. (1972) The apportionment of human diversity. *Evolutionary Biology* 6, 381–398.
Lewontin, R. (1982) *Human Diversity*. New York: Scientific American Books Inc.
Lipset, S.M. and Schneider, W. (1987) *The Confidence Gap*. Baltimore, MD: The Johns Hopkins University Press.
Loewwnstein, M.J. (2013) Benefit corporations: A challenge in corporate governance. *Business Lawyer* 68 (4), 1007–1038.
Lovelock, B. (2008) Ethical travel decisions: Travel agents and human rights. *Annals of Tourism Research* 35 (2), 338–358.
Lovelock, B. and Lovelock, K. (2013) *The Ethics of Tourism: Critical and Applied Perspectives*. London: Routledge.
Low, B. (1996) Behavioral ecology of conservation in traditional societies. *Human Nature* 7 (4), 353–379.
Lucas, R.C. (1964) Wilderness perception and use: The example of the Boundary Waters Canoe Area. *Natural Resources Journal* 3 (3), 394–411.
Lück, M. (2003) Education on marine mammal tours as agent for conservation: But do tourists want to be educated? *Ocean and Coastal Management* 46, 943–956.
Luo, Y. and Deng, J. (2008) The new Environmental Paradigm and nature-based tourism motivation. *Journal of Travel Research* 46, 392–402.
Lutchmaya, S., Baron-Cohen, S. and Raggatt, P. (2002) Foetal testosterone and eye contact in 12 month old human infants. *Infant Behavior Development* 25, 327–335.
Lykken, D. (1995) *The Antisocial Personalities*. Mahwah, NJ: Erlbaum.
Lykken, D. (2000) The causes and costs of crime and a controversial cure. *Journal of Personality* 68, 559–605.
Macbeth, J. (2005) Towards an ethics platform for tourism. *Annals of Tourism Research* 32 (4), 962–984.
MacCannell, D. (1992) *Empty Meeting Grounds: The Tourist Papers*. London: Routledge.
Machiavelli, N. (1513/1999) *The Prince* (George Bull, trans.). London: Penguin Books.
MacIntyre, A. (1981) A crisis in moral philosophy: Why is the search for the foundations of ethics so frustrating? In D. Callahan and H.T. Englehardt Jr. (eds) *The Roots of Ethics* (pp. 3–30). New York: Plenum Press.
MacLagan, P. (1996) The organizational context for moral development: Questions of power and access. *Journal of Business Ethics* 15, 645–654.
Maitland, A. (2004) Social conscience can be a tough sell. *National Post*, 8 March, SR5.
Malloy, D.C. and Fennell, D.A. (1998a) Codes of ethics and tourism: An exploratory content analysis. *Tourism Management* 19 (5), 453–461.
Malloy, D.C. and Fennell, D.A. (1998b) Ecotourism and ethics: Moral development and organizational cultures. *Journal of Travel Research* 36, 47–56.
Malloy, D.C., Ross, S. and Zakus, D.H. (2000) *Sport Ethics: Concepts and Cases in Sport and Recreation*. Buffalo, NY: Thompson Educational Publishing.
Malone, S. (2014) Ethical tourism: The role of emotion. In C. Weeden and K. Boluk (eds) *Managing Ethical Consumption in Tourism* (pp. 153–165). London: Routledge.
Markuson, A. (2002) Two frontiers for regional science: Regional policy and interdisciplinary research. *Papers in Regional Science* 81, 271–290.
Marnburg, E. (2000) The behavioural effects of corporate ethical codes: Empirical findings and discussion. *Business Ethics: A European Review* 9 (3), 200–210.

Marnburg, E. (2006) 'I hope it won't happen to me!' Hospitality and tourism students' fear of difficult moral situations as managers. *Tourism Management* 27, 561–575.
Martin, G. (1998) Once again: Why should business be ethical? *Business and Professional Ethics Journal* 17 (4), 39–60.
Mason, P. (1997) Tourism codes of conduct in the Arctic and sub-Arctic region. *Journal of Sustainable Tourism* 5 (2), 151–165.
Mason, P. (2005) Visitor management in protected areas: From 'hard' to 'soft' approaches. *Current Issues in Tourism* 8 (2&3), 181–194.
Mason, P. and Mowforth, M. (1995) *Codes of Conduct in Tourism*. Occasional Papers in Geography No. 1, Department of Geographical Sciences, University of Plymouth.
Mason, P. and Mowforth, M. (1996) Codes of conduct in tourism. *Progress in Tourism and Hospitality Research* 2 (2), 151–167.
Masson, J.M. (1999) *The Emperor's Embrace: Reflections on Animal Families and Fatherhood*. New York: Pocket Books.
Masson, J.M. and McCarthy, S. (1996) *When Elephants Weep: The Emotional Lives of Animals*. New York: Delta.
Masterton, A.M. (1992) Environmental ethics. *Island Destinations (a supplement to Tour and Travel News)*. November, 16–18.
Mathieson, A. and Wall, G. (1982) *Tourism: Economic, Physical and Social Impacts*. New York: Longman.
Maynard Smith, J. (1964) Group selection and kin selection. *Nature* 201 (4924), 1145–1147.
Maynard Smith, J. (1974) The theory of games and the evolution of animal conflicts. *Journal of Theoretical Biology* 47, 209–221.
Maynard Smith, J. and Price, G. (1973) The logic of animal conflict. *Nature* 246, 15–18.
Maynard Smith, J. and Szathmary, E. (1995) *The Major Transitions in Evolution*. Oxford: W.H. Freeman.
Mayo, M.A. and Marks, L.J. (1990) An empirical investigation of a general theory of marketing ethics. *Journal of the Academy of Marketing Science* 18 (2), 163–171.
Mayr, E. (1988) *Toward a New Philosophy of Biology: Observations of an Evolutionist*. Cambridge, MA: The Belknap Press.
McAneny, L. (1992) Pharmacists again top 'honesty and ethics' poll; ratings for Congress hit new low. *The Gallup Poll Monthly* 322, 2–4.
McAvoy, L. (1990) An environmental ethic for parks and recreation. *Parks & Recreation* 25 (9), 68–72.
MCB University Press (1998) MCB University Press Internet Conference on 'Tourism and Ethics'. See http://www.mcb.co.uk/services/conferen/jan98/eit/conhome.html (accessed 11 January 1999).
McChesney, R.W. (1999) *Rich Media, Poor Democracy: Communication Politics in Dubious Times*. Chicago, IL: University of Illinois Press.
McDonald, G. (1999) Business ethics: Practical proposals for organisations. *Journal of Business Ethics* 19, 143–158.
McGran, K. (2003) Taking tourists for a ride? *Toronto Star*, section B, 20 June.
McGregor, D. (1997) Exploring aboriginal environmental ethics. In A. Wellington, A. Greenbaum and W. Cragg (eds) *Canadian Issues in Environmental Ethics* (pp. 325–330). Toronto, ON: Broadview Press.
McHugh, F.P. (1988) *Keyguide to Information Sources in Business Ethics*. New York: Nichols Publishing.
McKay, B.J. and Acheson, J.M. (1990) Human ecology of the commons. In B.J. McKay and J.M. Acheson (eds) *The Question of the Commons: The Culture and Ecology of Communal Resources* (pp. 1–6) Tucson, AZ: The University of Arizona Press.
McKercher, B. (1999) A chaos approach to tourism. *Tourism Management* 20, 425–434.

McKercher, B. and Bauer, T. (2003) Conceptual framework for the nexus between tourism, romance, and sex. In B. McKercher and T. Bauer (eds) *Sex and Tourism: Journeys of Romance, Love, and Lust* (pp. 3–17). New York: The Haworth Hospitality Press.

McKibben, B. (1999) Consuming nature. In R. Rosenblatt (ed.) *Consuming Desires: Consumption, Culture, and the Pursuit of Happiness* (pp. 87–95). Washington, DC: Island Press.

McLain, D.L. and Keenan, J.P. (1999) Risk, information, and the decision about response to wrongdoing in an organization. *Journal of Business Ethics* 19, 255–271.

McNaughton, D. (1988) *Moral Vision: An Introduction to Ethics*. New York: Basil Blackwell.

Mead, M. (1935/1963) *Sex and Temperament in Three Primitive Societies*. New York: William Morrow.

Merchant, C. (1990) Environmental ethics and political conflict: A view from California. *Environmental Ethics*, Spring, 45–68.

Meghani, Z. (2011) A robust, particularist ethical assessment of medical tourism. *Developing World Bioethics* 11 (1), 16–29.

Mensch, J.R. (2015) *Levinas's Existential Analytic: A Commentary on Totality and Infinity*. Evanston, IL: Northwestern University Press.

Midgley, M. (1994) *The Ethical Primate: Humans, Freedom and Morality*. London: Routledge.

Mies, M. (1997) Do we need a new 'moral economy'? *Canadian Women Studies* 17 (2), 12–20.

Mihalic, T. (2000) Environmental management of a tourist destination: A factor of tourism competitiveness. *Tourism Management* 21, 65–78.

Mihalic, T. and Fennell, D.A. (2014) In pursuit of a more just international tourism: The concept of trading tourism rights. *Journal of Sustainable Tourism* 23 (2), 188–206.

Mill, J.S. (1861/1979) *Utilitarianism*. Indianapolis, IN: Hacket.

Millar, C. and Yoon, H.-K. (2000) Morality, goodness and love: A rhetoric for resource management. *Ethics, Place and Environment* 3 (2), 155–172.

Miller, A.S. (1991) *Gaia Connections: An Introduction to Ecology, Ecoethics, and Economics* (pp. 9–30). New York: Rowman & Littlefield Publishers.

Miller, G. (2001) Corporate responsibility in the UK tourism industry. *Tourism Management* 22, 589–598.

Miller, J. and Szekely, F. (1995) What is 'green'? *European Management Journal* 13 (3), 322–333.

Milne, S. and Ateljevic, I. (2001) Tourism, economic development and the global-local nexus: Theory embracing complexity. *Tourism Geographies* 3 (4), 369–393.

Minnaert, L., Maitland, R. and Graham, M. (2006) Social tourism and its ethical foundations. *Tourism, Culture & Communication* 7 (1), 7–17.

Mitchell, B. (1995) *Geography and Resource Analysis* (2nd edn). New York: John Wiley & Sons.

Mitchell, T., Daniels, D., Hopper, H., George-Falvy, J. and Ferris, G. (1996) Perceived correlates of illegal behavior in organisations. *Journal of Business Ethics* 15 (4), 439–455.

Moore, G.E. (1903) *Principia Ethica*. New York: Cambridge University Press.

Morgan, K.C. (n.d.) These Burmese women are taking off their restrictive neck coils. http://www.allday.com/these-burmese-women-are-taking-off-their-restrictive-neck-coils-2180791136.html (accessed 25 March 2017).

Moscardo, G. (2010) Tourism research ethics: Current considerations and future options. In D.G. Pearce and R.W. Butler (eds) *Tourism Research: A 20–20 Vision* (pp. 203–214). Oxford: Goodfellow Publishers.

Mostafanezhad, M. (2013) The geography of compassion in volunteer tourism. *Tourism Geographies* 15 (2), 318–337.

Mostafanezhad, M. and Hannam, K. (2014) (eds) *Moral Encounters in Tourism*. Burlington, VT: Ashgate.

Mowforth, M. and Munt, I. (1998) *Tourism and Sustainability: New Tourism in the Third World*. London: Routledge.
Mowforth, M. and Munt, I. (2008) *Tourism and Responsibility: Perspectives from Latin America and the Caribbean.* New York: Routledge.
Munt, I. (1994) The 'other' post-modern tourism: Culture, travel and the new middle classes. *Theory, Culture and Society* 11, 101–123.
Murphy, P.R., Smith, J.E. and Daley, J.M. (1992) Executive attitudes, organizational size and ethical issues: Perspectives on a service industry. *Journal of Business Ethics* 11, 11–19.
Myers, B. (2003) The case for volunteerism. *Ontario Out of Doors*, June, 4.
Myers, D. and Diener, E. (1995) Who is happy? *Psychological Science* 6, 10–18.
Myers, N. (1988) Threatened biotas: 'Hotspots' in tropical forests. *The Environmentalist* 8 (3), 187–208.
Myers, N., Mittermeier, R.A., Mittermeier, C.G., da Fonseca, G.A.B. and Kent, J. (2000) Biodiversity hotspots for conservation priorities. *Nature* 403, 853–858.
Narveson, J. (1997) Resources and environmental policy. In E. Soifer (ed.) *Ethical Issues: Perspectives for Canadians* (2nd edn; pp. 107–124). Peterborough, ON: Broadview Publishing.
Nash, R.F. (1989) *The Rights of Nature: A History of Environmental Ethics*. Madison, WI: The University of Wisconsin Press.
Naylon, J. (1967) Tourism: Spain's most important industry. *Geography* 52, 23–40.
Neulinger, J. (1974) *Psychology of Leisure: Research Approaches to the Study of Leisure*. Springfield, IL: Charles C. Thomas.
NGO Steering Committee Tourism Caucus (1999) NGO Paper on Tourism. http://www.earthsummit2002.org/toolkits/women/ngo-doku/ngo-comm/csd/commissi6.html#ngocsd (accessed 3 March 1999).
Nielson, R.P. (1987) What can managers do about unethical management? *Journal of Business Ethics* 6, 309–320.
Nisan, M. and Kohlberg, L. (1982) University and cross-cultural variation in moral development: A longitudinal and cross-sectional study in Turkey. *Child Development* 53, 865–876.
Noble, B.F. (2000) Institutional criteria for co-management. *Marine Policy* 24, 69–77.
Noland, J. and Phillips, R. (2010) Stakeholder engagement, discourse ethics and strategic management. *International Journal of Management Reviews* 12, 39–49.
Noronha, F. (1999) Ten years later, Goa still uneasy over the impact of tourism. *International Journal of Contemporary Hospitality Management* 11 (2–3), 100–106.
Nozick, R. (1974) *Anarchy, State and Utopia*. Oxford: Blackwell.
Nussbaum, M.C. (2006) *Frontiers of Justice: Disability, Nationality, Species Membership*. Cambridge, MA: Harvard University Press.
Oakeshott, M. (1962) *Rationalism in Politics*. London: Methuen.
O'Halloran, R. (1991) Ethics in hospitality and tourism education: The new managers. *Hospitality and Tourism Educator* 3 (3), 33–37.
Olen, J. and Barry, V. (1989) *Applying Ethics: A Text with Readings* (3rd edn). Belmont, CA: Wadsworth.
Orams, M. (1997) The effectiveness of environmental education: Can we turn tourists into 'greenies'? *Progress in Tourism and Hospitality Research* 3, 295–306.
O'Riordan, T. and Cameron, J. (1994) The history and contemporary significance of the precautionary principle. In T. O'Riordan and J. Cameron (eds) *Interpreting the Precautionary Principle* (pp. 12–30). London: Earthscan.
Orr, D.W. (1993) Love it or lose it: The coming biophilia revolution. In S.R. Kellert and E.O. Wilson (eds) *The Biophilia Hypothesis* (pp. 415–440). Washington, DC: Island Press.

Orr, D.W. (1999) The ecology of giving and consuming. In R. Rosenblatt (ed.) *Consuming Desires: Consumption, Culture, and the Pursuit of Happiness* (pp. 137–154). Washington, DC: Island Press.

Ostrom, E. (1990) *Governing the Commons: The Evolution of Institutions for Collective Action*. New York: Cambridge University Press.

Ostrom, E. (2003) Toward a behavioral theory linking trust, reciprocity, and reputation. In E. Ostrom and J. Walker (eds) *Trust and Reciprocity: Interdisciplinary Lessons from Experimental Research* (pp. 19–79). New York: Sage.

Ostrom, E., Walker, J. and Gardner, R. (1992) Covenants with and without a sword: Self-governance is possible. *American Political Science Review* 86 (2), 404–417.

Oxfam International (2017) Just 8 men own same wealth as half the world. https://www.oxfam.org/en/pressroom/pressreleases/2017-01-16/just-8-men-own-same-wealth-half-world (accessed 27 March 2017).

Paradis, J. and Williams, G.C. (1989) *Evolution and Ethics: T.H. Huxley's Evolution and Ethics with New Essays on its Victorian and Sociobiological Context*. Princeton, NJ: Princeton University Press.

Parker, S. (1999) Ecotourism, environmental policy, and development. In D.L. Soden and B.S. Steel (eds) *Handbook of Global Environmental Policy and Administration* (pp. 315–345). New York: Marcel Dekker.

Pattullo, P. (1997/8) Like an alien in we own land. *In Focus* 26, 7–9.

Payne, D. and Dimanche, F. (1996) Towards a code of conduct for the tourism industry: An ethics model. *Journal of Business Ethics* 15, 997–1007.

Peattie, K. (1999) Trappings versus substance in the greening of marketing planning. *Journal of Strategic Marketing*, June, 131–148.

Pellow, D.N., Weinberg, A. and Schnaiberg, A. (2001) The environmental justice movement: Equitable allocation of the costs and benefits of environmental management outcomes. *Social Justice Research* 14 (4), 423–439.

Penn, W. and Collier, B. (1985) Current research in moral development as a decision support system. *Journal of Business Ethics* 4, 131–136.

Pera, L. and McLaren, D. (1999) Globalization, tourism and indigenous people: What you should know about the world's largest 'industry'. St Paul, MN: Rethinking Tourism Project. See http://www.akha.org/content/tourismecotourism/globalizationtourismindigenouspeoples.html (accessed 22 July 1999).

Perkins, H.E. and Brown, P.R. (2012) Environmental values and the so-called true ecotourist. *Journal of Travel Research* 51 (6), 793–803.

Perrings, C., Turner, R.K. and Folke, C. (1995) *Ecological Economics: The Study of Interdependent Economic and Ecological Systems*. Beijer Discussion Paper Series No. 55. Beijer Institute, Stockholm.

Peters, P.E. (1990) The grazing lands of Botswana and the commons debate. In B.J. McKay and J.M. Acheson (eds) *The Question of the Commons: The Culture and Ecology of Communal Resources* (pp. 171–194). Tucson, AZ: The University of Arizona Press.

Peters, T.J. and Waterman, R.H. (1982) *In Search of Excellence*. New York: Harper & Row.

Peterson, J.A. and Hronek, B.B. (1992) *Risk Management for Park, Recreation, and Leisure Services* (2nd edn). Champaign, IL: Sagamore.

Petrinovich, L., O'Neill, P. and Jorgensen, M. (1993) An empirical study of moral intuitions: Toward an evolutionary ethics. *Journal of Personality and Social Psychology* 64 (3), 467–478.

Pigram, J. (1996) Best practice environmental management and the tourism industry. *Progress in Tourism and Hospitality Research* 2, 261–271.

Pigram, J. (2004) *The Ecological Management of Tourism Sites: Do Successful Examples Exist?* Centre for Water Policy Research, University of New England, Australia.

See http://www.une.edu.au/cwpr/Papers/chamonix.pdf (accessed 3 December 2004).

Pinker, S. (2002) *The Blank Slate: The Modern Denial of Human Nature*. New York: Viking.

Pinkerton, E. (ed.) (1989) *Co-operative Management of Local Fisheries: New Directions for Improved Management and Community Development*. Vancouver, BC: University of British Columbia Press.

Platt, J. (1973) Social traps. *American Psychologist* 28, 641–51.

Plomin, R., DeFries, J.C. and McClearn, G.E. (1990) *Behavioral Genetics: A Primer*. New York: W.H. Freeman.

Plummer, R. and Fitzgibbon, J. (2004) Co-management of natural resources: A proposed framework. *Environmental Management* 33 (6), 876–885.

Polis, G.A. (1981) The evolution and dynamics of intraspecific predation. *Annual Review of Ecology & Systematics* 12, 225–251.

Polonsky, M.J. and Rosenberger, P.J. (2001) Reevaluating green marketing: A strategic approach. *Business Horizons*, Sept–Oct, 21–30.

Pomeroy, R.S. and Berkes, F. (1997) Two to tango: The role of government in fisheries co-management. *Marine Policy* 21 (5), 465–480.

PPT (2002) What is pro-poor tourism? See http://www.propoortourism.org.uk (accessed 6 June 2002).

Prigogine, I., Nicolis, G. and Babloyantz, A. (1972) Thermodynamics of evolution. *Physics Today* 25 (11), 23–28.

Przeclawski, K. (1996) Deontology of tourism. *Progress in Tourism and Hospitality Research* 2, 239–245.

Pugh, D.L. (1991) The origins of ethical frameworks in public administration. In J.S. Brown (ed.) *Ethical Frontiers in Public Management: Seeking New Strategies for Resolving Ethical Dilemmas* (pp. 9–33). San Francisco, CA: Jossey-Bass.

Rachels, J. (1989) Morality and moral philosophy. In J. Rachels (ed.) *The Right Thing to Do* (pp. 3–32). New York: Random House.

Rachels, J. (1990) *Created from Animals: The Moral Implications of Darwinism*. New York: Oxford University Press.

Rand, A. (1957) *Atlas Shrugged*. New York: Random House.

Rand, A. (1964) *The Virtue of Selfishness: A New Concept of Egoism*. New York: Signet.

Rapaport, A. and Chammah, A.M. (1965) *Prisoner's Dilemma: A Study in Conflict and Cooperation*. Ann Arbor, MI: The University of Michigan Press.

Raphael, D.D. (1989) *Moral Philosophy*. Oxford: Oxford University Press.

Ravinder, R. (2007) Ethical issues in collaboration in the aviation industry. *Tourism Review International* 11(2), 175–185.

Rawls, J. (1971) *A Theory of Justice*. Cambridge, MA: Belknap Press.

Rawls, J. (2000) *Lectures on the History of Moral Philosophy*. Cambridge, MA: Harvard University Press.

Ray, R. (2000) *Management Strategies in Athletic Training* (2nd edn). Champaign, IL: Human Kinetics.

Redclift, M. (1987) *Sustainable Development: Exploring the Contradictions*. London: Methuen.

Redclift, M. (1997) Sustainability and theory: An agenda for action. In D. Goodman and M. Watts (eds) *Globalising Food: Agrarian Questions and Global Restructuring* (pp. 333–343). London: Routledge.

Reed, M. and Harvey, D.L. (1992) The new science and the old: Complexity and realism in the social sciences. *Journal for the Theory of Social Behaviour* 22 (4), 353–380.

Rees, W. (1992) Ecological footprints and appropriated carrying capacity: What urban economics leaves out. *Environment and Urbanization* 4 (2), 121–130.

Regan, T. (1981) The nature and possibility of an environmental ethic. *Environmental Ethics* 3, 34.

Regan, T. (1983) The case for animal rights. In J. Rachels (ed.) *The Right Thing To Do: Basic Readings in Moral Philosophy* (pp. 211–225). New York: Random House.
Reidenbach, R.E. and Robin, D.P. (1988) Some initial steps toward improving the measurement of ethical evaluations of marketing activities. *Journal of Business Ethics* 7, 871–879.
Reidenbach, R.E. and Robin, D.P. (1990) Toward the development of a multidimensional scale for improving evaluations of business ethics. *Journal of Business Ethics* 9, 639–653.
Reis, A.C. and Shelton, E. (2011) The nature of tourism studies. *Tourism Analysis* 16, 375–384.
Rest, J.R. (1979) *Development in Judging Moral Issues*. Minneapolis, MN: University of Minnesota Press.
Rest, J.R. (1986) *Moral Development: Advance in Research and Theory*. New York: Praeger Publishers.
Richards, D. (2001) Reciprocity and shared knowledge structures in the Prisoner's Dilemma game. *Journal of Conflict Resolution* 45 (5), 621–635.
Riddell, R. (1981) *Ecodevelopment*. New York: St Martin's Press.
Ridley, M. (1998) *The Origins of Virtue: Human Instincts and the Evolution of Cooperation*. Toronto, ON: Penguin Books.
Ridley, M. (2003) *Nature via Nurture*. Toronto, ON: HarperCollins.
Ritchie, J.R.B. (1999) Crafting a value-driven vision for a national tourism treasure. *Tourism Management* 20, 273–282.
Rivers, P. (1973) Tourist troubles. *New Society* 23, 250.
Robertson, C. and Fadil, P.A. (1999) Ethical decision making in multinational organizations: A culture-based model. *Journal of Business Ethics* 19, 385–392.
Robinson, F. (1997) Globalizing care: Ethics, feminist theory, and international relations. *Alternatives* 22, 113–133.
Robson, J. and Robson, I. (1996) From shareholders to stakeholders: Critical issues for tourism marketers. *Tourism Management* 17 (7), 533–540.
Roe, D. and Urquhart, P. (2002) *Pro-poor Tourism: Harnessing the World's Largest Industry for the World's Poor*. London: International Institute for Environment and Development. See http://pubs.iied.org/pdfs/11007IIED.pdf (accessed 27 June 2002).
Rogers, M.F., Sinden, J.A. and De Lacy, T. (1997) The precautionary principle for environmental management: A defensive-expenditure application. *Journal of Environmental Management* 51, 343–360.
Rohr, J.A. (1978) *Ethics for Bureaucrats: An Essay on Law and Values*. New York: Marcel Dekker.
Rokeach, M. (1973) *The Nature of Human Values*. New York: Free Press.
Rolston, H. (1986) *Philosophy Gone Wild: Essays in Environmental Ethics*. Buffalo, NY: Prometheus Books.
Rolston, H. (1988) *Environmental Ethics: Duties to and Values in the Natural World*. Philadelphia, PA: Temple University Press.
Romp, G. (1997) *Game Theory: Introduction and Applications*. Oxford: Oxford University Press.
Rosenau, J.N. (1992) Governance, order and change in world politics. In J.N. Rosenau and E.O. Czempeil (eds) *Governance without Government: Order and Change in World Politics* (pp. 1–30). Cambridge: Cambridge University Press.
Rosenberg, N. and Birdzell, L.E. (1986) *How the West Grew Rich: The Economic Transformation of the Industrial World*. New York: Basic Books.
Rosin, P. (1997) Moralization. In A.M. Brandt and P. Rosin (eds) *Morality and Health* (pp. 379–402). New York: Routledge.
Ross, G.F. (2003) Workstress response perceptions among potential employees: The influence of ethics and trust. *Tourism Review* 58 (1), 25–33.

Rousseau, J.-J. (1755/1964) *The First and Second Discourses* (Roger and Judith Masters, trans.). New York: St Martin's Press.

Ruse, M. and Wilson, E.O. (1986) Moral philosophy as applied science. *Philosophy* 61, 173–192.

Russell, B. (1957) *Why I Am Not a Christian and Other Essays on Religion and Related Subjects.* New York: Simon & Schuster.

Russell, R. (1982) *Planning Programs in Recreation*. St Louis, MO: Mosby.

Russell, R. and Faulkner, B. (1999) Movers and shakers: Chaos makers in tourism development. *Tourism Management* 20, 411–423.

Ryan, C. (2002) Equity, management, power sharing and sustainability: Issues of the 'new tourism'. *Tourism Management* 23, 17–26.

Ryan, C. (2005) Ethics in tourism research: Objectives and personal perspectives. In B.W. Ritchie, P. Burns and C. Palmer (eds.) Tourism research methods: Integrating theory with practice (pp. 9–19). Wallingford: CABI.

Ryan, C. and Kinder, R. (1996) Sex, tourism and sex tourism: Fulfilling similar needs? *Tourism Management* 17 (7), 507–518.

Ryle, G. (1949) *The Concept of Mind*. London: Penguin.

Sachs, A. (1995) Individuals: The traditional human rights focus. In J.A. Peterson (ed.) *Eco-justice: Linking Human Rights and the Environment* (pp. 18–34). Washington, DC: World Watch Paper 127.

Sartre, J.-P. (2003) *Being and Nothingness: An Essay on Phenomenological Ontology*. London: Routledge.

Saul, J.R. (1995) *The Unconscious Civilization*. Toronto, ON: Anansi.

Saul, J.R. (2001) *On Equilibrium*. Toronto, ON: Viking.

Sautter, E.T. and Leisen, B. (1999) Managing stakeholders: A tourism planning model. *Annals of Tourism Research* 26 (2), 312–328.

Scace, R.C., Grifone, E. and Usher, R. (1992) *Ecotourism in Canada*. Consulting report prepared for the Canadian Environmental Advisory Council. Hull, PQ: Minister of Supply and Services.

Schein, E. (1985) *Organizational Culture and Leadership*. San Francisco, CA: Jossey-Bass.

Scheyvens, R. (2007) Exploring the tourism-poverty nexus. *Current Issues in Tourism* 10 (2&3), 231–254.

Schmitter, P.C. (1974) Still the century of corporatism? *Review of Politics* 36 (1), 85–131.

Schor, J.B. (1998) *The Overspent American: Upscaling, Downshifting, and the New Consumer.* New York: Basic Books.

Schumann, P.L. (2001) A moral principles framework for human resource management ethics. *Human Resource Management Review* 11, 93–111.

Schwartz, S.H. (1970) Moral decision making and behavior. In J. Macauley and L. Berkowitz (eds) *Altruism and Helping Behavior* (pp. 127–141). New York: Academic Press.

Schwepker, C.H. and Ingram, T.N. (1996) Improving sales performance through ethics: The relationship between salesperson moral judgement and job performance. *Journal of Business Ethics* 15, 1151–1160.

Scoones, I. (1999) New ecology and the social sciences: What prospects for a fruitful engagement? *Annual Review Anthropology* 28, 479–507.

Searle, M.S. and Brayley, R.E. (1993) *Leisure Services in Canada.* State College, PA: Venture Publishing.

Selinger, E. (2009) Ethics and poverty tours. *Philosophy & Pubic Policy Quarterly* 29(1/2), 2–7.

Shackleford, P. (1985) The World Tourism Organisation: 30 years of commitment to environmental protection. *International Journal of Environmental Studies* 25, 257–263.

Shani, A. and Pizam, A. (2008) Towards an ethical framework for animal-based attractions. *New Zealand Management* 20 (6), 679–693.

Sharp, L. (1952) Steel axes for stone-age Australians. *Human Organization* 11, 17–22.

Sharp Paine, L. (1991) Ethics as character development: Reflections on the objective of ethics education. In R.E. Freeman (ed.) *Business Ethics: The State of the Art* (pp. 67–86). New York: Oxford University Press.

Shaw, R. (2000) Stacked decks: Big ships belch most pollution. *Environmental News Network*, 26 July. See http://www.cnn.com/2000/NATURE/07/26/dirty.ships.enn/ (accessed 14 March 2001).

Sheldon, P.J., Fesenmaier, D.R. and Tribe, J. (2011) The Tourism Education Futures Initiative (TEFI): Activating change in tourism education. *Journal of Teaching in Travel & Tourism* 11, 2–23.

Shweder, R.A. (1990) In defense of moral realism: Reply to Gabennesch. *Child Development* 61, 2060–2067.

Silver, I. (1993) Marketing authenticity in Third World countries. *Annals of Tourism Research* 20 (2), 302–318.

Simon, G.L. and Alagona, P.S. (2009) Beyond Leave No Trace. *Ethics, Place and Environment* 12 (1), 17–34.

Simpson, G.G. (1969) Biology and ethics. In G.G. Simpson (ed.) *Biology and Man* (pp. 130–148). New York: Harcourt, Brace.

Sinclair, A. (1993) Approaches to organizational culture and ethics. *Journal of Business Ethics* 12, 63–73.

Singer, M.G. (1961) *Generalization in Ethics*. New York: A.A. Knopf.

Singer, P. (1981) *The Expanding Circle: Ethics and Sociobiology*. New York: Farrer, Straus & Giroux.

Singer, P. (1997) Rich and poor. In E. Soifer (ed.) *Ethical Issues: Perspectives for Canadians* (2nd edn; pp. 208–221). Peterborough, ON: Broadview Publishing.

Singh, T.V. (2002) Altruistic tourism – another shade of sustainable tourism: Case of Kanda community in the Himalaya. *Tourism* 50 (4), 361–370.

Singh, T.V. (2015) *Challenges in Tourism Research*. Bristol: Channel View Publications.

Singhapakdi, A. and Vitell, S.J. (1991) Selected factors influencing marketers' deontological norms. *Journal of the Academy of Marketing Science* 19 (1), 37–42.

Sizer, S.R. (1999) The ethical challenges of managing pilgrimages to the Holy Land. *International Journal of Contemporary Hospitality Management* 11 (2–3), 85–90.

Slobodkin, L. (1993) The complex of questions relating evolution to ethics. In M.H. Nitecki and D.V. Nitecki (eds) *Evolutionary Ethics* (pp. 337–347). Albany, NY: State University of New York.

Slote, M. 1992. *From Morality to Virtue*. New York; Oxford University Press.

Smith, A. (1759/1966) *The Theory of Moral Sentiments*. New York: Kelly.

Smith, A. (1776/1964) *The Wealth of Nations*. London: Dent & Sons.

Smith, H. (2002) A dog's life in Greece. *New Statesman* 131, 15–16.

Smith, J.Q. (1993) *The Moral Sense*. New York: Macmillan.

Smith, M. and Duffy, R. (2003) *The Ethics of Tourism Development*. London: Routledge.

Sober, E. (1988) What is evolutionary altruism? *Canadian Journal of Philosophy* 14, 75–99.

Sober, E. (1992) The evolution of altruism: Correlation, cost, and benefit. *Biology and Philosophy* VII, 177–87.

Sober, E. and Wilson, D.S. (1998) *Unto Others: The Evolution and Psychology of Unselfish Behavior*. Cambridge, MA: Harvard University Press.

Soifer, E. (1997) Introduction. In E. Soifer (ed.) *Ethical Issues*. Peterborough, ON: Broadview Press.

Soper, K. (2008) Alternative hedonism, cultural theory and the role of aesthetic revisioning. *Cultural Studies* 22 (5), 567–587.

Sowell, T. (1987) *A Conflict of Visions*. New York: William Morrow & Co.

Spears, T. (2004) Saving Santorini. *The Welland Tribune*, 12 June, C2.

Speed, C. (2008) Are backpackers ethical tourists? In K. Hannam and I. Ateljevic (eds) *Backpacker Tourism: Concepts and Profiles* (pp. 54–81). Clevedon: Channel View Publications.

Spinage, C. (1998) Social change and conservation misrepresentation in Africa. *Oryx* 32 (4), 265–276.

Sprigge, T. (1987) *The Rational Foundations of Ethics*. London: Routledge & Kegan Paul.

Sridhar, B.S. and Camburn, A. (1993) Stages of moral development of corporations. *Journal of Business Ethics* 12, 727–739.

Stark, J.C. (2002) Ethics and ecotourism: Connection and conflicts. *Philosophy and Geography* 5 (1), 101–113.

Steadman, D.W. (1995) Prehistoric extinctions of Pacific Island birds: Biodiversity meets zooarchaeology. *Science* 267, 1123–1131.

Stearman, A. (1994) Only slaves climb trees: Revisiting the myth of the ecologically noble savage in Amazonia. *Human Nature* 5, 339–357.

Stefanovic, I.L. (1997) A code of ethics for Short Hills Park. In A. Wellington, A Grenbaum and W. Cragg (eds) *Canadian Issues in Environmental Ethics* (pp. 246–258). Toronto, ON: Broadview Press.

Steiner, C.J. and Reisinger, Y. (2006) Understanding existential authenticity. *Annals of Tourism Research* 33 (2), 299–318.

Steiner, G. (1978) Heidegger. Sussex: The Harvester Press.

Stern, P.C. and Dietz, T. (1994) The value basis of environmental concern. *Journal of Social Issues* 50 (3), 65–84.

Stevens, B. (1994) An analysis of corporate ethical code studies: 'Where do we go from here?'. *Journal of Business Ethics* 13, 63–69.

Stevens, B. (1997) Hotel ethical codes: A content analysis. *International Journal of Hospitality Management* 16 (3), 261–271.

Stevens, B. and Fleckenstein, A. (1999) Comparative ethics: How students and human-resources directors react to real-life situations. *Cornell Hotel and Restaurant Quarterly* 40 (2), 69–75.

Stevenson, L. and Haberman, D.L. (1998) *Ten Theories of Human Nature*. New York: Oxford University Press.

Stewart, P. (2001) Complexity theories, social theory, and the question of social complexity. *Philosophy of the Social Sciences* 3 (1), 323–360.

Stilwell, F. (2002) *Political Economy: The Contest of Ideas*. Oxford: Oxford University Press.

Stone, C.D. (1979) Should trees have standing? Toward legal rights for natural objects. In T.L. Beauchamp and N.E. Bowie (eds) *Ethical Theory and Business* (pp. 563–567) Englewood Cliffs, NJ: Prentice-Hall.

Stonehouse, B. (1997) Tourism codes of conduct in the Arctic and sub-Arctic region. *Journal of Sustainable Tourism* 5 (2), 151–165.

Stonehouse, B. (2001) Polar environments. In D.B. Weaver (ed.) *The Encyclopedia of Ecotourism* (pp. 219–234). Wallingford: CABI.

Storr, A. (1983) *The Essential Jung*. Princeton, NJ: Princeton University Press.

Strassberg, B.A. (2003) 'The plague of blood': HIV/AIDS and ethics of the global health-care challenge. *Zygon* 38 (1), 169–184.

Suzuki, D. (1994) *Time to Change: Essays*. Toronto, ON: Stoddart.

Tangley, L. (1988) Who's polluting Antarctica? *BioScience* 38 (9), 590–594.

Tapper, R. (2001) Tourism and socio-economic development: UK tour operators' business approaches in the context of the new international agenda. *International Journal of Tourism Research* 3, 351–366.

Taylor, C. (1991) *The Ethics of Authenticity*. Cambridge, MA: Harvard University Press.

Taylor, P.W. (1986) *Respect for Nature: A Theory of Environmental Ethics*. Princeton, NJ: Princeton University Press.

Tearfund (2000a) *Tourism – An Ethical Issue: Market Research Report*. London: Tearfund.
Tearfund (2000b) *A Tearfund Guide to Tourism: Don't Forget your Ethics*. London: Tearfund.
Tearfund (2001) *Tourism: Putting Ethics into Practice*. London: Tearfund.
Tearfund (2002) *Worlds Apart*. London: Tearfund.
Tetlock, P.E. (1999) Coping with tradeoffs: Psychological constraints and political implications. In A. Lupia, M. McCubbins and S. Popkin (eds) *Political Reasoning and Choice*. Berkeley, CA: University of California Press.
ThaiMed (2009) Women of the long neck Karen tribe removing rings. http://www.thaimedicalnews.com/medical-tourism-thailand/long-neck-karen-tribe-thailand-burmese-border-remove-rings/ (accessed 25 March 2017).
Theerapappisit, P. (2003) Mekong tourism development: Capital or social mobilization? *Tourism Recreation Research* 28 (1), 47–56.
Terry, W.C. (2009) Working on the water: On legal space and seafarer protection in the cruise industry. *Economic Geography* 85 (4), 463–482.
Thomas, M. (1766–1834) An essay on the principle of population as it affects the future improvement of society, with remarks on the speculations of Mr. Goodwin, M. Condorcet and Other Writers (1st ed.). London: J. Johnson in St Paul's Church-yard, 1798 (Retrieved 25 June 2016).
Tichy, N.M. and Devanna, M.A. (1986) *The Transformational Leader*. New York: John Wiley & Sons.
Tickner, J. and Raffensberger, C. (1998) The precautionary principle: A framework for sustainable business decision-making. *Environmental Policy* 5 (4), 75–82.
Tierney, N.L. (1994) *Imagination and Ethical Ideals: Prospects for a Unified Philosophical and Psychological Understanding*. New York: State University of New York Press.
Tisdell, C.A. (1989) Environmental conservation: Economics, ecology, ethics. *Environmental Conservation* 16 (2), 107–112, 162.
Todaro, M. (1983) *The Struggle for Economic Development: Readings in Problems and Policies*. New York: Longman.
Tooby, J. and Cosmides, L. (1990) On the universality of human nature and the uniqueness of the individual: The role of genetics and adaptation. *Journal of Personality* 58, 17–67.
Toulmin, S. (1986) How medicine saved the life of ethics. In J.P. DeMarco and R.M. Fox (eds) *New Directions in Ethics: The Challenge of Applied Ethics* (pp. 265–281). New York: Routledge & Kegan Paul.
Tourism Concern (1992) Marketing tourism responsibly. In S. Eber (ed.) *Beyond the Green Horizon: A Discussion Paper on Principles for Sustainable Tourism* (pp. 31–33). Surrey: WWF-UK.
Tourism Industry Association of Canada (TIAC) (1991) *Code of Ethics and Guidelines for Sustainable Tourism*. In association with the National Roundtable on the Environment and Economy, Ottawa, Canada.
Tourism Operators' Initiative (2002) Mission and Objectives. See http://apps.unep.org/redirect.php?file=/publications/pmtdocuments/-Sustainable%20Tourism_%20The%20Tour%20Operator%27s%20Contribution-2003647.pdf (accessed 29 May 2002).
Townsend, P. (1987) Deprivation. *Journal of Social Policy* 16 (2), 125–146.
Towsend, R. and Wilson, J.A. (1990) An economic view of the tragedy of the commons. In B.J. McKay and J.M. Acheson (eds) *The Question of the Commons: The Culture and Ecology of Communal Resources* (pp. 311–326). Tucson, AZ: The University of Arizona Press.
Travel Wire News (2004) *Lazy Brits are Hopeless Holidaymakers*. See travelwirenews@travelwirenews.com (accessed 27 January 2004).
Trevino, L.K. (1986) Ethical decision making in organizations: A person–situation interactionist model. *Academy of Marketing Review* 11, 601–617.
Trevino, L.K. (1990) A cultural perspective on changing and developing organizational ethics. *Organizational Change and Development* 4, 195–230.

Tribe, J. (2002) Education for ethical tourism action. *Journal of Sustainable Tourism* 10 (4), 309–324.
Tribe, J. (2009) (ed.) *Philosophical Issues in Tourism*. Bristol: Channel View Publications.
Tribe, J., Font, X., Vickery, R. and Yale, K. (2000) *Environmental Management for Rural Tourism and Recreation*. London: Cassells.
Trivers, R. (1971) The evolution of reciprocal altruism. *Quarterly Review of Biology* 46, 35–57.
Trivers, R. (1985) *Social Evolution*. Menlo Park, CA: The Benjamin/Cummings Publishing Co.
Tuan, Y. (1976) Geopiety: A theme in man's attachment to nature and place. In D. Lowenthal and M.J. Bowden (eds) *Geographies of the Mind: Essays in Historical Geosophy* (pp. 11–39). New York: Oxford University Press.
UNEP (1995) *Environmental Code of Conduct for Tourism*. Technical Report No. 29. United Nations: Paris, France.
Upchurch, R.S. and Ruhland, S.K. (1995) An analysis of ethical work climate and leadership relationship in lodging operations. *Journal of Travel Research* 34 (2), 36–42.
Urry, J. (1990) *The Tourist Gaze: Leisure and Travel in Contemporary Society*. London: Sage.
Vail, D. and Hultkrantz, L. (2000) Property rights and sustainable nature tourism: Adaptation and mal-adaptation in Dalarna (Sweden) and Maine (USA). *Ecological Economics* 35, 223–242.
Vallen, G. and Casado, M. (2000) Ethical principles for the hospitality curriculum. *Cornell Hotel and Restaurant Quarterly* 41 (2), 44–51.
Van Buren, H.J. (1996) Why business should help save the rainforest. *Business and Society Review* 95, 22–24.
VanderZwaag, D. (1994) *CEPA and the Precautionary Principle/Approach*. Hull, QP: Minister of Supply and Services.
Veatch, R.M. (2003) *The Basics of Bioethics* (2nd edn). Upper Saddle River, NJ: Prentice Hall.
Veblen, T. (1899/1953) *The Theory of the Leisure Class*. New York: Mentor.
Vegan Magazine (2013) Ecuador's Tropic Journeys in Nature Opens Window on Ancient Ethnic People Threatened by Oil Interests in Amazon Rainforest. http://www.vegan-magazine.com/2013/06/18/ecuadors-tropic-journeys-in-nature-opens-window-on-ancient-ethnic-people-threatened-by-oil-interests-in-amazon-rainforest/ (accessed 27 March 2017).
Velasquez, M.G. (1992) *Business Ethics: Concepts and Cases* (3rd edn). Englewood Cliffs, NJ.: Prentice Hall.
Verschoor, C.C. (2003) New evidence of benefits from effective ethics systems. *Strategic Finance* 84 (11), 20–21.
Victor, B. and Cullen, J. (1988) The organizational bases of ethical work climates. *Administrative Science Quarterly* 33, 101–125.
Vitell, S.J. and Ho, F.N. (1997) Ethical decision making in marketing: A synthesis and evaluation of scales measuring the various components of decision making in ethical situations. *Journal of Business Ethics* 16, 699–717.
Vitell, S.J. and Singhapakdi, A. (1991) Factors influencing the perceived importance of stakeholder groups. *Business and Professional Ethics Journal* 10 (3), 53–72.
Von Neumann, J. and Morgenstern, O. (1944) *Theory of Games and Economic Behaviour*. Princeton, NJ: Princeton University Press.
Wagar, J.A. (1964) The carrying capacity of wildlands for recreation. *Society of American Foresters*, Forest Service Monograph 7, 23.
Wahn, J. (1993) Organizational dependence and the likelihood of complying with organizational pressures to behave unethically. *Journal of Business Ethics* 12, 245–251.
Waitt, G. and Cook, L. (2007) Leaving nothing but ripples on the water: Performing ecotourism natures. *Social & Cultural Geography* 8 (4), 535–550.

Wakefield, S. (1976) Ethics and the public service: A case for individual responsibility. *Public Administration Review* 36, 661–666.

Walker, B., Carpenter, S., Anderies, J., Abel, N., Cumming, G., Janssen, M., Lebel, L., Norberg, J., Peterson, G. and Pritchard, R. (2002) Resilience management in social-ecological systems: A working hypothesis for a participatory approach. *Conservation Ecology* 6 (1), 14–26.

Walker, L. (1984) Sex differences in the development of moral reasoning: A critical review. *Child Development* 55, 677–691.

Wallach, M.A. and Wallach, L. (1983) *Psychology's Sanction for Selfishness: The Error of Egoism in Theory and Therapy*. San Francisco, CA: W.H. Freeman & Co.

Walle, A.H. (1995) Business ethics and tourism: From micro to macro perspectives. *Tourism Management* 16 (4), 263–268.

Walster, E., Berscheid, E. and Walster, G.E. (1973) New directions in equity research. *Journal of Personality and Social Psychology* 25, 151–176.

Watson, A.E., Hendee, J.C. and Zaglauer, H.P. (1996) Human values and codes of behavior: Changes in Oregon's Eagle Cap Wilderness visitors and their attitudes. *Natural Areas Journal* 16 (2), 89–93.

Watson, J.B. (1924/1998) *Behaviorism*. New Brunswick, NJ: Transaction.

Watson, R. (1996) Risk management: A plan for safer activities. *CAHPERD Journal*, Spring, 13–17.

Wearing, S. (2001) *Volunteer Tourism: Experiences that Make a Difference*. Wallingford: CABI.

Wearing, S. and Wearing, M. (2014) On decommodifying ecotourism's social value: Neoliberal reformism or the new environmental morality? In M. Mostafanezhad and K. Hannam (eds) *Moral Encounters in Tourism* (pp. 123–136). London: Routledge.

Weaver, D.B. (1998) *Ecotourism in the Less Developed Countries*. Wallingford: CABI.

Weaver, G.R. (1995) Does ethics code design matter? Effects of ethics code rationales and sanctions on recipients' justice perceptions and content recall. *Journal of Business Ethics* 14, 367–385.

Weber, M. (1927/2003) *General Economic History*. (F. Knight, Trans.) Mineola, New York: Dover Publications, Inc.

Weeden, C. (2001) Ethical tourism: An opportunity for competitive advantage? *Journal of Vacation Marketing* 8 (2), 141–153.

Weeden, C. (2011) Responsible tourist motivation: how valuable is the Schwartz value survey? *Journal of Ecotourism* 10 (3), 214–234.

Weeden, C. and Boluk, K. (2014) (eds) *Managing Ethical Consumption in Tourism*. London: Routledge.

Weggeman, M.C.D.P. (1997) *Kennismanagement: Inrichting en besturing van Kennisintensieve Organisaties*. Schiedam.

Wellington, A., Greenbaum, A. and Cragg, W. (1997) Introduction. In A. Wellington, A. Greenbaum and W. Cragg (eds) *Canadian Issues in Environmental Ethics* (pp. 1–34). Peterborough, Ontario, Broadview Press.

Wenger, E., McDermott, R. and Snyder, W. (2002) *Cultivating Communities of Practice: A Guide to Managing Knowledge*. Boston, MA: Harvard Business School Press.

Wheat, S. (1999) *Ethical Tourism*. Tourism Concern. See https://www.tourismconcern.org.uk/ethical-travel-guide/ (accessed 11 January 1999).

Wheeler, M. (1995) Tourism marketing ethics: An introduction. *International Marketing Review* 12 (4), 38–49.

Wheeller, B. (1994) Egotourism, sustainable tourism and the environment: A symbiotic, symbolic or shambolic relationship. In A.V. Seaton (ed.) *Tourism: The State of the Art*. Chichester: John Wiley and Sons.

White, L. Jr (1971) The historical roots of our ecologic crisis. In Robert M. Irving and George B. Priddle (eds) *Crisis* (pp. 5–17). London: Macmillan and Co.

Whitehill, J. (2000) Buddhism and the virtues. In D. Keown (ed.) *Contemporary Buddhist Ethics* (pp. 17–36). London: Routledge.
Whitehead, A.N. (1953) Dialogues of Alfred North Whitehead. In J. Kaplan (ed.) (1992) *Bartlett's Familiar Quotations* (p. 584). Boston, MA: Little, Brown and Company.
Whitney, D.L. (1990) Ethics in the hospitality sector: With a focus on hotel managers. *International Journal of Hospitality Management* 9 (1), 59–68.
Wight, P.A. (1993a) Sustainable ecotourism: Balancing economic, environmental and social goals within an ethical framework. *The Journal of Tourism Studies* 4 (2), 54–66.
Wight, P.A. (1993b) Ecotourism: Ethics or eco-sell? *Journal of Travel Research* 21 (3), 3–9.
Wight, P.A. (2001) Environmental management tools in Canada: Ecolabelling and best practice benchmarking. In X. Font and R. Buckley (eds) *Tourism Ecolabelling: Certification and Promotion of Sustainable Management* (pp. 141–164). Wallingford: CABI.
Wiig, K.M. (1997) Knowledge management: An introduction and perspective. *Journal of Knowledge Management* 1 (1), 6–14.
Wilde, O. (1895/2001) *The Soul of Man Under Socialism*. New York: Penguin Classics.
Wilkins, A.L. and Ouchi, W.G. (1983) Efficient cultures: Exploring the relationship between culture and organizational performance. *Administrative Science Quarterly* 28, 468–481.
Wilkinson, G.S. (1984) Reciprocal food sharing in the vampire bat. *Nature* 308, 181–184.
Wilkinson, R. (2000) *Mind the Gap: Hierarchies, Health, and Human Evolution*. London: Weidenfeld & Nicolson.
Williams, B. (1985) *Ethics and the Limits of Philosophy*. London: Fontana.
Williams, F. (2004) Child labour costs financial burden on developing countries. *National Post*, 8 March, SR 4.
Williams, G.C. (1966) *Adaptation and Natural Selection: A Critique of some Current Evolutionary Thought*. Princeton, NJ: Princeton University Press.
Williams, G.C. (1988) Huxley's evolution and ethics in sociobiological perspective. *Zygon* 23 (4), 383–407.
Williams, M. (2000) Sokal, chaos and the way forward? *Sociology* 34 (2), 341–346.
Williams, P. (1993a) Environmental business practice: Ethical codes for tourism. In S. Hawkes and P. Williams (eds) *The Greening of Tourism: From Principles to Practice* (pp. 81–93). Burnaby, BC: Simon Fraser University Central Duplicating.
Williams, P.A. (1993b) Can beings whose ethics evolved be ethical beings? In M.H. Nitecki and D.V. Nitecki (eds) *Evolutionary Ethics* (pp. 233–239). Albany, NY: State University of New York.
Williams, P.A. (1999) Challenges and lessons on the route to sustainable tourism practice. In P. Williams and I. Budke (eds) *On Route to Sustainability: Best Practices in Canadian Tourism* (pp. 1–5). Ottawa, ON: The CTC Distribution Centre.
Williamson, M. (2003) Space ethics and protection of the space environment. *Space Policy* 19, 47–52.
Williamson, P.J. (1989) *Corporatism in Perspective: An Introductory Guide to Corporatist Theory*. London: Sage.
Wilson, D.S. and Sober, E. (1994) Reintroducing group selection to the human behavioral sciences. *Behavioral and Brain Sciences* 17, 585–654.
Wilson, E.O. (1975/2000) *Sociobiology: The New Synthesis*. Cambridge, MA: Harvard University Press.
Wilson, E.O. (1984) *Biophilia*. Cambridge, MA: Harvard University Press.
Wilson, E.O. (1999) *Consilience: The Unity of Knowledge*. New York: Vintage Books.
Wilson, E.O. (2000) *The Future of Life*. New York: Vintage Books.

Wilson, J.Q. (1993) *The Moral Sense*. New York: Free Press.
Winpenny, J.T. (1982) Issues in the identification and appraisals of tourism projects in developing countries. *Tourism Management* 3 (4), 218–221.
Winterhalder, B. and Lu, F. (1997) A forager-resource population ecology model and implications for indigenous conservation. *Conservation Biology* 11 (6), 1354–1364.
Wise, J. (1999) How cruise ships shortchange the Caribbean: Does Kathie Lee know about this? [Electronic version]. *Fortune* 139 (6), 44–45.
Wolf, H.A. (1990) Dorrence Corporation trade-offs. In W.M. Hoffman, R.E. Frederick and M.S. Schwartz (eds) *Business Ethics: Readings and Cases in Corporate Morality* (4th edn; pp. 123–128). Boston, MA: McGraw Hill.
Wolfe, J. and Fuller, T. (1983) Geography and community in collaborative research: Prospects for mutual learning. Paper presented to the 6th Annual Applied Geography Conference, Toronto, Ontario, October.
Woller, G.M. and Patterson, K.D. (1997) Public administration ethics. *American Behavioral Scientist* 41 (1), 103–118.
Wollstonecraft, M. (1792) *A Vindication of the Rights of Women*. Boston, MA: Peter Edes.
Wonders, N.A. and Michalowski, R. (2001) Bodies, borders, and sex tourism in a globalized world: A tale of two cities – Amsterdam and Havana. *Social Problems* 48(4), 545–571.
Wood, R. E. (1999) Caribbean cruise tourism: Globalization at sea. *Annals of Tourism Research* 27 (2), 345–370.
World Commission on Environment and Development (WCED) (1987) *Our Common Future*. Oxford: Oxford University Press.
World Tourism Organization (WTO) (2001) *Global Code of Ethics for Tourism*. WTO: Madrid, Spain.
World Wildlife Fund (WWF) (2000) *Tourism Certification: An Analysis of Green Globe 21 and other Tourism Certification Programmes*. See http://www.eldis.org/document/A10691 (accessed 17 June 2002).
World Wildlife Fund (WWF) (n.d.) *Arctic Tourism Guidelines*. See http://wwf.panda.org/what_we_do/where_we_work/arctic/what_we_do/tourism/ (accessed 4 June 2002).
Wotruba, T.R. (1990) A comprehensive framework for the analysis of ethical behavior, with a focus on sales organizations. *Journal of Personal Selling and Sales Management* 10, 29–42.
Wright, J.K. (1966) *Human Nature in Geography: Fourteen Papers, 1925–1965*. Cambridge, MA: Harvard University Press.
Wright, R. (2000) *Nonzero: The Logic of Human Destiny*. New York: Pantheon Books.
Yaman, H.R. (2003) Skinner's naturalism as a paradigm for teaching business ethics: A discussion from tourism. *Teaching Business Ethics* 7, 107–122.
Yaman, H.R. and Gurel, E. (2006) Ethical ideologies of tourism marketers. *Annals of Tourism Research* 33 (2), 470–489.
Yeung, S. (2004) Hospitality ethics curriculum: An industry perspective. *International Journal of Contemporary Hospitality Management* 16 (4), 253–262
Yeung, S.Y., Wong, S.C. and Chan, B.M. (2002) Ethical beliefs of hospitality and tourism students towards their school life. *International Journal of Contemporary Hospitality Management* 14 (4), 183–192.
Yudina, O. and Fennell, D.A. (2013) Ecofeminism in the tourism context: A discussion of the use of other-than-human animals as food in tourism. *Tourism Recreation Research* 38 (1), 55–69.
Yudina, O. and Grimwood, B.S.R. (2016) Situating the wildlife spectacle: Ecofeminism, representation, and polar bear tourism. *Journal of Sustainable Tourism* 24 (5), 715–734.
Zahn-Waxler, C., Radke-Yarrow, M., Wagner, E. and Chapman, M. (1992) Development of concern for others. *Developmental Psychology* 28, 126–136.
Zuckerman, M. (1979) *Sensation Seeking: Beyond the Optimal Level of Arousal*. Hillsdale, NJ: Erlbaum.

Index

aboriginal communities
  human rights and exploitation 103–104, 105, 350
  staged authenticity for tourists 156–157
  stone axe trading (Yir Yoront, Australia) 118–119
  traditional knowledge and ethics 233–234, 328
accidents, legal responsibility 97–98, 110
accreditation 97, 213–217, 224
Ache people, Paraguay 117
actions
  interaction of motivations and outcomes 92, 94–95, 187–189, 304–309
  judgement according to principles/rules 69, 70, 73
  moral basis, moderating factors 260, 267–268, 290–292
  moral evaluation by consequences 66, 67, 108
  teleological approach in codes of ethics 232–233
Addo Elephant National Park, South Africa 164
advertising *see* marketing
Agenda 21, Earth Summit (Rio, 1992) 9, 218
airports, bag-toting by children 298–300
akrasia 287
Alaska, cruise line tourism 309, 310–311
all-inclusive tourism 295–301
alternative tourism (AT) 4–5, 39, 101, 136, 157–158
altruism
  biological basis 23–29, 38, 68
  related to values 40, 46
  and volunteerism 304–309
  see also reciprocal altruism
Amazon basin, local communities 163, 167–168
animals
  altruistic/cooperative behaviour 23, 25–26, 116–117, 305–306
  behavioural survival strategies 29–30, 34–35, 168
  capacity for emotion 37–38
  disturbance during tourist visits 289–295
  human impacts, hunting and extinction 167–168, 171
  recognition of rights 102, 185–187
Annals of Tourism Research (journal) 12
antagonism, biological 21–25
Antarctic tourism 210, 288–295
anthropocentric viewpoints 178, 186, 190, 194, 250
anthropology 17, 19, 21, 117
antisocial behaviour 43, 179, 335
Arctic tourism 197–198, 232–233
Aristotle 51, 148, 149, 341
  phronesis (practical wisdom) concept 343
  virtue ethics and the golden mean 61–62, 78, 273
Ascent of Man, The (Bronowski) 17
auditing
  audit culture, in research 134–135, 323, 340
  ethical audit of organisations 217, 218, 348
authenticity, tourist/community encounters 156–157, 158, 186, 188, 300
autism 258

backpacking 11, 188
Bad Boll Conference (Germany, 1986) 345
balance, in Eastern decision-making 273–275, 302–304

Balearic Islands, eco-taxes 96
Banff National Park, Canada 58
bats, vampire, food-sharing behaviour 25–26, 33, 334
bees, behaviour 23, 25
behaviourism 18–19, 87, 207
benchmarking, best practice (BPB) 218
Benedict, Ruth 86
benefit corporations (BCs) 143–144
benevolence 78, 147, 246
  motivations for 74–75, 79
Bentham, Jeremy 65, 66, 67, 185
Bermuda National Trust 310, 311
best practice 217–220
billfishing 188–189, 192
biocentric viewpoints 173, 190, 197
biodiversity
  duty of conservation 185, 233
  'hotspot' areas 178
  human threats to 167
biophilia concept (Wilson) 183
biopiracy 187
Birmingham, Alabama, church bombing 139
blackfly control, Adirondacks 154
'blank slate' *(tabula rasa)* concept 18–19, 37, 65–66
Blue Flag awards 214
Bluewater (cruise line watchdog) 310
Blyth's Bridge (knowledge model) 329
British tourists 12, 13–14
Brundtland Report, WCED *(Our Common Future)* 179–180, 228, 233
Buddhist ethics (Theerapappisit model) 271–275, 302–304
business ethics, theoretical
  attitudes to social responsibility 140–142
  challenges of cultural relativism 88
  definition 53
  development history and prospects 148–152
  framework for ethical conduct (Martin) 262–263
  religious and ideological views 124
business organisations
  codes of ethics 9, 145, 227, 236–238
  common pool equity 204
  comprehensive ethical training programmes 347–348
  conflict with environmentalists 180–181
  cultural change, transition mechanisms 315–318
  encouragement of consumerism 154–155
  knowledge management 328–330
  organisational culture 145–148, 252, 260–261, 276
  power and communication with stakeholders 77, 310–311
  pro-poor partnerships 163, 164
  public perceptions of management intentions 103
  responsibility and accountability 138–144
  see also transnational corporations
business schools 149–150

Canada, sustainable tourism
  codes of ethics 228–230, 234–235, 240, 248
  implementing responsible marketing 160
Canadian Centre for Ethics and Business Policy 151
canoeing 97–98, 197
capitalism
  history of Western development 122, 123–125, 138, 152
  related to moral reasoning 286, 314, 317, 350
  understanding, Hayek's 'fatal conceit' 121
care ethics 83–85, 257–258, 269, 287
Caribbean islands
  cultural dependency 128
  economic implications of cruise tourism 311
  impacts of all-inclusive resorts 296–299
Carson, Rachel 136, 190
case method, for ethics education 343–345
casuistry (case study guidance) 244, 250
categorical imperative (Kant) 72–73, 74, 286
Center for Environmental Policy (Texas) 196
certification schemes 214, 217, 316
cheating 26–27, 39, 93
Chibcha Indians, Colombia 119–120
children
  child labour 284, 298–300
  development of empathy and ethics 38, 109
  development of moral reasoning 257, 258
  exploitation in sex trade 284–285

human rights 106–107
inheritance of personality 43–44
chimpanzees
behaviour 37, 116–117
genetic similarity to humans 43
Christianity 63, 115, 124, 184, 303
circle of morality (Singer), expansion of 111–113, 191–192, 350
circumscription, theory of (Carneiro) 76–77
class structure of society 61, 124–125, 133–134
based on race 174, 297–298
codes of ethics
as basis for tourist behaviour 53, 228–229, 232, 236, 247
construction and implementation 246–247, 248, 347
as example of deontological approach 69, 232, 246
history and scope 224–230
influence on consumer choices 12, 230
pros and cons 237–243
related to organisational culture 145, 227, 236–238, 252–254
research on tourism industry codes 231–237, 243
Rules for Antarctic Visitors 288–289, 293
self-regulatory systems 210, 214, 231
WTO Global Code 7, 241, 247–250, 353–359
colonialism 125–126, 160, 174, 197
commerce *see* trade
commitment, theory of 39–42, 144–145, 334
common pool resources 200–208, 333–334
communities of practice (shared knowledge) 328–329
community-based conservation (CBC) 202
community-based tourism 163, 164, 165
complexity 324–327, 331
compliance, with codes of ethics 231, 237, 247
conferences 11, 215, 347
conflict resolution 47–48, 104, 212–213
cross-cultural strategies 235–236
conservation
biodiversity hotspots, role of ecotourism 178
community-based 202
conservation tourism, fair trade principles 161
ethic, in traditional societies 167–168, 169
funded by eco-taxes, Balearic Islands 96
of natural resources 201
self-interested motives in volunteers 305
consilience 323, 337
consumers
destination choice factors 13–14, 152–153, 158
ethical purchasing trends 12, 131
growth of consumerist culture 134–136, 153–155
cooperation
human reciprocal altruism 26–29, 46, 117–118
interactive resource management 211–213
between non-relatives, game theory 29–34, 38, 111
observed in animal behaviour 25–26
between research disciplines 321–324
core values
among nature-based tourists 196–197
in human relationships 56–57
importance in companies/institutions 137, 263, 348
objectives in codes of ethics 227–228
corporate social responsibility 108, 138–144, 149, 263
corporations *see* business organisations
corporatism 133–136
Country/Countryside Code (UK) 225–226
cruise tourism 109, 207, 280, 288, 309–318
Crusades, impact on trade 121–122
culture
commodification, in tourism 106, 127–128, 157
cultural exchange in tourism 332–333
cultural relativism 86–89, 90, 279, 299
in evolution of ethics 46–48, 50
individualistic and collectivist types 144–145
role in individual human development 17–21, 24, 44
WTO code of ethics on cultural heritage 355
cynicism 62

Darwinism 21, 45, 121
Dawkins, Richard 24, 31–32, 207
De Waal, Frans 116–117, 132

decision-making
  codes of ethical conduct 252, 253–254
  comprehensive ethical model (Malloy *et al.*) 268, 270–271, 272, 290–295
  ethical, individual and social norms 53, 252
  existential freedom in 82–83
  influence of emotion 38–39
  levels of ethical discourse (Veatch) 243–246
  moral methods, in business/marketing 277–282
  participation and justice 175, 211–213
  person-specific interactions model (Trevino) 259–262
  pros and cons of utilitarianism 66–67, 83
  stages of process 271, 292–295
deep ecology 191
Defining Issue Test (DIT) 278
dehumanisation 112, 285
democracy
  democratic values in social systems 56, 135–136, 304
  participatory and representative 175
  related to capitalism 125
demonstration effect 127, 128
deontology 69–77, 160, 232, 265, 279
dependency
  cultural 128–129
  economic 125–128
  and technology 136–137
deprivation, definition (Townsend) 155–156
Descartes, René 20, 37, 185
determinism 22, 110, 213, 327
development, role of tourism 125–131, 202, 273, 327
dilemmas (ethical), recognition of 253, 292–293
Diogenes 62, 183–184
discourse ethics (Habermas) 77, 151
distributive justice 100–101, 130, 210, 269, 286
division of labour 119, 120–121
dualism
  mind–body, in human nature 20, 193
  'two tribes,' socioeconomic division 125–126
duty
  of care, legal standards 97–98
  sense of, as behaviour motivation 40, 41
  toward moral principles (Kantian ethics) 69, 72–75
  toward natural world 185, 191, 195

eco-taxes, legislation in Spain 96
ecological consciousness 191, 317, 318
ecology, definition and scope 169–170
economics
  history of Western development 121–125
  importance of tourism 1, 324
  influence on civic value systems 135–136
  of lesser developed countries 125–128, 172, 179–180
  linkage stimulation, in pro-poor tourism 162, 164
  Nobel Prize 138, 150
  value of externalities 9, 173–174
ecosystem services 170
ecotourism
  codes of ethics, effectiveness 240
  ecolabelling of tourism products 197, 214, 215–216
  educational aims 345
  ethical principles 9–10, 131, 147–148, 187
  evaluation of ethical practices 280, 287, 290–295
  local community benefits and control 202, 210
  role in biodiversity conservation 178
  tourist enterprises, case studies 163, 188, 288–290
Ecotourism Summit (2002), Quebec City 213
education
  case method approach 343–345
  ethics teaching in business schools 149–150, 341, 342
  as part of visitor management strategies 345
  in tourism, terms of success 150, 340, 342–343, 346
  training in ethical development 257, 261, 341–342
  used to escape social traps 208
efficiency
  as aim of corporatist culture 134–135, 136
  economic, as driver of growth 123–124, 179
  international application 174

egalitarianism 100, 101, 117, 286
egocentric moral reasoning 193, 255, 257
egoism 68–69, 147, 207, 337
elephant-back tourism 186–187
emissions
  hazardous waste incineration 176–177
  travel and transport 182–183, 310
emotion
  in attitudes to natural world 183–185
  influence on tourist purchasing choices 38–39
  in moral reasoning 264, 300–301, 335
  role in human cooperation 36–42
empathy-induced altruism 28, 40–41, 267
energy consumption 173, 182
environment, influence on genes 42
Environmental Code of Conduct for Tourism (UNEP) 246–247
environmental ethics
  classes of concern/value approaches 173, 189–190, 192–195
  definitions 192
  gap between values and behaviour 171–172
  green consumer demands 159
  inequity and justice 174–177
  precautionary principle 220–222
  in tourism, responsibility for impacts 143, 181, 310–311, 312–318
  in traditional societies 167–168
Ethical Climate Questionnaire 278
ethical climate types, in organisations 147–148
Ethical Purchasing Index 12
ethics
  absolutist and subjectivist theories 51, 148–149
  as constraint to business 150–151
  definition and scope 52–53, 55, 186–187, 349
  ethical systems development 54, 87, 226–230
  relativism and universality 86–90, 268
  relevance to tourism 85, 158, 330–336, 338–340
  see also codes of ethics; morality
eudaimonia (Aristotle) 61, 67, 78, 79
evolution
  of altruism and morality 27, 45–48, 76, 168–169
  and culture, differences and connections 17, 21
  of differing body plans 43
  as far-from-equilibrated system 325–326
  mind and behaviour functions 44–45, 49, 89–90
  natural selection as basis for 22, 45
  social cooperation and trade 116–121
evolutionary stable strategies (ESSs), for survival 29–30, 32
existentialism 79–83, 184, 272
external goods 238–239
extremism 113

Fair Trade movement 160–161
fairness
  as basis for justice 99, 100
  in tourism exchanges 132
  in treatment of employees 73
famine workers (volunteers) 305
fashion, corporate exploitation of 134, 158
feminist perspectives 84, 142
Firm-but-fair programme (game theory) 33
First World War truces 32–33
fisheries management 34, 200–201, 212, 221
food, production and safety 134
Francis (Saint) 115
Frankl, Viktor 82–83, 107, 306–307
free will
  and consent, in social contract theory 75–76
  moral constraints on freedom 47, 73–74, 81, 351
  responsibility for, in existentialism 79, 82, 110–111
Friedman, Milton 138, 140, 141, 142

Galapagos Islands tourism 163, 207, 290
game theory 29–34, 38, 204
genes, role in human behaviour 23–25, 42–44, 335
geopiety concept (Wright) 183–184
'ghost in the machine' doctrine (Descartes) 20
'giraffe women,' Burma/Thailand 104
Global Code of Ethics for Tourism (WTO) 7, 241, 247–250, 353–359
globalisation 121, 155, 330, 350
  in tourism 7, 101, 136
Goa, India 2, 3
golden mean (Aristotle) 61, 62, 78–79, 273
Golden Rule (of human conduct) 57, 72, 73, 306

golf courses 106
'good,' philosophical meaning 52–53, 60, 63–64
gorilla tourism, Central Africa 330
Gould, Stephen J. 323
governance
Antarctic Treaty System (1959) 288
definition 209
strategies in tourism 129–131, 209–213
governments
consequences of institutional thinking 135
environmental policy reform 172, 177, 221
relationship with businesses 139, 310–311
and social contract theory 75–76, 91
tourism manipulation for political ends 301–302, 304
tourism policy 107, 129–130, 162, 180
Greece
ancient, philosophy and ethics 58–63, 85, 148, 224–225
conflicts over turtle protection 180–181
pre-Olympics destruction of stray animals 185–186
tourism impacts on Santorini 324
Green Globe 21 certification 214, 215
green marketing 159–160
groups
collective behaviour, biological basis 17–18, 34–35
evolution of ethical systems 46–48
factors aiding social cohesion 92, 118–119
group socialisation theory 43–44
marginalisation and exclusion 66–67, 101, 172
temporary relationships in tour groups 36
threats to common good 28, 205–208

Hadza people, Tanzania 117–118
Haida people, Watchman Program (Canada) 234–235
Hamilton, W. D. 23, 25, 34, 307
Hardin, Garrett 190, 201, 205
Hawaii
bird diversity and extinctions 167
tourism development 158, 181
'hawks' and 'doves,' behaviour strategies 30, 32
Hayek, Friedrich 121
hedonism 10, 39, 67, 68, 351
Heidegger, Martin 79, 80, 84, 222
high-risk activities
reasons for participation 44
risk management and responsibility 97–99
Hippocratic Oath (medicine) 224–225
Hobbes, Thomas 19, 36–37, 75, 207, 208
holistic viewpoints 80, 85, 140–141, 192, 195
Holy Land pilgrimage tourism 301–304
hospitality industry
carbon footprint measurement 143
ethical standards 8, 147, 346
survey of ethical issue perceptions 251–252
hotel complex development 67, 103, 297
human genome project 42–43, 44
human nature
cooperation and competition 35, 207–208, 304
emotion and reason 36–42
importance of morality 14, 35–36, 110
individual uniqueness and freedom 81
involvement in trade 115, 118–120, 148
the need to travel 14, 290
philosophical understanding 59, 321, 323
self-interest and altruism 22–29, 49–50, 68–69, 140, 334–335
universal social instincts 89–90
virtues as character traits 78
human resource management 268
human rights
abuses associated with tourism 7, 103–107, 285–286, 301
individual, interactions with state/society 75, 81, 101–102
right for healthy environment 175–176
Universal Declaration, articles 105–107, 357
humanism 63, 79, 122, 170–171
Hume, David 37, 48, 100
hunter-gatherer communities 46, 121
cooperation and food sharing 117–118
optimal hunting group size 35
overkill of animal species 167–168
Huxley, Thomas 169, 201

incest, and moral intuition 263–264
inclusive fitness theory (kin selection) 23, 46, 49, 307–309, 334–335
indigenous people

co-management of tourism 234–235
dispossession *vs.* involvement in tourism 301–304
environmental stewardship 166–168, 169, 202
see also aboriginal communities
individualism 135–136, 137, 144
Industrial Revolution 64–65, 167
instrumental rationality 56, 77, 113, 137
applied to environmental ethics 177–183, 240
restriction of free thought 213
intentions, related to ethics 173, 187–189, 267, 294
interactive governance 211–213
interdisciplinarity, in research 321–324, 326, 329, 338
internal goods 238–239
International Hotel and Restaurant Association 346
International Institute for Quality and Ethics in Service and Tourism (IIQUEST) 8, 346
intrinsic value 73, 185
definition 56
nature and environment 177–183, 190, 204, 240
people and relationships 141, 185
intuition
as basis for moral judgement 70, 239, 254
ecological consciousness 191
Haidt's social intuitionist model 263–265, 299–301
Inuit people, Canada 168, 197

Japan Association of Travel Agents 231
Jerusalem, history of pilgrimage 14
Josephson Institute of Ethics 251–252
'just' tourism (Hultsman) 10
justice
application in society 99–101, 255, 257
environmental equity 174–177
preferred over charity (Wollstonecraft) 299, 305
relevance in tourism 101, 131
as social exchange 29, 99
theories, deontological nature 69
see also human rights

Kant, Immanuel 37, 70–71, 72–75, 240, 286
Kierkegaard, Sören 79–80
kin selection (Hamilton) *see* inclusive fitness theory
knowledge
division between science and humanities 319–321
experience or reasoning as basis 37, 48
management 328–330, 338–339
and moral competence 54, 60
philosophical theory (epistemology) 52
related to understanding and utility 327–328, 336
uncertainty 349
knowledge value chain (Weggeman) 330, 338

labelling schemes 197, 214, 215–216
labour conditions, cruise lines 311–312
land management, standards and rights 102–103
laws 92, 96–99, 108, 119
'Leave No Trace' principles 198
leisure
consumer choices and status 152–153
as human right 107
meaningful use 83, 306
Leopold, Aldo 10, 190
Lesser Developed Countries (LDCs)
common property management 203
economic and cultural dependency 125–129
environmental regulation, and reform 172, 174
human rights restrictions 105–106, 107
pro-poor tourism strategies 162–165
prosperity increase and inequity 111–112, 155, 161
Leviathan (Hobbes) 19, 75
Levinas, Emmanuel 80–81
liberal arts 59
in education 150, 340, 341
libertarianism 100–101, 286
Lindblad, Lars-Eric 288
locally unwanted land uses (LULUs) 174, 176–177
Locke, John 37, 75–76, 107
logging industry 171–172
logotherapy (Frankl) 82, 107
love
attachment to place 184
duty of care for others (Kierkegaard) 80, 211
in human relationships 39, 41–42, 91, 185, 213
as ultimate human goal 82

Machiavelli, Niccolò 122–123, 132
Makong region (Thailand), tourism development 273
Malawi, eviction of local community 103
Malthus, Thomas 21, 124
Manila Declaration (WTO) 107
Marcy's Woods, Ontario 103
marginalised groups
  moral condemnation of 112
  subsistence lifestyles 172, 202
  tourism development impacts 66–67, 101, 296–298
market culture 314, 315
marketing
  destination image, foreign control 126–127
  ethical evaluation, MES scale 277–279
  ethics, Hunt and Vitell model 265–268
  green 'enviropreneurial' approach 159–160, 181
  intensity, and growth of consumerism 155
  pressure for tourism development 158, 327
  truth and reality in advertising 342
mass tourism
  alternatives and balanced growth 5
  consumer choices 152–153
  effects on local (host) culture 127–128, 295–298, 333
  international development, 20th C. 2, 296
  public awareness of impacts 103, 158
MCB University Press, tourism ethics conference 11
meaning, awareness of in life 82–83, 112–113, 306–307
medical ethics 148–149, 224–225
medical tourism 130
men, compared with women 258
merchants, medieval Italy 122
meta-ethics (Veatch) 245, 246, 250
Mill, John Stuart 65–66
minority groups 66–67, 112
mission statements 139, 316, 318
models
  Buddhist ethics (Theerapappisit) 271–275, 302–304
  comprehensive ethical decision-making (Malloy) 268, 270–271, 272, 290–295
  cooperative behaviour (game theory) 29–34
  educational, for ethics-based tourism 342–345, 346
  ethic of care (Gilligan) 257–258
  ethical decision-making (Trevino) 259–262
  ethics knowledge diffusion in tourism (Fennell) 338–339
  framework for ethical conduct (Martin) 262–263
  marketing ethics (Hunt and Vitell) 265–268
  moral development stages (Kohlberg) 255–257, 313
  moral principles framework (Schumann) 268, 269, 285–287
  organisational culture, 'peeled onion' (Schein) 312–313, 314
  psychological, of social relations (Fiske) 129
  social intuitionist, of moral judgement (Haidt) 263–265, 299–301
  tourism development 6, 327
  tourism organisational framework (Malloy & Fennell) 313–315
Mohonk Conference (2000), New York 215
monoamine oxidase A (MAOA) gene 335
moral reasoning
  applied to wrongful acts 108, 275–277
  classes of moral intention 173
  as conscious activity 254, 264–265, 270
  evaluation methods 277–282
  gender differences 257–258
  Kohlberg's model of development 255–257, 292, 313
  levels of moral discourse (Veatch) 243–246
  see also decision-making
morality
  adherence to rules 66, 74, 90–92
  application in tourism 10–11, 216, 351–352
  evolution of 45–48, 168–169
  expansion of scope 111–112
  importance as human characteristic 14, 27, 35, 38
  moral fortitude in the workplace 252
  moralisation, perceptions of 13, 112–113
  role of sentiment (emotion) 37, 41, 335
  Schumann's moral principles framework 268, 269, 285–287
  standards, in society 53–54, 59, 111

motivations
  of tourism stakeholders 332–333
  in volunteerism 306–309
mountain climbing 183
multidimensional ethics scale (MES)
    278–282
multinational enterprises *see*
    transnational corporations
myths
  narratives in ancient past 59, 208–209
  stewardship of nature by traditional
    societies 166–169
  of Third World marketing 160

NASA shuttle flight (1986) disaster 239
nationalisation 201–202
natural selection concept (Darwin)
    21–22, 45, 146
natural world
  common pool resource management
    200–208
  legal recognition of rights 102
  reverence for 183–185, 209
  stewardship by indigenous people
    166–168, 169
naturalistic fallacy (Hume) 48
nature *vs.* nurture debate 16, 18–21, 42, 110
negative entropy 325–326
negligence, legal definition 97
neoliberalism 130–131, 134–135,
    186–187, 241
neuroscience, recent findings 45
Niagara-on-the-Lake (Canada),
    tourism 333
Nietzsche, Friedrich 79, 83, 113, 350
nihilism 63, 113
NIMBY (not in my backyard) attitudes
    174, 176–177
'noble savage' concept (Rousseau) 19–20,
    125, 166
non-governmental organisations (NGOs)
    103, 130–131, 240
  codes of ethics development 232
  NGO Steering Committee Tourism
    Caucus 104–105, 220–221
norms
  conflict with individual freedom 81, 179
  and cultural relativism 87–88
  definition 58
  normative ethics in decision-making
    244–246, 255
  role in structure of society 90–91, 144

objectivism in ethics (Rand) 337
off-road activities 65, 98
organisational culture 145–148, 252,
    260–261, 276
  Schein's 'peeled onion' model
    312–313, 314
'Other' concept (Levinas) 80–81, 160
Our Common Future (Brundtland Report,
    WCED) 179–180, 228, 233
outdoor recreation 58, 196–198, 345

Paduang tourist model village,
    Thailand 104
Palestinians, and Holy Land tours 301–304
Papua New Guinea, tourism 158
participation, in decision-making 175,
    316, 317
paternalism 164–165, 212
peer groups, influence on behaviour
    44, 294
personality
  assumed by tourists abroad 127
  heritability 43–44
philosophy
  characteristics and applications 15, 52,
    85, 149, 282
  in classical antiquity, schools of
    thought 58–63
  paradigms of environmental ethics
    192–195
  terminology 51–58
phronesis (practical wisdom) concept 343
pilgrimage 14, 301–302
Plato 59, 60–61, 85, 342
pleasure
  happiness measurement (Bentham/
    Mill) 65, 66, 67
  neurological basis 45
  pursuit related to happiness 83
  seeking, morality of 10, 112–113, 287
  self-image and consumerism 154
poaching, wildlife (Africa) 202
politics
  choices and freedom of citizens 137,
    154–155
  definition and social role 116
  of environmental issues 172, 196
  instability, impacts on tourism 163
  Israeli/Palestinian tensions 301–304
  related to morality, Machiavelli's
    views 123
  social contract (Rousseau) 75

pollution 173–174, 310
possessions *see* property ownership
poverty, global incidence 155, 156, 172
power
 decentralisation 212
 imbalances and communication 77
 Machiavellian strategies 122–123
 of multinational firms in tourism 125, 126–129
precautionary principle 220–222
Priestley, Joseph 64–65
'primitive' cultures and tourism 156–157, 160
Prince, The (Machiavelli) 122–123
Prisoner's Dilemma (game theory) 30–32, 33–34, 205–206, 236–237
private sector *see* business organisations
pro-poor tourism 162–165, 175
profit, as focus of business 135, 139–141, 146
property ownership
 as basis for class structure 124, 133–134
 as human right 75, 106
 related to happiness 153–154
 and responsible management 102–103, 201, 202–203
 rights to common pool resources 203–204
prostitution 49, 119, 130, 283–287
psychology
 evolutionary 44–45
 models of social relations (Fiske) 121, 129
 social, reciprocal altruism theory 26–28
public service
 bureaucratic ideals 55–56
 care ethics 84–85
 corporatist tendencies 134–135
 development of ethical codes 238
punishment
 and moral responsibility 108, 110–111, 240
 in moralistic strategy 34
Pythagorean philosophy 59

race relations 128, 174, 297–298, 311–312
rainforest clearance 109, 171–172
Rand, Ayn 15, 55, 337
ranking of subordinates 69–70
reason
 communicative, for a just society 77
 and cultural relativism 88–89
 in Kantian ethics 72, 73, 286
 role in human understanding 37, 39, 41, 350
reciprocal altruism (Trivers)
 constraints, for tourists 334
 observed in animals 25–26, 116–117
 systems of human behaviour 26–29, 38, 39, 46, 49
 in volunteer tourism scenario 307–309
reciprocity, indirect 91–92, 93
recreational activities
 ethical comparison 65
 legal standards of care 97–98
 see also outdoor recreation
Red Light Districts (TV show) 283–284
reductionism, in scientific approach 320, 325, 326
regulations, role in governance 209–210, 227
relativism, cultural 86–89, 90, 279, 299
religion
 attitudes to business ethics 124
 diversity and conflict 301, 302, 303, 304
 moral codes and commandments 71–72, 208
 reasons for questioning 59, 80
Renaissance, and humanism 63, 122
research, tourism
 content analysis of codes of ethics 231–233
 growth of interest in ethics 8–12, 196
 incorporation of values 57–58, 336, 339–340
 methods for ethical behaviour analysis 243, 280–282
 need for interdisciplinary approach 322, 323–324, 326–327, 329, 338
 quality and autonomy 134–135
responsibility 48–49, 81, 83, 107–110
 civic 135–136, 138
 education for, case method approach 343–345
 individual, substituted by codes of ethics 239–240
 legal 97–98
 for unintended consequences 189
responsible tourism 80, 109–110, 161
 operators/organisations 131, 140, 315
rights, human *see* human rights
Rio Earth Summit (1992) 9, 218, 220
risk management planning 98–99, 220–222

rituals, cultural importance 47, 236
Rousseau, Jean Jacques 19, 75, 184–185
rule violation in sports 272
rules and rights (in decision-making) 244, 245
rules of society 90–92, 208, 211
Ruskin, John 181–182

Sandals La Toye resort, St. Lucia 297
Sartre, Jean-Paul 79, 81
scenarios, for ethics evaluation 278, 280, 281–282, 307–309, 318
scepticism 37, 62–63
sciences
  differences between natural and social 320–321
  scientific method 320, 325
  systems complexity and modelling 326
self-interest
  beneficial role in social progress 120
  biological basis 22–25, 34–36
  egocentric environmental ethics 173, 193
  moral judgement of egoism 68–69, 337–338
  as nature of humanity (Hobbes) 36–37
self-regulation 210, 231, 255
'selfish gene' concept (Dawkins) 24–25, 49, 118, 146, 207
sex tourism 112, 283–287
Smith, Adam
  definitions of 'value' 55
  division of labour concept 120–121
  on moral sentiments 37, 91, 100, 121
  opinion of businessmen 148
  Wealth of Nations (economics) 124
Snow, C. P. 320
snowmobiling 204
social contract theory 75–77, 185, 313
social exchange theory 28–29, 99
social sciences 320–321, 326, 336
social tourism 131, 357
social trap theory (Platt) 204–208
socialism (in justice ethics) 286
sociobiology 34
sociobureaucratic culture (tourism) 314–315
Socrates 51, 59, 60
Sophist philosophy 59, 342
space tourism 250
stakeholder theory 141–142
status, social
  awareness and stress 112
  establishment in cooperative groups 116–119
  and rules of etiquette, Brunei 90
  shown in wealth and leisure choices 152–153
stewardship, of natural world 166–169, 184
stoicism 62
Stonehenge, site protection 70, 74
sufficiency concept, tourism planning 273, 302, 303–304
sustainable development
  applied in tourism 5–6, 9, 101, 152, 204
  green marketing strategies 159
  institutional and moral basis 211
  mismatch with current economic systems 121, 172, 179
  resource use in traditional societies 167, 168
  in WTO Code of Ethics 354–355
systems dynamics 325–326

taboos 92, 96
taxation 96, 203, 208, 311
technology, development impacts 136–137, 222
teleology 63–69, 232–233, 265–267
testosterone, and empathy 258
Thailand
  elephant-back tourism 186, 187
  rural tourism development 104, 197, 273
  sex tourism industry 283–284
theology 61, 63, 71–72
thermodynamics 325–326
thinking
  awareness and existence (Descartes) 37
  calculative and meditative (Heidegger) 80, 84
  critical (rational) 52, 56
  moral cognitions 259, 260, 335
  see also moral reasoning
tipping (for services) 39–40
Tit-for-tat (strategy game) 32–33
Toronto (Canada), rickshaw fares 8, 100
tort law 97
Tourism Concern 105, 159, 161
Tourism Education Futures Initiative (TEFI) 346
tourism industry
  codes of ethics 227–230, 240–241
  corporate social responsibility 142–144

tourism industry (*Continued*)
dependency in Third World economies 125, 126–129
environmental inequity and justice 174–175
ethical evaluation of operators 280–282
governance strategies 129–131, 136, 140–142, 209–213
impacts (benefits and costs) 1–2, 3, 5–6, 309–310
implementation of ethical practices 7–8, 12–14, 54, 218, 221–222
interactions between stakeholders 2, 4, 28–29, 332–333
mass–alternative development continuum 4–5, 157–158
organisational framework model (Malloy & Fennell) 313–315
quality standards 213–220
resort development in sensitive areas 79, 103, 180–181, 281
Tourism Industry Association of Canada (TIAC) 228–230, 240
Tourism Operators' Initiative (2002) 219
trade
evolutionary origins 116–118
Fair Trade movement 160–161
history, in Western civilisations 121–125
in prehistory/primitive society 118–120
tradable tourism rights 101
'tragedy of the commons' (Hardin) 190, 200–201, 205–206, 208
transnational corporations
environmental resource management 9, 203
market dominance 106, 136
in tourism, power in less developed countries 125, 126–129, 161
transport
comparisons of fuel consumption 182
cruise ship pollution 310
rickshaws, control in Toronto 8, 100
'trapped tourists' scenario 40–41
travel agents 231, 267
travel experience, compared with virtual reality 45
triangulation, ethical 271, 295, 296
Trivers, Robert 25–27, 38, 43, 307
Tropic Ecological Adventures, Ecuador 163
trust 36, 39–40, 144–145, 207–208
turtles, conservation 103, 180–181

United Kingdom (UK)
Country (/Countryside) Code 225–226
Tourism and Human Rights (Tourism Concern) 105–107
tourist attitudes and motives 12, 13–14
United States (US)
constitutional acknowledgement of rights 102
corporate fraud and business ethics 149
environmental racism 174
labour–management relations 145
September 11th (2001) attack 324
Universal Declaration of Human Rights 105–107, 357
universal laws of ethics 70–71, 73
see also Golden Rule
universalism 89–90
universities, educational scope 341
utilitarianism 64–67, 83, 186, 269, 285
utopian ideals/beliefs 61, 137, 139, 157, 183

values
in changing economic systems 124–125
definitions and applications 55–58, 84, 145–146, 226–227, 246
evolution in hunter-gatherer communities 46
globally shared (universality) 89–90
internalisation, and moral judgement 112, 312–313
as motivations for empathy 40–41
types, in environmental ethics 173, 192–195
virtual reality 45
virtue
displayed in actions 269
internal goods of organisations 238–239
moral, Aristotlean definition 61–62, 78
related to self-knowledge 60
virtue ethics, social utility 77–79, 84, 246, 286
vision (in theory development) 57–58
volunteer tourism 131, 304–309

waste disposal 174, 176–177, 188, 310, 312–318
water sports, safety issues 97–98, 109
water supply
contamination by agrochemicals 106
local use management, Valencia 203
pollution from cruise ships 310
Watson, John 18–19

wealth inequality 155
welfare, animal 186, 250, 295
whales
  code of ethics for whale-watching 69, 230–231, 233
  held in captivity 66
  hunting by Inuit peoples 168
  value, philosophical approaches 178
whistleblowing 149, 275–277
Wilde, Oscar 299, 305
wilderness management 198, 233, 288–290
Wilson, E. O. 183, 323
'Wise Use' lobby group (US) 176
women
  human rights restriction and exploitation 105, 106
  involved in sex tourism 283–287
  moral outlook and care ethics 84, 257–258
workplace ethics 252, 259–262, 275–277, 279
World Tourism Organization (WTO)
  Global Code of Ethics 7, 241, 247–250, 353–359
  Manila Declaration 107
  origins 2
World Travel and Tourism Council 240
World Wildlife Fund (WWF) 171–172, 202, 214, 232–233
wrongdoing, in organisations
  definition 275–276
  risks and responses 277

Xenophanes 59

Yir Yoront people (Australia), trading for axes 118–119
youth hostelling 36